ENERGY RESOURCE, SLAVE, POLLUTANT
A Physical Science Text

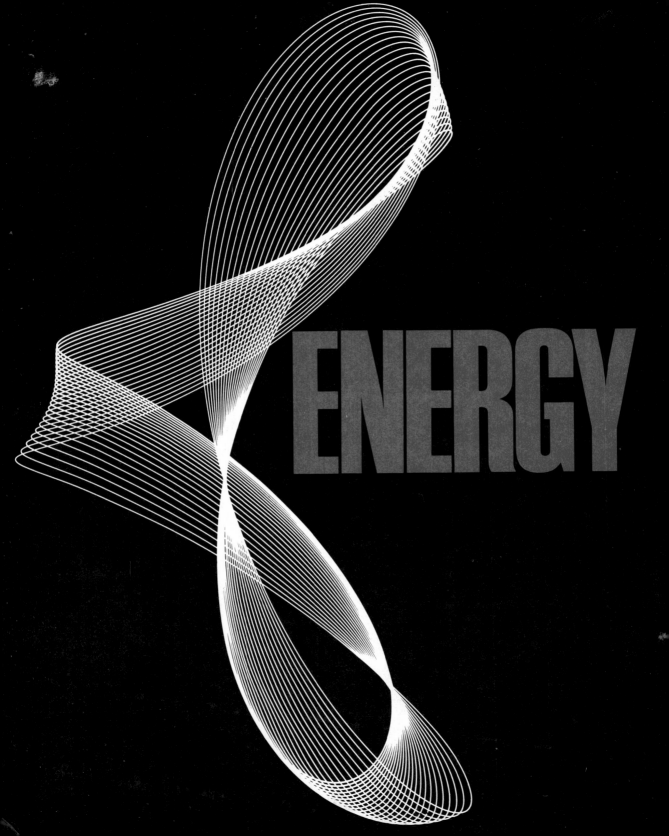

ENERGY

Robert S. Rouse
Monmouth College

Robert O. Smith
Monmouth College

RESOURCE
SLAVE
POLLUTANT

A Physical Science Text

Macmillan Publishing Co., Inc.
New York

Collier Macmillan Publishers
London

Macmillan Publishing Co., Inc.
866 Third Avenue, New York, New York 10022

Collier-Macmillan Canada, Ltd.

Library of Congress Cataloging in Publication Data

Rouse, Robert S
  Energy: resource, slave, pollutant.

  1. Power resources. 2. Pollution. 3. Science.
I. Smith, Robert Owens, joint author. II. Title.
TJ153.R69   338.4   74-3793
ISBN 0-02-404000-2

Printing: 1 2 3 4 5 6 7 8   Year: 5 6 7 8 9 0

Preface

This is a textbook for a one-semester physical science course for the student whose main interests lie outside the field of science. Several years ago, we faced the issue that we as scientists need to answer the question posed by students—"Is science relevant to my life?"

In the search for a meaningful answer to that question, we asked ourselves what single scientific concept or principle might best relate the activities of a scientific-technological world to the student whose main interests lie outside the field of science. We sought a balance between relevance and scientific principle by the choice of energy as our theme. Energy is a term that is increasingly a key word in our daily lives. Our civilization and our economy are predicated on the prodigious consumption of energy and energy resources. We thus could present scientific topics in such a way as to have readily identifiable meaning to the student. Most physical science texts present numerous topics for their importance to science or for their intrinsic beauty. Neither reason is truly meaningful to the uninitiated.

The choice of topics and organization of this physical science text are nontraditional in that we have selected principles which are directly related to the central theme of energy and energy use by society. Certain questions about energy occur and recur in various ways throughout the text: (1) What is energy? (2) What are our energy sources? (3) What basic laws of the natural sciences govern the generation and use of energy? (4) What are the consequences of consuming very large amounts of energy? (5) Are there any limits to continued rapid growth in energy consumption? To illustrate these relevant questions, we survey in some depth one aspect of energy, namely the generation of electricity or electrical energy.

The text has three introductory chapters: Chapter 1 interrelates science, technology, and the economy with emphasis on electrical energy generation; Chapter 2 presents the astronomical rationale for the popular metaphor, spaceship earth; Chapter 3 discusses the flow of solar energy through the ecosphere. We discuss the production of electricity by moving water (Chapter 4), the production of electricity by burning fossil fuels (Chapters 5, 6, 7, and also 8 and 9), and the production of electricity by nuclear fission (Chapters 10 and 11). The concluding chapters of the text, Chapters 12, 13, and 14, discuss various types of pollution of the environment—radiological, thermal, and air, respectively. These chapters review material presented in earlier chapters and apply scientific principles to the environmental problems deriving from electrical energy generation. The prospects for alternate and supplemental sources of energy, along with projections for future usage, are discussed in both Chapters 8 and 15.

We have found that our students get most out of the course when all chapters are covered. Each chapter has a specific purpose and is a necessary part of the total presentation. Parts of chapters may be omitted without loss of continuity. While no supplemental reading material is required, references for additional reading have been included.

The text has been class-tested for three years at Monmouth College in both large (up to 200) and small classes (7 to 20). The backgrounds of our students have varied widely, from those who have had physics and chemistry in secondary school to those who have had no secondary school science at all. Basic arithmetic skills have been expected. Over 98 per cent of these students have completed our course successfully. Student response to the course, by a variety of measures, has varied from over 50 to 100 per cent favorable, depending on class size, in contrast with only about 10 per cent favorable to what we call a traditional physical science course.

This text is unique in that a unifying, relevant theme has been presented in such a way as to unfold a reasonably complete story when the entire set of chapters is covered. The text is written so that the course may be very qualitative. However, many calculations are included in the body of the text. These examples are clearly set off from the text. The narrative is written so that either a qualitative or a quantitative reading of the chapter is possible. In the text, significant questions for the reader are recorded and set off to encourage thought about the concepts presented and about how these concepts may affect one's daily life. At the end of each chapter, there are both questions and numerical exercises to review the material of the chapter and encourage the reader to relate the chapter material to the energy theme. For us, the course and the text, even in its embryonic stages, caused our students to become meaningfully involved with science, scientific thinking, and the related technology and at the same time to understand the role and the limits of the physical world described by science. Overall, our students have come to understand the balanced picture of energy and energy consumption, which includes the concept of trade-offs, i.e. the benefits of various types of energy versus the hazards or costs. Science, technology, and society have become real to our students. We hope that our approach will be as successful with your students as it has been with ours.

R. S. R.
R. O. S.

Acknowl-
edgments

We are pleased to record our sincere appreciation for the help and encouragement given during the preparation of this manuscript. We can acknowledge here but a few. Colleagues at Monmouth College who provided ready answers to questions and who effectively contributed to our approach include D. Barnes, R. Teeters, W. Hieslmair, L. Kijewski, D. Shick, J. Sweer, and L. Spiegel. Our most severe and most helpful critics have been the many students who used our manuscripts. We are certain that the book is much the better for their perceptive comments. The preparation of manuscripts for our classes was initially supported by a grant to the authors from the Grant-in-Aid-for-Creativity Committee of Monmouth College. We are most grateful for this encouragement and support.

This book owes much to the diligence, efficiency, cooperativeness, and enthusiasm of the staff of the Macmillan Publishing Co., Inc. In particular, we appreciate the guidance and suggestions throughout the project from J. Walsh, Executive Editor, and the coordination by L. Malek of the many inputs necessary to make this a finished text. We are indebted to the many reviewers who contributed at various stages, and in particular for the helpful, substantive reviews by the following in the final stages of the project: M. Goldberg, Westchester (N.Y.) Community College; J. Howell, Western Michigan University; K. Mendelson, Marquette University (Wisconsin); and L. Painter, University of Tennessee.

We are eternally indebted for the supportive help and typing of Mrs. Carol Krebs, who has done and redone many portions of the text with unexcelled rapidity, accuracy, and enthusiasm. We thank R. Binder for the preparation of many glossy prints and J. Tardiolo for the hundreds of slides prepared for the course.

Our wives, Nancy R. and Carol S., have been generous with encouragement and patience during this project. We dedicate this book to our children with hope—hope that we have made a contribution that will in some large or small way insure that they will live in the best of all possible worlds.

R. S. R.
R. O. S.

Contents

xi

6 Reactions Provide Heat 163

7 Heat Is Transformed and Transferred 185

8 Do You Consume Power? 227

9 Entropy and Disorder 265

10 Our Nuclear Age 291

11 Peaceful Utilization of Nuclear Energy 323

12 Is Your Geiger Counter Working?— Pollution from Radiation 349

13 We May Be in Hot Water— Thermal Pollution 375

14 Our Air Is Unclean— Air Pollution 411

15 What Next? 445

Appendixes 481

Glossary 497

Index 507

A Note to the Student

This book has been written for you. Although you may have had some science in your educational experience, we hope that this book will present a somewhat different and more meaningful approach to the place of science in society. We have attempted to meld the fundamental quantitative way of looking at the world of the scientist with some of the problems facing you today and tomorrow.

A key feature of any analytical activity is involvement. You have been involved in many different things in your lives; however, the involvement we urge you to undertake is one of separating fact from emotion. This takes effort. We believe that the results of the effort will prove valuable to you. Our world-wide society, but particularly society in the United States, is a scientific-technological society. Much of what we take for granted, men of a few hundred years ago would have found incomprehensible—the automobile, permanent-press fabrics, self-winding watches, transmission of electricity, movement of foods over long distances, frozen foods, and so on. We routinely use concepts such as speed, acceleration, molecules, and energy in our daily lives. How we have gotten to this state of sophistication is very important and will be some of our story. However, our attention in this book focuses on a problem that is found wherever men have a moderate or a high standard of living— the developed countries such as Australia, Canada, England, France, Germany, Italy, Sweden, the United States. This standard of living is possible because we have "energy slaves" doing much of our work for us. This work would occupy much of our lives if energy were not available to us. Buckminster Fuller calculated that the amount of energy consumed per capita in the United States in 1940 was the equivalent of 153 slaves working full time, "energy slaves." He estimates that in 1970 the typical American used energy at a rate equivalent to some 400 "energy slaves." Most of us use this energy without thinking (other than paying the bill for fuel and/or electricity). The authors believe that all of us must do some hard thinking about our use of energy.

Critical to an understanding of energy in our society, and some of the consequent pollution, are certain fundamental principles of chemistry and physics. A basic aspect of all chemistry and physics is quantitative reasoning. All significant quantitative reasoning can be handled with simple arithmetic. These principles of chemistry and physics, the derived technology, and the implications for us today and tomorrow comprise our approach to the subject of electrical power generation and use.

We invite your attention and vigorous participation because the issues surrounding society's use of electrical power will influence your life-style today and tomorrow. You will be called upon as intelligent citizens to make judgments that will determine future developments. You will vote. We hope you will vote cognizant of significant facts. In our physical world, man has uncovered a number of laws that govern the behavior of matter. The laws of physics and chemistry are important to you. Solutions to society's problems cannot defy nature's laws. Come with us and become aware of our limitations—the limitations of rational men, the limitations of our being together at a high standard of living.

At the end of each chapter, you will find exercises pertinent to the chapter. These exercises are designed to assist you to understand not only the laws and principles of the chapter but how these may affect you and me. Most of us who have pursued science as a career believe in the educational philosophy—"learning by doing." We invite you to participate with us. We further hope that you really want to participate in living rather than be on the sidelines as a spectator. Let's go!

R. S. R.
R. O. S.

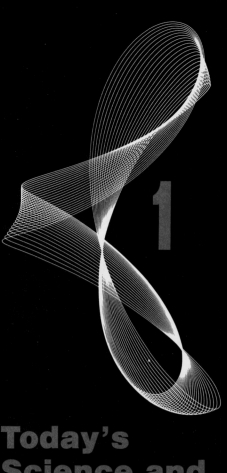

Today's Science and Technology

You have been making observations for years. You size things up. You look over the opposite sex. You gather as much information as you need in order to draw a conclusion or to make a decision. Isn't that so? Do you remember trying something just to see what happens? Nearly all of us have. In these ways we are all scientists.

At this point in your life, you may have mixed feelings about science and technology. Do you see science (and the resulting technology) as potentially beneficial? Do you see science as potentially harmful? Do you find science interesting, or do you find science complex, confusing, dull, or just not understandable. Many of you have probably stereotyped scientists (Figure 1-1)—white lab coat, glasses, fancy equipment, noises, colors, flashes.

Perhaps you have formed some attitude about science and scientists via the products deriving from scientific endeavor (technology): the atomic bomb and other weapons, the Apollo program, the transistor. Some of these accomplishments have been beneficial to mankind, some have been harmful, some have been both beneficial and harmful. Many technological advances that are the result of scientific work are beneficial; yet some of the products of technological society are disturbing.

Figure 1-1. Is this your stereotype of a scientist?

Questions for the Reader: How do you feel about the Salk polio vaccine, the atomic bomb, paper mills, textile mills, electric power stations, aluminum cans, polyethylene bottles, rubber tires? How do you feel about the people, including the scientists associated with these advances?

All of these products contribute toward your attitude about science and technology, and your attitude toward a product and its use may have been transferred into an attitude toward the men and women we call scientists and technologists (engineers, for example).

Both the vast array of products and the wealth of knowledge surrounding you can easily leave you confused and bewildered about what science is, what technology is, how they interact with your life, and whether or not science and technology really are beneficial to society. We do not attempt here to change your attitude, but we will try to show you many ways in which science and technology affect your daily lives, particularly through society's need for **energy.**

To understand your place in our scientific, technological society, you must put aside your preformed opinions and prejudices and try to understand some of the subtle differences between science and technology. You may have been surprised when we said in the first paragraph that, in a sense, you actually are a scientist. As a child, you had a natural curiosity about things and happenings. We hope you still have this curiosity! You have been making observations for years. You have been gathering information (whether you have liked it or not) and have been drawing conclusions and making decisions. At times you experiment just to see what happens. All of these characteristics or attributes are common to scientists and technologists, too!

Scientists have chosen to make use of these attributes more directly in their careers, in contrast to many other members of our society. You may further be surprised that so many conveniences that you now enjoy have their origin in scientific endeavor. Glenn Seaborg, Nobel Laureate in chemistry and a former Director of the United States Atomic Energy Commission, has observed that the problems posed by science enter into the fabric of our whole society in many ways. He further observed that man in the Middle Ages could not ignore the Church, that Renaissance man could not ignore the arts, that eighteenth century man could not ignore political thought, and that modern man cannot ignore science. We believe that modern man cannot ignore either science or technology. And so we feel that it behooves you, a member of our modern society, to understand how science and technology interact with you. We further submit to you that one basis for your understanding of this interaction is through the subject of energy.

In this chapter, we examine what science is and who these people are that we call scientists. Then, after discussing some of the various sciences, we describe technology and attempt both to differentiate science and technology and to show how science and technology interact in modern society. Lastly, we survey some of the implications of modern technology for us today as an introduction to the fuller discussions of these implications later in the book.

What Is Science?

Science is many things. Each person you ask, including scientists, will give you a different answer. Let us note some of the things we feel that science is. Science is made up of facts. Science is one of the bodies of knowledge that man has accumulated over many centuries. Man, in his attempt to make life more comfortable and pleasant, has gathered information about natural processes and material objects. Furthermore, gathering these facts requires observations. In other words, science is the body of knowledge obtained by methods based on observation. However, as history has shown, the way in which we make the observations and the inferences that we draw from them are critical as to how they contribute to scientific knowledge. Following are two examples.

From earliest times through the Middle Ages, what we now call science was part of philosophy or natural philosophy because man's observations focused on nature, which man has always regarded with a mixture of fear, reverence, and curiosity. The history recorded before Jesus Christ showed that natural philosophy was studied and practiced by only a few. These early philosophers were able to use their observations (for example, for predicting eclipses of the sun and the moon) to gain political power (particularly back in the time of the ancient Egyptian, Babylonian, and Assyrian societies). Probably the best documented early natural philosophy was that of Aristotle (384–322 B.C.) who relied only on observation and human reasoning.

An excellent example of Aristotle's reasoning was his philosophy of the nature of motion. He thought that if two bodies of different weights were dropped from the same height and at the same time, the heavier object would reach the ground first. Those who, on their television screens, saw Astronaut David Scott in 1971 drop the feather and the hammer on the moon (Figures 1-2 and 1-3) know that objects fall at a rate without regard to their weights. Here is your chance to be a scientist, too. Choose a cork and a rock, about the same size but differing in weight, and drop them. Was Aristotle's reasoning about this aspect of motion right or wrong?

Unfortunately, Aristotle's reasoning dominated the thoughts of man for nearly 2000 years. Put simply, observations without carefully designed checks (experiments) are often insufficient or even incorrect. Thus, scientific fact is based on carefully designed experiments to prove validity of inferences drawn from observation. Although the importance of experimentation to science and to what is often referred to as the scientific method seems fairly obvious to us today, the central role of experimentation to the scientific enterprise only dates from Galileo (1584–1642). Despite Galileo, it has only been since the end of the seventeenth century that both observation and experimentation have been fundamental to all science. Thus, a more complete definition of **science** is that science is *the body of knowledge obtained by methods based on observation, including careful experimentation.*

Science ultimately depends on observation along with experiments for its validity, but science is also public knowledge and, as such, is a consensus discipline. A central concern of science is achieving a consensus, that is, general agreement (at least by scientists) on the implications of certain observations. Scientific observations, experiments, and laws are available

Figure 1-2. Astronaut David Scott (Apollo 17) on the moon, standing on the slope of the Hadley Delta. [*Courtesy of NASA.*]

Figure 1-3. As in the sketch, Astronaut Scott did do an elementary physics experiment by dropping a hammer and a feather. Those who saw this experiment on TV saw that both objects are accelerated equally by the moon's gravity, despite their differences in mass and weight.

for perusal and checking not only by the scientific community but by the informed general public.

However, to leave you with the impression that science is only facts based on observation would reinforce that all too popular stereotype of science as mere facts to be memorized and the scientist as an impersonal machine. Science also includes concepts and relations that are only indirectly products of observation and experiment. Directly, they are products of the human mind and, as such, represent a high degree of creativity. Some of the greatest scientists of our time should appropriately be called artists because of their creative minds. Albert Einstein is but one of many examples. History is loaded with examples of the influence of science on the fine arts and of the fine arts on science. The poet, Goethe, was a scientist, too [1].[1] We recognize that he was a far better poet than he was a scientist. Our point is that science is very human and scientists are people. Furthermore, scientific laws are descriptions of how nature behaves.

Finally, science is dynamic and never completely final. For example, by the late seventeenth century, enough information had been gathered for Sir Isaac Newton to perceive the relations that we now know as Newton's laws of motion. Although these laws are as valid today as they were in Newton's day, more recent studies of very small particles and of the heavens have shown that Newton's laws of motion are really a part of a larger view of the movement of bodies—Einstein's theories of relativity. The modern ideas in any area of science have developed to their present forms by incorporating successively the ideas developed over centuries. In this way, science is cumulative, too. Strangely enough, the more the scientist learns the more he realizes that he knows rather little indeed about nature's complex processes.

The Practice of Science

One of the practices of science is observation. Who makes observations? Observations are made by people. Experiments are designed and performed by people. The person who considers himself a scientist (and many who do not, such as yourself) believes that the natural world is orderly and that definite patterns of behavior in nature exist. Such a belief, however, has not always been held by man. Early man attributed natural phenomena to moods of anger or benevolence of the "gods." The philosopher, Alfred North Whitehead, in pondering the fact that the scientific revolution occurred within the past 400 years, suggested this breakthrough could be attributed to the Judeo-Christian idea of a rational creator. The monotheistic belief in a rational creator then led to the firm expectation that every

[1]Numerals in square brackets refer to references at the end of each chapter.

occurrence can be related to earlier events in a very definite way through general principles. All scientists today believe in the orderliness of nature. In this respect, the laws of nature are very different from the laws of society as developed through a judicial system. The laws of nature do not vary. Man's laws, however, change as society evolves. Because nature is orderly, behavior is predictable from past observation. Thus, scientists believe in an organized gathering of knowledge, beginning with accurate observation.

The scientist does not dispute accurate observation, but he studies carefully the accuracy of observations. That accurate observations depend on the observer and his equipment can be illustrated as follows. In all the three cases below, a flower garden is being observed. **Case 1:** You are color-blind and are on the roof of a house. What do you see in the garden? Shape, patterns, light and dark. **Case 2:** You have normal vision, but you are still on the roof of the house. You certainly see all that the observer in Case 1 sees, but now you see all the colors and the patterns of the colors in the garden in addition. **Case 3:** You have normal vision, and you are now in the garden. You may not see the larger patterns of the garden quite as well, but you see the differences in shadings of the flowers, the different greens, the bare spots, the bugs, the eaten leaves, and so on. Your closer view more highly differentiates the components of the garden; however, in seeing the many little things, you do not see the overall effect on the pattern of the garden that one gets at a distance! On presentation of their observations, these three people would give different pictures of the same garden—all "correct" and "accurate" in some way. Which is truth? What do you think?

Accurate observation in scientific work also requires reproducibility of the observation under a common set of circumstances. When we turn an ordinary glass of water upside down, the water spills out and falls to the ground. This observation has been reported so many times over—in restaurants by children, at kitchen tables—that we expect this will happen whenever we try it. Most of you are familiar with many other examples of the law of gravity, but most of you probably are not familiar with the larger statement, namely, that any two bodies of matter attract each other in a way that depends on the product of the masses of both bodies and also on the distance between the two bodies.[2] The dependence on distance is such that the mutual attraction of the two bodies decreases rapidly with increasing distance.

The precise statement[2] of the law of gravity is the result of numerous, careful experiments and observations. Some of the observations, for exam-

---

[2]To be precise

$$\text{force of attraction} = \frac{\text{mass}_1 \times \text{mass}_2}{(\text{distance})^2}$$

ple, those on falling objects, are easily made with the naked eye. However, other experiments showing the effect of distance on attraction between bodies require more intricate methods of observations. For example, the orbits of the planets are described well by this law of gravity. Such experiments require precision tools and equipment along with careful design of experiments. Using our precision equipment and making accurate observations will, in most cases, yield reproducible results.

Ultimately, even our most sophisticated tools may be inadequate for some observations. For example, our largest telescope, the 200-in. reflector telescope (Figure 1-4) on Mt. Palomar, is limited to observations out to a few billion-trillion miles. Effectively, this is the limit to the visible universe. Here there is an illustration of the inherent limitations to science because intricate observations often depend on sophisticated equipment. Anything that is outside or beyond the senses (including the extension of our senses by various tools or devices) is outside the bounds of today's science. *Today's* science is limited to those observations made possible with the tools and equipment of today. Today's tools and equipment are made possible by today's technology. Tomorrow's science will depend on whatever better techniques will become available—tomorrow's technology.

Observations of the heavens by the eye alone provided enough information for mariners to develop techniques for navigation. First, the simple telescope and now very powerful telescopes allow man to reach out into space far beyond the dreams and imaginations of a few hundred years ago. The practice of science evolves and becomes more sophisticated as more detailed observations are made. More detailed observations are made possible by more advanced technology. Today's science and technology has allowed us to view the earth from the moon. Color photographs show the earth to be blue and small when compared to the vastness of space. This is another view of our only habitat made possible with today's equipment. Equipment of the future may allow us to see other habitats.

Although in the preceding discussion we have stressed the importance of observations to the practice of science, we have also seen that experiment is necessary in the practice of science. The experimental aspect of the scientific method is an excellent and widely agreed upon technique for achieving a consensus. In general, there are four steps to the scientific method: (1) gathering data, (2) setting a hypothesis (or model) on the basis of the data, (3) taking more data (usually experimentally) to check out the predictions of the hypothesis, and (4) finally formulating a reasonable theory. If the experiments do not support the hypothesis, then the hypothesis must be reevaluated. In reality, only the last two steps belong uniquely to science.

The practice of science involves the subjection of any hypothesis to further tests, frequently experimental tests performed by colleagues and

Figure 1-4. **(A)** 200-in. Hale telescope dome with shutter open. **(B)** 200-in. telescope showing observer in the prime focus and the reflecting surface of the 200-in. mirror.

(A)

(B)

peers. Because a scientist believes in the necessity of peer acceptance and agreement with his ideas, he acts as his own devil's advocate in order to establish the validity of his work beyond any doubt. Again, we should mention that these steps are simply a way of characterizing the method of science and are not hard and fast rules.

Thus, who are these people who practice science? Observations are made by people. The experimental method of science is used by people. There are many personal attributes of scientists. Accurate observation is important. Curiosity is important. Albert Einstein observed that there exists in children a passion for comprehension that gets lost in most people later on. Design of experiments is important. The capacity to develop hypotheses or develop models is important. Absolute honesty is important. All of these attributes are important for the ideal practitioner of science. A scientist combines curiosity with imagination to obtain answers by experimentation. But wait a minute. We said above that observations are made by people. The scientist is really a person with a particular kind of training and experience.

You can understand how scientists may have conflicting opinions. Although scientists rarely dispute accurate observations (the scientist will study carefully the accuracy of the observations), the interpretation of the observations is very much another matter. All scientists do not agree about the building of nuclear power stations, the use of DDT, the development of the supersonic transport (SST), the off-shore drilling for oil, the fluoridation of certain water supplies, and so on. Why? Because given the factual data or observations, different conclusions or hypotheses may be made. Thus, because interpretation involves decisions other than purely scientific, the scientist as a fellow human being develops his own human point of view. He has left the domain of science because science does not extend itself to include "laws" about what is right or what is wrong. Right and wrong are outside the orderliness and invariance of the natural world.

You should note that observation and curiosity alone are not enough to make a scientist. Thus, Aristotle was not a true scientist. Yes, Aristotle made observations. However, he never experimented to check them, and he used only his human reasoning powers to interpret his observations in order to formulate his natural philosophy. Alas, Aristotle's hypotheses fell apart with the investigations of Galileo, Newton, and others who were real practitioners of science.

The Sciences

The vast reservoir of knowledge known as science is divided by its practitioners into several groupings, for example, physics, chemistry, astronomy, biology, and geology. The growth of information within any one

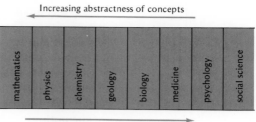

Figure 1-5. A suggested relationship among the sciences. (Auguste Comte).

of these sciences is so great that, today a field such as chemistry must not only be divided into physical chemistry, organic chemistry, and biochemistry but also into groupings such as pharmaceutical chemistry, radiochemistry, and fluorine chemistry. The most recent man considered to have been an authority in all fields of science was Leonardo da Vinci (1452–1519). Most scientists today are considered an expert if they know one portion of a single science thoroughly.

For our descriptive purposes, the branches of science are organized in a sequence as in Figure 1-5. This useful organization for all of science was first suggested in the midnineteenth century by the French mathematician-philosopher, Auguste Comte. Mathematics at the left is the science whose concepts are most abstract and at the same time the simplest. At the far right, the concepts of the social science are not abstract but are very complex. Let us now turn from the sciences to a consideration of technology.

What Is Technology?

Science is not technology. Technology is not science. Many people confuse these two terms and use them incorrectly. As in the case of science, technology is many things. For example, we can define **technology** as *the means we have to control or manipulate nature.* The practical man has always sought ways to achieve more and better food, better supplies of fuel, more suitable clothing, more comfortable housing, and numerous other improvements in his way of life. Furthermore, craftsmen down the centuries have been moved by necessity and inspiration to improve their products. The motivating impulse came from the desires and needs of the societies that the craftsmen sensed. Limitations were set by the quality and quantity of the raw materials available to them. The goals of technology are solutions to these practical problems.

In our society, many technologies are based on science. Thus, in addition to the satisfaction of man's curiosity about nature, science is the basis for various technologies which have practical value. Sir Francis Bacon (1561–

1620) was among the first to believe that science could be put to such practical use. That practical use is what we today call technology. Since Bacon's time, the scientist's knowledge of nature has come to be joined effectively with the trial-and-error developed skills of craftsmen and the development of the "scientific method." The machine-dominated environment of the Western world dates from Bacon's time, even though the extensive practice of science and the development of technology did not occur until after 1800, for example, as in the development of the steam engine and the burning of coal as a fuel.

The essential impact of science on technology derives from the idea that, if we know nature's laws, we can manipulate nature in accordance with these laws to achieve certain valued ends. Herein lies the confusion in the terms of science and technology. Most of us today recognize science primarily through the fantastic array of goods and services utilized by man. These goods and services are, however, directly a product of technology. Indirectly though, scientific endeavor is often responsible. The products (Figure 1-6) available from our science-based technology are clearly visible around us: rubber, plastics, glass, automobiles, fuels, airplanes, drugs, satellites and pesticides. Most contemporary technological achievements are based on a logical array of scientifically measured and checked facts and data—all determined in accord with the laws of nature as defined by science.

We should reemphasize that advancing technology does not free man in any ultimate sense from the limitations imposed by nature. For example, the signals in our computers cannot pass from one point to another with a speed exceeding that of light. This sets a natural limit to the number of signals that can be transmitted per second. As we will later see, pollution in its various forms is the result of energy consumption. The environmental impact of technology in a different sense is one of the limits definable by science. Thus, we have added to Seaborg's observations that modern man cannot ignore technology.

Technology, or the demands of practical problems, has had a great stimulating effect on science. That is, to solve practical problems we must appeal to knowledge of the situation at hand. For example, when energy resources become scarce, we look to scientists for the discovery and development of new sources (Chapter 15). Further, when pollution becomes extensive in an area, we appeal to the scientific community to study the nature of the pollution and to make recommendations as to how technologists might solve the problem. In turn, science has a stimulating effect on technology. For example, scientific study of energy sources and their best methods of utilization serves as the basis on which large scale technology has been developed and will continue to be developed for the use of these

(A)

(B)

Figure 1-6. Montage of modern technologies and products of technology. **(A)** shows the Bayway refinery of the Exxon Company. Crude petroleum is refined to give gasoline, kerosene, lubricating oil, heating oils, greases and asphalts. In addition, materials are obtained which serve as raw materials for the chemical industry. [*Courtesy of the Exxon Company.*] The construction industry annually uses millions of feet of laminated foam board as insulation in buildings. **(B)** shows urethane chemicals, derived ultimately from petroleum or coal, being poured on the bottom skin, or membrane. The top skin is placed over the chemicals and the whole system, when heated, expands to give the laminated board stock. In **(C)** the cut pieces of laminated board stock are being stacked and wrapped for shipping. With a k-factor (heat conductivity—Chapter 7) of 0.14, urethane is the most efficient insulating material now on the market. Interestingly, although the urethane foam is a "non-energy" use of fuel, the end-product—insulation board—is an excellent conserver of energy. [*Courtesy of the Mobay Chemical Company, which makes the chemicals used in manufacturing urethane foam insulation.*] **(D)** shows fibers of polyester which is obtained from petrochemicals. These fibers are used to reinforce the rubber of tires. About one-half of all new cars are equipped with such polyester cord tires. [*Courtesy of Goodyear Tire and Rubber Company.*]

(C)

(D)

Figure 1-7. Electrical power generating station. The big Homer City station in central Pennsylvania is one of several giant mine-mouth power plants which consume coal from nearby mines to generate electricity for use many miles away. Electric utilities consume more than half of the nation's coal output. [*Courtesy of National Coal Association. Photograph: New York State Electric and Gas Corp., 1970.*]

energy sources and the methods of control of pollution. For another example, scientists need better and more precise tools and instruments to make observations (including experiments). Technology provides these. In summary then, we add the cliché, "Necessity is the mother of invention."

The human side to technology (and to science) is the part that man (and woman) plays in the advancement in both areas. Using nature's laws as guidelines, man must make decisions as to what practical problems are to be solved and which of the possible solutions are to be used by society. Consider an electrical power generating station (Figure 1-7). Man decides what land will be used. Man decides what will be burned to provide heat for generating power. Man designs and builds the plant. Man decides what will be done with the waste materials from these plants. In this sense, man is the originator and controller of technology. He is not technology's slave. However, man must abide by the laws of nature and to this extent he may be considered as "enslaved" by these laws (for our purposes these are laws of chemistry and physics). In summary, *the laws of nature limit both man and his technologies, whereas man in turn controls technology.*

The Development of Modern Technology

In our time, the progressive technological improvement in material standards of living (we often refer to this as the quality of life), as in medical techniques and in transportation and communication, has transcended technology's earliest goal of liberating man from drudgery. In fact, the goal of recent technology seems to be the provision of all the conveniences that man's wildest dreams can conjure! Let us look at this concept further.

Modern technology demands energy—large quantities of inexpensive energy. Each day from the moment we rise and until we retire, everything we do requires some form of energy. What is energy? The definition of energy is really dependent on the form that energy takes. For example, water above a fall or a rapidly moving stream has energy because such water is capable of producing work (Figure 1-8), whether it be work used in producing electricity or grinding wheat. Our autos use the chemical energy from gasoline. The rate at which we can obtain work from these various sources of energy is called power. We often use these terms interchangeably in our society. However, at this point, we will not attempt a more formal definition of energy other than to note that energy is the prime necessity for our modern technological processes.

Technology in preclassical and classical antiquity was characterized by the small workshop run by a single craftsman who might have one or two slaves to assist him in producing goods on a modest scale. As viewed from today, the dominant characteristic of ancient technology was the lack of machinery. Those few machines, treadle cranes or water-raising devices, that were developed necessitated harnessing humans or animals. Classical engineers made a breakthrough when they began utilizing moving water to perform tasks. The water wheel driven millstone for grinding grain could do work 6 to 7 times as fast as previous donkey driven devices. That is, the wheel was 6 to 7 times as powerful.

Gradually the use of water wheels spread so that the great power sources of the Middle Ages became the water wheel and the windmill. The *Domesday Book* mentions that by 1086 A.D. there were 5624 water driven mills in 3000 communities in England south of the rivers Trent and the Severn. By this time, the water wheel not only ground grain but was widely used to activate other machinery, such as hammer mills and stamping mills for crushing rocks and ores. In Central Europe during the fourteenth and fifteenth centuries, powerful water driven hammer forges and bellows made possible blast furnace processes that yielded large masses of cast iron for the first time. The increasing tendency to substitute machines for hand tools soon demonstrated the grave limitations of the existing power sources. Wind was cheap but unreliable. Water power was strictly limited by topographical conditions. Thus, inventors turned to steam, which suffers none of these disadvantages.

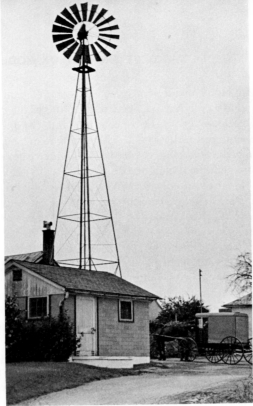

(A)   (B)

The invention of the steam engine must be regarded as the crucial factor for the Industrial Revolution. At the same time the introduction of the steam engine marks the beginning of use of large quantities of energy, particularly energy that could be produced in the location where it was needed and in the desired amounts, to produce goods and services for man. The onset of the Industrial Revolution was by no means as sudden as is often claimed. Although its roots lay in the important technological advancements of the sixteenth century, it did not gain momentum until about 1800. Water power remained the main source of power of eighteenth century England. The first steam engines merely pumped water back into a reservoir for the operation of a water wheel when natural water flow rates were low.

James Watt (1736–1819) developed a steam engine of greater capabilities and thus opened the way for this new "slave" to work for man in myriad ways. For fuel the steam engine depended on coal, which quickly became the essential fuel of the Industrial Revolution. Coal and iron were the symbols of the new machine age, and the steam engine was its "prime mover." Improvements in and enlarging of the steam engine's capacity to do work increased the demand for coal. In the production of iron, coke obtained by heating coal was needed. As a by-product of this heating, combustible gases were produced that found use in gas-illuminating lamps (methane).

Figure 1-8. Montage of energy converters. (Figure 1-8B, The Birkenhead Powder Mills, courtesy The Hagley Museum)

(C)

Gas-fired engines were developed in the middle of the nineteenth century, followed shortly by internal combustion engines using gasoline, which then was a by-product of the refining of petroleum to obtain lubricants and kerosene. Subsequent design of engines as small, mobile power plants has freed industry and man of the traditional restrictions of geography. Thus, the Industrial Age has spread far beyond the limits of its European origin, and no part of the world is now physically beyond its reach.

As late as 1800, man derived most of the energy he needed for heating, cooking, and manufacturing from farm refuse and wood. This situation still prevails in many underdeveloped countries. Against that reference point, the energy consumed by the industrial nations has increased 20- to 200-fold or more. Unfortunately, the result of industrialization is waste (Figure 1-9). The technological processes, including conversion of energy from one form to another and the production and transportation of fuels, waste about half the energy put in and use the other half to supply man with the necessities and amenities of life. Later, we talk about this "waste" as pollution.

Thus, the history of technology is the story of the mechanization of labor by virtue of consumption of large quantities of energy. This is the basis for referring to energy consumption as using "energy slaves." Put another way, the difficulties nature put in the way of Western man have been overcome to a great extent by harnessing large energy resources: water, coal, oil, and natural gas. Man's material needs have largely been met, and many of his dreams have become realities as a result of such consumption of energy. In the United States, we often refer to advanced technology; advanced technology, in this context, means that fewer men can do more in less time because more extensive means are available to manipulate nature more rapidly and more completely. However, our environment and our resources (in particular, energy resources) must "bear the brunt" of this mechanization and industrialization.

Technology, Energy, and the Economy

Energy has largely been viewed as the essential resource to stimulate and support economic growth. The United States in the mid and late twentieth century is the world's greatest economic power, has the highest standard of living, and is the world leader in energy consumption. In the 1970s, the United States used annually about 35% of the world's energy, although it has only 6% of the world's population. For over a century, the annual growth in energy consumption has been about 3%, but the rate of increase in consumption has been rising. Energy use grew from an average annual increment of 1.2% for 50 years beginning in 1880 to an average annual increment of 2% for the next 30 years, then to an annual average of 2.7% for the early part of the 1960–1970 decade, and to a 4.9% annual

Figure 1-9. **(A)** Disposal of waste by burning is a common practice by many towns. Such practices contribute to air pollution. [USDA-SCS photo by S. R. Flourney.] **(B)** Glass bottle waste cannot be disposed of by burning. The cost of new glass is about the same as recycled glass, but where do we put all these bottles if they are not recycled?

(A)

(B)

growth in the last 5 years of that decade. For comparison, in Figure 1-10, the United States' gross national product has been plotted along with the total United States energy consumption. The curves are essentially parallel, which stresses the intimate interaction among technology, energy consumption, and the economy as measured by the gross national product.

Growth in population has had a significant impact on our needs for energy. However, Barry Commoner (Professor of Botany, Washington University, St. Louis, Missouri) also identifies the critical impact of new and highly energy consuming technologies as having an impact on our ecosystem (insofar as resource depletion and pollution) far beyond mere population growth. For example, recently the world's population has been growing at 2% per annum, whereas world energy consumption has been growing at 5% per annum. Commoner pinpoints and accuses modern technologies that have replaced older technologies (which are less energy consuming) such as, the displacement of natural fibers (cotton and wool) by synthetic fibers (nylon, dacron, and so on); of lumber by plastics; of steel by aluminum and/or cement; of railroad freight by truck freight; of animal

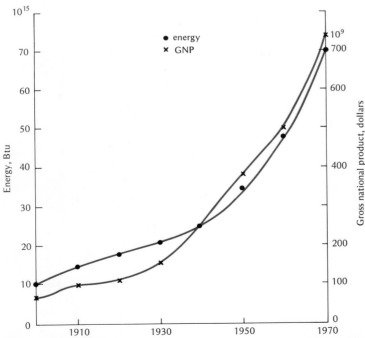

Figure 1-10. Change in gross national product (United States) and energy consumption with time. [From Bureau of the Census, Long Term Economic Growth, 1860–1965.]

wastes for fertilizer by synthetic, chemical fertilizers; of returnable bottles by nonreturnable bottles.

All of these changes in our way of life mean a higher per capita energy consumption. Thus, we need to ask where is our technological society headed. Does this change mean a better quality of life? Should fewer men do more in less time? If so, what should we do with our increasing spare time? Some people have suggested that growth be slowed or that technological progress be stopped. Some have suggested that we place a ceiling on the gross national product. Others have insisted that we stop using the gross national product as a measure of desirable growth. Is this reasonable? Is this necessary?

Quantity versus Quality

A general problem we must all be concerned with is the demand for quantity or quality. Are these the same? Many people feel that technology has improved the quantity of life at the expense of the quality. Instead of the clean country air of a past rural society, there is now the pollution in the cities of the new urban society. Traditions and family life have been disrupted by the telephone and television. There is still starvation in the midst of plenty.

Certainly the most easily measurable aspect of the quantity of life is the length of time that the average person lives. This can be measured (Figure 1-11) by the infant mortality rate (lower), the average life span (longer), and the life expectancy as a function of age. However, measures of the quality of life are very subjective. Quality means something different to most of us. Healthier people living a longer life may be more alert and capable, but they may also be enjoying life less because of technological degradation of the whole social millieu. Discussions of the quality of life have two dominant features: one is a worry about the pleasure of living, the other is about the relative satisfaction deriving from various technological improvements, for example, the convenience of air conditioning versus the consequent danger of thermal pollution, the availability of physical conveniences that relieve drudgery versus the consequent competitive rat race of keeping up with the Joneses, and so on. Such concerns are not new, but are more difficult to quantify because quality involves a subtle value judgment or subjective judgment. What do you think?

The Technology of Electric Power Generation

Finally, the subject of the generation of electric power brings into focus many of the problems facing man with regard to the interactions of science, technology, and society: the growth in demand for electric service versus

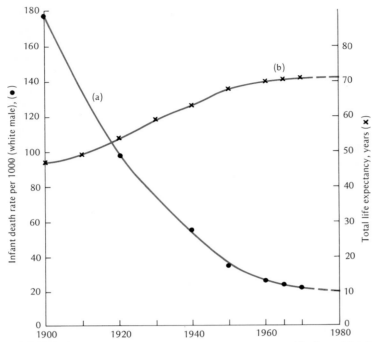

Figure 1-11. **(a)** Change in infant mortality rate with time (dots). Note the sharp decrease in the first half of the twentieth century. **(b)** Life span/mortality rate changes with time (crosses). The longer life span of Americans is due largely to good nutrition and good medical care.

consumption of natural resources, the growth in demand for electrical service versus pollution, and the question of quantities in life versus quality of life. The most recent figures show that electrical power generation takes about 25% of the fuel energy consumed in the United States. Furthermore, it is predicted (Chapter 8) that by the beginning of the twenty-first century, electrical power may consume 50% of the fuel energy while needing 100% of the nation's stream flow or runoff for cooling.

Each year many in both the private and industrial sectors are turning to the cleanliness and convenience of "electrical energy." Thus, electrical power generation is now and will continue to be the single most important technology in our society. As we mentioned previously, the general public—you—are ultimately responsible for the choice of plant siting, the fuel to be used (fossil fuel, nuclear, or some alternate source), and the method of cooling (river, lake, ocean, or cooling tower). You must weigh the benefits of convenient, plentiful electrical power against the costs and hazards involved in its generation. A choice that maximizes benefits and minimizes the hazards can only be made if you understand the basic scientific principles involved—the laws and limitations imposed by nature. The laws pertinent to the generation of electricity are found in the sciences of chemistry and physics. We present these sciences at a level appropriate for you, the liberal arts student.

In the course of the following chapters, we examine the subject of energy: energy as a resource, energy as a slave to society, and energy as a pollutant. We believe that you, who are the users of this technology, should be aware of some of the critical data and information necessary in the planning and maintenance of this technology. We begin by taking a critical look at our small, marbled sphere called "Earth."

Questions

1. What is science?

2. Does accuracy in an observation make that which is observed valuable to science?

3. On what assumption do all experiments rest?

4. How can one tell whether or not a certain field of endeavor is a science. Which of the following are sciences? Explain.
   (a) astronomy
   (b) astrology
   (c) navigation
   (d) collecting butterflies
   (e) art
   (f) mixing of colored paints
   (g) psychology
   (h) religion
   (i) geology

5. What happens to science if we give up the belief in God?

6. (a) Can an advertisement set forth scientific observations? Examples? (b) Can an advertisement merely imply that it is based on scientific observations? Give several examples.

7. A Federal regulation known as the Delaney Amendment (1958) requires the Food and Drug Administration (FDA) to prohibit the addition to foodstuffs of any substance that has been found to produce cancer in any species at any dose. Thus, if a compound induces cancer in infant mice after massive injections, that compound cannot be added to foods for humans even in trace amounts. Is this regulation consistent with careful scientific analysis? Why?

8. What is meant when we say that a person is "scientific minded"? What are some of the characteristics of a scientific attitude? Do you have some of these attributes?

9. Why has science continued to develop throughout the ages instead of repeating itself as has occurred in some other areas of man's activities and achievements?

10. Is a man a true scientist if he studies nature in secret and never puts together any of his data or theories?

11. Suggest a basic difference between a scientific law and a civil law.

12. In a court of law, evidence may be eye witness testimony or it may be circumstantial, such as fingerprints, foot prints, blood tests, drunkometer readings, and so on. Which do you consider the more reliable? Why?

13. What is technology?

14. How does technology differ from science?

15. Is science responsible for technology and its effects? Why? Illustrate your answer from your experience.

16. Does technology affect primarily the quantity or the quality of life? Why?

17. Which of the following is science and which is technology?
    (a) The reduction of visible emissions from power plant smokestacks.
    (b) Research to elucidate the chemical reactions that produce smog.
    (c) The discovery of the dulling of your sense of smell by "Airwick."
    (d) The development of a nuclear powered submarine.
    (e) The study of the heat produced in burning various kinds of wood.
    (f) The correlation of cigarette smoking with lung cancer.
    (g) The killing of fish in winter when heated water coming from a power plant is stopped.

18. It has been suggested that the development of the cotton gin (1794) led to a resurgence of cotton production and, hence, of slavery. Thus, is science and technology to blame for the Civil War?

Reference

[1] See, for example, D. Schroeer: *Physics and Its Fifth Dimension: Society.* Addison-Wesley Publishing Co., Inc., Reading, Mass., 1972.

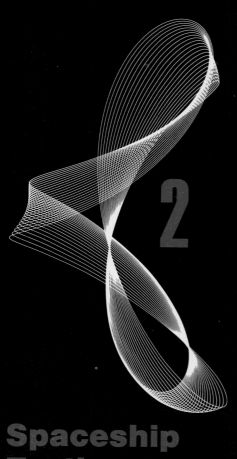

2

Spaceship Earth

We have considered the relationship between our society and technology and have indicated that the level of our technology and the use we make of that technology influences not only the "quantity" but the quality of life. Historically, the earliest interaction between society and technology was made through the "science and technology" of astronomy, which was then pervaded by the mystique of astrology. Astronomy is based on scientific fact, whereas astrology is based on myths and nonscientific interpretations of celestial events, for example, eclipses, the passing of comets, and the positions of stars and planets at the time of one's birth.

After sorting through the many facts, observations, theories and myths collected through the ages, one dominant thought or fact remains; early man needed to develop a technology that he could use to tell time for planting crops and for harvesting them, which aided him to schedule his tasks throughout the year. Man had observed that the stars and planets moved in a systematic manner. He soon used this fact to his advantage for telling time. He also observed the "heavenly bodies" because he was curious both about their nature and how he related to them. Thus, he hoped that his observations and the development of a technology based on these observations would provide the basis for a better way of life in his more "primitive" society.

In this respect, we have not changed very much. However, today our society and its needs are rather more complex, thus necessitating our digging a little further hoping to find answers to perplexing problems. To answer some of these problems, such as our search for supplies of energy, it is appropriate that we, too, begin with a study of our more distant environment, our solar system, the Milky Way galaxy, and the universe, in an attempt to understand how we relate to the whole. In so doing, we (Figure 2-1) hope to find some answers that may improve the quality of our life in "modern" society, much as did early man.

Even though it is not immediately obvious as to how a study of the universe might help us to understand contemporary problems of our immediate environment, historically, the science of astronomy has made man aware of the orderliness and simplicity of nature. The relative simplicity

Figure 2-1. Viewing the heavens. Although 2000 or more years apart in time, both early man and modern man have looked to the heavens, hoping to find answers to the many perplexing questions of life.

Early man

Modern man

of the physical models used to explain the motions of bodies and matter in the universe was probably in Albert Einstein's mind when he said, "The most incomprehensible thing about the whole universe is that it is comprehensible." With the confidence that nature is understandable, man has been more willing to tackle the complex problems closer at hand, such as the interactions of the land, the oceans, and the plant and animal life existing therein.

Out of man's curiosity about the heavens (and also out of competition with other nations) grew various efforts to reach our next nearest body in space, the moon. Modern man became spaceman with the success of Russia's Sputnik in 1957 followed by the United States' Mercury, Gemini, and Apollo missions. It was during one of the Apollo flights that the American astronaut, James Lovell, looked back at earth and exclaimed, "the earth from here is a grand oasis in the great vastness of space." From those words and from pictures showing a tiny marbled sphere developed the popular metaphor—**spaceship earth.** Our tiny spaceship carries its own life support system: atmosphere for supporting respiration, vegetation for sustaining life, and resources that can be transformed into goods and services. We have learned that earth is traveling through space at high speeds and the chance of collision with another planet or star is very, very remote because of our extreme isolation in space; that is, it is 24 trillion (24×10^{12}) miles to our next nearest star, Proxima Centauri.

Such knowledge has led man to ask questions such as: Where are we going? How long will it take us to get there (with our level of technology)? How long will our life support system last? No one really knows the answers to such questions. However, we hope that studies of our past history (back as far as 5 billion years) as to the way in which earth has evolved to its present state will help us in predicting our future course.

In this chapter, we present some basic astronomical observations that develop the concept of earth as a spaceship. We view the members of our solar system and discuss how they relate to the whole—the universe. Included are cosmological theories about the creation of the universe and our solar system. Earth's early history and the way it may have evolved to its present appearance is discussed. We develop the broad concept of earth as an ecosphere (ecological system) and include some speculations pertaining to the long range future of earth as an inhabitable planet.

Basic Astronomical Observations

Early man had little or no appreciation for the concept of earth as a lonely traveller because, to him, high speed might mean a fast camel. On a clear summer's eve, the stars seemed to be just beyond his finger tips. It is interesting to note that the ancients did distinguish between planets and stars. They referred to the planets as "wanderers" because the planets seemed to wander when compared to the fixed, "starry" background. At least 1400 years of recorded history elapsed before man began to realize that earth was not the center of the universe, as had been suggested by Ptolemy in 150 A.D., but that earth was one of nine planets, all revolving about the sun.

The forerunner of our modern view of the solar (sun centered) system was proposed by Nicolaus Copernicus (1473–1543). Copernicus wrote:

> In the midst of all, the sun reposes, unmoving. Who, indeed, in this most beautiful temple would place the light giver in any other part than whence it can illumine all other parts? . . . In this orderly arrangement there appears a wonderful symmetry in the universe and a precise relation between motions and sizes of orbs [orbits] which is impossible to attain any other way.

Later, the accurate observations of planetary motions made by Tycho Brahe (1546–1601) and the mathematical description by Johannes Kepler (1571–1630) did much to support Copernicus' views. But the final blows to the Ptolemaic model were not made until Galileo Galilei (1564–1642) developed the telescope and made observations consistent with Copernicus' model and inconsistent with certain aspects of the Ptolemaic model. Two of the most notable inconsistencies discovered by Galileo were that the planets in general were not perfectly spherical (a holdover from Aristotelian philosophy) and that Venus has a full set of phases similar to the moon's.

Then, Sir Isaac Newton (1642–1727) published his *Principia* (1701), setting forth his laws of motion and law of gravitation. In his works, Newton also made other contributions to astronomy, chemistry, and mathematics. The mathematician, Riemann, said about Newton, "never again will there be a way of the world to be established." Albert Einstein said, "no one must think that Newton's great creation can be overthrown by relativity or any other theory. His clear and wide ideas will forever retain their significance as the foundation on which our modern conceptions of physics have been built." These great men and many others have contributed to that body of knowledge called "science." We draw the following observations that are pertinent to our concept of spaceship earth from that body of knowledge.

Motions of Earth

Planet earth undergoes several motions, two of which are particularly important for our discussion: the **rotational** or daily motion and the **revolutional** or yearly motion. Most of us are familiar with the daily motion because it is responsible for day and night. The yearly motion is responsible, in part, for our seasons. Other movements of earth include the **precessional** motion, or wobbling, of the earth's axis (like a top) and the motion of our entire solar system about the center of our group of stars which we call a galaxy. The wobble is very slow, with a complete revolution of the wobble taking about 26,000 years, whereas a complete revolution of our solar system about the galaxial center requires about 200 million years. Our immediate concern is with those motions that affect the energy received by the earth over a short time period, so we will restrict our attention to the daily and yearly motions of the earth.

One way to understand the rotational and revolutional motions of earth is to imagine ourselves sitting on Polaris, our north polar star, looking down at our solar system as shown in Figure 2-2. We see that the earth is spinning on its axis in the counterclockwise (CCW) direction. Because the earth has a circumference at the equator of about 25,300 mi (miles) and makes one rotation in 24 hr (hours), a person standing at the equator moves in space at a speed[1] of about 1050 mi/hr (miles per hour) or 1690 km/hr (kilometers per hour). A person standing at 40° (degrees) latitude-north (New York City) would be moving at 800 mi/hr or 1280 km/hr. As we approach the poles, the rotational speed approaches zero.

We also observe that the earth revolves in the CCW direction as it orbits the sun, making one complete revolution in about 365 days or 8760 hr

[1]The details of determining speed will be discussed in Chapter 4.

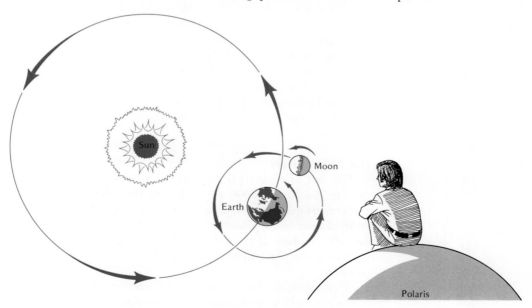

Figure 2-2.   Observation of the motions of planet earth. The motions of earth are best visualized by imagining ourselves sitting on Polaris, the north polar star. The rotation motions—daily, revolutional, and yearly—are all counterclockwise (CCW). The moon also revolves and rotates in the CCW direction with the same period, $27\frac{1}{3}$ days.

EXAMPLE 2-1

(a) Express the distance of 580,000,000 mi in powers of ten.

$$580,000,000 \text{ mi} = 5.8 \times 10^1 \times 10^1 \times 10^1 \times 10^1 \times 10^1 \times 10^1 \times 10^1 \times 10^1$$

$$= 5.8 \times 10^8 \text{ mi}$$

The exponent indicates the number of times 5.8 has been multiplied by ten.

(b) Express the number 0.0000006 m (meters) in powers of ten.

$$0.0000006 = 6 \times \frac{1}{10} \times \frac{1}{10} \times \frac{1}{10} \times \frac{1}{10} \times \frac{1}{10} \times \frac{1}{10} \times \frac{1}{10}$$

$$= 6 \times 10^{-1} \times 10^{-1} \times 10^{-1} \times 10^{-1} \times 10^{-1} \times 10^{-1} \times 10^{-1}$$

$$= 6 \times 10^{-7}$$

(365 days $\times$ 24 hr/day). Because the average earth-sun distance is about 93,000,000 mi (1.50×10^8 km), the earth travels about 580,000,000 mi in one complete revolution (which consequently yields an average orbital speed of 66,000 mi/hr). It is customary when using large numbers to write them as a number between one and ten multiplied by a power of ten. Thus, the distance traveled by the earth in one revolution about the sun would be written as 5.8×10^8 mi. The mechanics of this number notation is shown in Example 2-1. For a more complete development of powers of ten notation, the reader is referred to Appendix A.

The moon, likewise, moves about the earth and spins on its axis in the CCW direction. The moon's periods of revolution and rotation have been found to be the same, $27\frac{1}{3}$ days. All of the planets rotate counterclockwise (CCW), except Venus and Uranus which rotate in the clockwise (CW) direction.

Our Solar System

The sun, the nine planets, the 32 known satellites (moons) of these planets, the (about) 2000 asteroids, and the other bodies such as meteors and comets make up our solar system, as partly shown in Figure 2-3. The orbits of most of the planets lie nearly within the same plane in space which we call the **plane of the ecliptic.** Pluto is the exception. This great plane in the sky, the plane of the ecliptic, is 7.5×10^9 (billion) miles in diameter (across).

Even with the large number of bodies within our solar system, it is still very empty as these bodies are so far apart. Light, traveling at 186,000 mi/sec (miles per second), takes about half a day to move all the way across our solar system. Interestingly, it takes light about 4 years to arrive from our next nearest star, Proxima Centauri. Thus, a planetary system associated

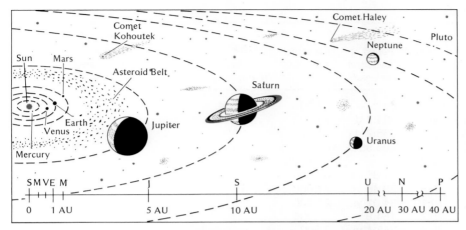

Figure 2-3. The solar system. The nine planets of the solar system, along with the collection of asteroids between Mars and Jupiter, are shown. When the earth-sun distance is defined as 1 AU (astronomical unit), the mean distance of the planets from the sun varies from 0.39 AU for Mercury out to 40 AU for Pluto.

with even the nearer stars, all of which are outside our own solar system, would be difficult to see because planets shine only by reflected light. A star, such as the sun, has its own intense source of light and other forms of energy. Even our own outer planets are hard to observe and were discovered only by searching the heavens for "wanderers" against the fixed starry background.

The distances to the planets along with selected physical characteristics are given in Table 2-1. For convenience, the average earth-sun distance, 93×10^6 mi or preferably 9.3×10^7 miles, is often referred to as an astronomical unit (AU). The average sun-to-planet distance is then given in astronomical units in Table 2–1.

Mercury is the planet closest to the sun with an average planet-sun distance (orbital radius) of 0.39 AU or 3.6×10^7 miles and a revolutional period of 88 days. For the conversion between astronomical units and miles refer to Example 2-2.

EXAMPLE 2-2

If Mercury is 0.39 AU from the sun, what is the distance in miles?

$$0.39 \text{ AU} \times \frac{9.3 \times 10^7 \text{ mi}}{\text{AU}} = 3.63 \times 10^7 \text{ mi}$$

$$= 3.6 \times 10^7 \text{ mi or 36 million miles.}$$

Table 2-1.   The Planets

Planets	Mean Distance from Sun (10⁶ mi)	(A.U.)	Period of Revolution (sidereal)	Equatorial Diameter (mi)	Mass (Earth = 1)	Density (g/cm³; water = 1)	Gravity at Surface Compared to Earth	Number of Moons
Mercury	36.0	0.39	88 d	3,000	0.054	5.4	0.36	0
Venus	67.2	0.72	224.7 d	7,600	0.815	5.1	0.85	0
Earth	92.9	1.00	365.2 d	7,927	1.00	5.52	1.00	1
Mars	141.6	1.52	686.98 d	4,200	0.11	3.97	0.38	2
Asteroids	260	(2.8)	—	1 to 480	—	—	—	—
Jupiter	483.3	5.20	11.86 yr	88,600	318	1.3	2.6	12
Saturn	886.2	9.54	29.46 yr	74,000	95	0.72	1.1	10
Uranus	1783.0	19.19	84.01 yr	29,400	14.5	1.6	0.92	5
Neptune	2794.0	30.7	164.8 yr	28,000	17.3	2.2	1.42	2
Pluto	3671.0	39.4	248.4 yr	3,900	0.18	5 ?	0.4 ?	0
Moon	*	—	27.33 d	2,160	0.01	3.4	0.2	—

*The mean distance of the Moon from the Earth is 238,000 mi.

The next planet is Venus, which is often referred to as our sister planet because its orbital radius is 0.72 AU (6.7×10^7 miles) and it has a revolutional period of 225 days (both rather similar to earth's). Earth is the third planet from the sun with an average orbital radius of 1 AU and a rotational period of about 365 days. As we continue outward, the distances between the sun and the planets become very large, for example, Pluto at 39.4 AU is 3.67×10^9 miles from the sun.

Mars, the "red" planet, is the fourth planet and has an average orbital radius of 1.52 AU and a revolutional period of 687 days. As can be seen from Figure 2-2 and Table 2-1, a very clear break occurs in the orbital spacing between Mars and Jupiter, Jupiter having an orbital radius of 5.20 AU. Hence, with regard to distance, the planets appear to form two distinct groups; we call them the **terrestrial** planets (earthlike), and the **jovian** (jupiterlike) planets, those farther away.

Certain physical properties such as size, mass, and density, make the grouping of the planets even more distinct. We find that the terrestrial planets have average densities similar to the earth, about 5 g/cm³ (grams per cubic centimeter)[2], as compared to 1–2 g/cm³ for the jovian planets.

[2] The density is defined as the mass of a sample of material divided by the volume of the sample ($\rho = m/v$). For example, a sample of earth containing both crust and core constituents in the proper proportions, 1 cm (0.4 in.) on a side would have a volume of 1 cm³ and a mass of about 5 g. Since there are 454 g of mass per pound of weight on earth, our 1-cm³ sample would weigh 0.011 lb (pounds) or 0.18 oz (ounces).

For comparison, water, ice, and some of the densest woods have densities of about 1 g/cm³. Besides water and ice, other lighter substances that make up the earth's crust, such as the silicates (compounds containing silicon and oxygen: for example, quartz, feldspar, granite, agate) have densities of about 2.5 g/cm³, whereas some heavier substances like iron and nickel metal, which are believed to make up the earth's core, have densities of about 8 g/cm³. However, when the total mass of the earth is divided by the volume of the earth, the average density is about 5 g/cm³ as given above.

As to size, the terrestrial planets are similar to earth, with earth the largest. The jovian planets are much larger than the terrestrial planets but similar in size to Jupiter, whose diameter is 11 times that of earth. A comparison of sizes of planets (also with that of the sun) is shown in Figure 2-4.

It is interesting to note that a band of **asteroids** occupies the space between Mars and Jupiter. About 2000 asteroids have been discovered, with

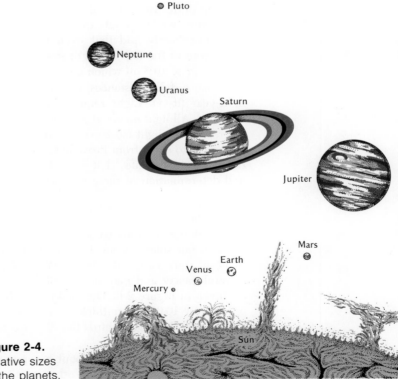

Figure 2-4.
Relative sizes
of the planets.

EXAMPLE 2-3

If Proxima Centauri is 4 lt-yr from the sun, and if a light-year corresponds to a distance of 6×10^{12} mi, what is the distance to Proxima Centauri from the sun expressed in miles and in astronomical units?

$$4 \text{ lt-yr} \times 6 \times 10^{12} \frac{\text{mi}}{\text{lt-yr}} = 24 \times 10^{12} \text{ mi}$$

$$24 \times 10^{12} \text{ mi} \times \frac{1 \text{AU}}{93 \times 10^{6} \text{ mi}} = 2.6 \times 10^{5} \text{ AU}$$

---

the largest being about 500 mi in diameter. However, most have diameters of less than a mile. One speculation is that the matter in space that formed our solar system might have condensed into smaller bodies in that region instead of forming a planet or planet plus satellites as was the case for the other regions.

In viewing our solar system, it has been convenient to measure distance in terms of the astronomical unit, 9.3×10^{7} miles. However, in the following section, we "step out" of our solar system into the emptiness of space and immediately find that even the astronomical unit is too small in discussions of the universe. For example, from earth to Proxima Centauri (star) is 24×10^{12} mi which converts to 2.6×10^{5} AU. Thus, as we begin to measure stellar distances, it is convenient to use a still larger unit, the light-year (lt-yr). Light requires 4 years to travel to us from Proxima Centauri. All light moves at a speed of 186,000 mi/sec (or 3×10^{8} m/sec). Therefore, the light has traveled a distance of 24×10^{12} (24 trillion) miles in coming to earth from Proxima Centauri. This means that light in 1 lt-yr travels 6×10^{12} miles (1 lt-yr = 6×10^{12} mi). The relationship between and determination of these numbers are shown in Example 2-3.

Our Galaxy

Across the heavens we observe a starry band or plane of stars that includes our solar system, Proxima Centauri, and many other stars. This plane of stars, called the **Milky Way galaxy** (Figure 2-5), is classified as a spiral galaxy and may be described as being disc or pinwheel shaped as shown in Figure 2-6. The Milky Way is about 100,000 lt-yr across but only about 10,000 lt-yr thick. Our solar system is located in one of the spiral arms about two thirds the distance from the center to the edge of the disc.

The spiral arms revolve about a common center, and each spiral has a different speed. The sun and other stars in our spiral are revolving about

Figure 2-5. Mosaic of the Milky Way, from Sagitarius to Cassiopeia.
[*Composite photograph, courtesy of Hale Observatories.*]

Figure 2-6. Probable structure of the Milky Way galaxy. Two views along with the sun's location are shown: **(A)** edge, **(B)** top. The arms are revolving about the center.

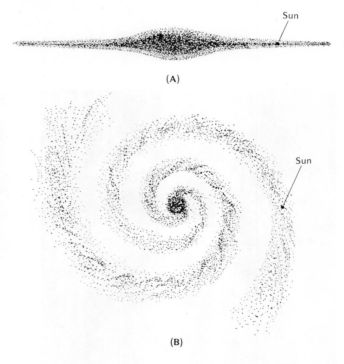

(A)

(B)

the galactic center at a rate of 180 mi/sec or 648,000 mi/hr or about 10 times the earth's revolutional speed around the sun. Even at this speed it will take 200,000,000 years for our solar system to make one complete revolution about the galactic center. The revolutional or swirling motion will take 200,000,000 years for our solar system to make one complete revolution about the galactic center. The revolutional or swirling motion typically observed in galaxies is believed necessary to avoid the collapse of the outer stars and matter toward the center of the galaxy due to **gravitational attraction.** Gravitational forces are those forces of mutual attraction between bodies described by Sir Isaac Newton in his *Principia.* The result of mutual attraction of bodies is collapse or falling toward a common point. The forces are sometimes described as the inward forces that keep us on the surface of the earth and keep the earth revolving about the sun.[3]

Figure 2-7 and Figure 2-8 show actual photographs of edge and top views of spiral galaxies considered to be similar to our own Milky Way. The fuzzy appearance, particularly in the galaxial plane and toward the center, is thought to be due to the presence of large quantities of interstellar dust along with many stars. An average size galaxy (like ours) is thought to be made up of about 100×10^9 (100 billion) stars. It is interesting to note that we have never "seen" the center of our galaxy because of the

[3]Gravitational forces and the concept of weight are discussed more fully in Chapter 4.

Figure 2-7. Spiral galaxy in Coma Berenices seen edge-on. [*Photographed using the 200-in. telescope (see Figures 1-4A and B), courtesy of Hale Observatories.*]

Figure 2-8. Spiral galaxy in Canes Venatici, a top view. [*200-in. telescope photograph, courtesy of Hale Observatories.*]

Figure 2-9. Great Spiral in constellation Andromeda. Also shown are two satellite galaxies. [*48-in. telescope photograph, courtesy of Hale Observatories.*]

large quantities of interstellar matter. The shape of our galaxy has been inferred by observations of other spirals. The closest spiral—the Great Spiral in constellation Andromeda—is 2×10^6 lt-yr from us (Figure 2-9). In general, galaxies are widely varying as to their size and shape, but the regular shaped galaxies are divided into two broad classes, **spirals** and **ellipticals,** with spirals accounting for about 77% of the total.

The Universe

As we view the heavens, we observe a very large number of stars, galaxies, and clouds of interstellar dust and matter. Some regions of the sky reveal few stars whereas other regions are clustered with galaxies, each perhaps containing 100 billion stars (Figure 2-10). Estimates based on how far we can see with our best telescopes have set the number of galaxies at 100 billion. Thus, there are at least 10^{22} stars in the heavens. Furthermore, many stars are not sources of visible energy (light) but are only sources of radio waves and x rays. Stars close to us appear to be moving away from us at speeds of 100 mi/sec (3.6×10^5 mi/hr), whereas those farther away are traveling at speeds nearly the speed of light, 186,000 mi/sec.

What are the limits to the universe? Although various estimates of the limits have been made, the best current estimate sets the outer limits of the universe as 10 billion lt-yr or 6×10^{22} mi from earth. We must again

Figure 2-10. Cluster of galaxies in Hercules. [*200-in. telescope photograph, courtesy of Hale Observatories.*]

emphasize that this is only as far as we can see. Future technology—better telescopes or other means of observation—may permit us to see farther into space, thus allowing us to update our knowledge of the universe.

Hence, from our basic astronomical observations, we conclude that earth is a spaceship, a tiny traveler carrying its life support system through the vastness of space. Where is it going? Perhaps it is going somewhere or perhaps nowhere. Pictures sent back by satellites and the Apollo missions showed to everyone the tiny blue marbled sphere against a background of emptiness (Figure 2-11). These pictures of earth were unsettling to many. However, such pictures spurred a new interest and concern, first about the isolation and finiteness of our tiny planet and second about ways to preserve both life and a quality of life on planet earth. Before considering ideas about the origin of the earth, let us consider briefly a hypothesis about the origin of our solar system.

The Origin of Our Solar System (Protoplanet Hypothesis)

A basic scientific and philosophical question that perplexed early man was how did it all begin? Modern man is no less perplexed even after several thousand years of observations, some of which we have discussed in the previous section. However, using the observations and records of the

Figure 2-11. Earth. Earth appears as a blue marbled sphere when viewed from space. [*Courtesy NASA—Apollo 17*]

centuries past, we are able to make a more intelligent guess (hypothesis) about the beginning. Some pertinent ideas about the creation of our solar system are as follows.

Earlier, we mentioned that space is not empty but contains clouds of gas (probably mostly hydrogen) and dust of varying densities. These clouds are called **nebulae.** Some nebulae are hot, heated by nearby stars, and emit their own light. Some shine by reflected light. Some are cool and dark (see Figures 2-12 and 2-13).

Due to the mutual attraction of these particles (law of gravity), the nebula or regions of the nebula collapse. When the gravitational energy has been transformed into sufficient heat energy, the nebula's "nuclear power plant" is triggered, and a star (such as our sun) is born. The large quantities of hydrogen present in the average star are used for fuel, releasing large amounts of energy in a way similar to the large energy releases from the H-bomb (hydrogen bomb).

Our solar nebula, in the beginning, was probably cool and dark and occupied space about the size of our present solar system, that is, about 80 AU across. The particles were swirling about a common center, but the speed was not sufficient to avoid gravitational collapse. Like a spinning skater who brings in his limbs in order to spin faster and faster (Figure

Figure 2-12. The Lagoon Nebula in Sagittarius. This emission nebula was photographed in red light with the 200-in. telescope. [*Courtesy Hale Observatories.*]

Figure 2-13. Horse head Nebula. The cool dark nebula obscures the starlight coming from behind. [*200-in. telescope photograph, courtesy of Hale Observatories.*]

Figure 2-14. Spinning skater. The skater's speed of rotation is faster when arms are alongside the body than when arms are outstretched.

2-14), the cloud swirled faster and faster as it collapsed (Figure 2-15A). The spinning or swirling motion caused the cloud to bulge and flatten until it became disc shaped (Figure 2-15B), the plane of the disc being our present ecliptic plane.

Evidence for the bulging effected on matter due to spinning is observed for the planets themselves, and particularly for the low density, jovian planets. For example, both Jupiter and Saturn are spinning rapidly, which causes the observed flattening at the poles and bulging at the equators.

Eventually, further contraction of the outer regions in the plane of the disc ceased. Only the center region, the red-cool sun, continued its collapse and gradually heated up. About 95% of the total matter in our nebula collapsed into the center to form the sun. In this final process the sun's interior temperature (about 20×10^{6} °C (degrees Celsius or centigrade)) and pressure (about 10^9 times the pressure of the earth's atmosphere at sea level) became high enough to fuse hydrogen into helium with the release of large quantities of energy—the sun began to shine—a star was born!

The planets are thought to be the result of smaller regions of material condensing in the gaseous disc at about their present positions. As more material was attracted to the condensation regions, these began to swirl about their own common centers, becoming stable against forces tending to tear these regions apart. Early earth was probably a thousand times larger in diameter than it presently is. As the sun's nuclear power plant began to operate, it spewed forth high speed ionized particles and electromagnetic energy in sort of a "solar wind," which swept away much of the remaining dust and debris surrounding the new planets, or **protoplanets** (Figure 2-15C). This whole description is referred to as the **protoplanet hypothesis.**

The heat from the young sun boiled away much of the lighter, gaseous content of the closer planets, the terrestrial planets, to leave mainly the heavier elements and materials. This loss caused the planets to shrink to about their present size (Figure 2-15D). The present thought about the

Figure 2-15. The formation of the solar system. **(A)** Collection of gas and dust undergoing gravitational collapse. **(B)** The swirling motion during collapse causes the matter to become disc shaped. **(C)** Smaller regions of condensation occur at various distances from the young sun. These were the protoplanets. **(D)** The "solar wind" cleared most of the debris and boiled away much of the gaseous content of the planets. About 95% of the mass of the solar system collapsed to form the sun.

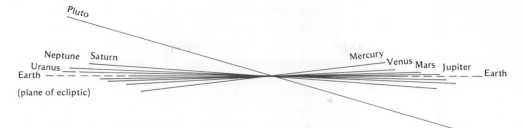

Figure 2-16. The plane of the ecliptic. The orbits of the planets are viewed edgewise; with the exception of Pluto, they lie nearly in the same plane.

planets' moons or satellites is that they were even smaller regions of condensation that were captured by the gravitational attraction of the larger, more massive protoplanets. Because the jovian planets are so much farther out, the young sun would have boiled away less of their gaseous content, which would account for their being much less dense than the terrestrial planets. Similarly, the solar wind would have been much less intense at Saturn's distance (9.54 AU), thus leaving the gas and dust that are thought to make up Saturn's rings.

Another related observation is that the planets' orbital planes are within a few degrees (angular) of the plane of the ecliptic (Figure 2-16). This is consistent with the theory that the nebula collapsed into a disc due to the swirling motion. Hence, most of the observations about our solar system seem to agree well with the protoplanet hypothesis. However, the one notable exception is Pluto, whose orbital plane lies well outside the disc. It has been suggested that Pluto was a wandering rock that became trapped by the sun's gravitational attraction.

The Origin of the Universe (Big Bang Theory)

We have looked at a theory of the origin of our solar system which starts with the idea that matter is widely spread out in space and condenses by gravitational attraction to form a star with a planetary system. Let us now consider a theory of the universe that begins with all matter concentrated in a region about the size of our solar system. This matter would have been ultradense and, consequently, at very high temperatures and pressures. Any suggestion as to how matter appeared in the universe in the beginning and to how it got into such a localized region at such high density is very much left to the reader's imagination. However, it is of interest to note that many theories have been proposed to explain the evolution (or lack of) into the ultradense ball, but none, based on scientific fact, have been proposed to explain the origin of the matter.

A Question for the Reader: Can you suggest reasons why it is not practical or profitable of science to explain why matter exists in the heavens at all?

Thus, according to the "Big Bang" theory it is believed that time (so far as we know it) began about 10 billion years ago when this ultradense mass blew apart in a primeval fireball.[4] The theory proposes that the fireball is still expanding with the outer layers expanding faster than the inner layers. Most of the energy of the original fireball would have been in the form of x rays and gamma (γ) rays (very high energy electromagnetic radiation) because of the extreme heat (at least $10^{6\circ}$C). As the fireball expanded, the high energy radiation flowed out into empty space, thus decreasing in intensity and in energy. Our nebula, which later (4.5×10^9 years later) became our solar system, was a part of that huge fireball.

One of the most significant observations favoring the Big Bang comes from the Hubble red shift. The red shift is, in principle, the shift of the colors making up a star's visible energy emissions due to its motion relative to us. If a source (a star) is moving away from an observer (on earth), the star's red light shifts toward deeper red, yellow toward orange, and blue toward green.[5] A similar effect occurs with the pitch of a train whistle as it approaches or recedes from a standing observer. The pitch increases (becomes higher) for an approaching train and decreases (becomes lower) for a receding train. Red shifts for stars and galaxies are observed and are thus consistent with the principal ideas of the theory.

Relative to our system, which is somewhere in the fireball, the stars are moving away and their speed is faster the farther away the star is. We think that stars moving close to the speed of light (radio stars called quasars) have been observed. With this speed, they would be located at about 10 billion lt-yr from us, which means that the light we are observing today left those stars 10 billion years ago. If this were true, we would in fact be looking back in time to the beginning of the universe. This is one of the most speculative and controversial ideas in modern astronomical science and cosmology.

Further evidence for the Big Bang theory was discovered by A. Penzias and R. Wilson (1965) of Bell Telephone Laboratories, Holmdel, New Jersey, when they observed radio waves coming uniformly from all directions in space. According to the late G. Gamow and R. Dicke (Princeton University) the intense, high energy radiation from the primeval fireball

---

[4] We call this explosion a fireball for the lack of something better to call it. However, after having observed a hydrogen bomb exploding, I (ROS) can think of no better description for it than a huge primeval fireball.

[5] Actually, superimposed in the continuous energy spectrum of a star is a series of dark lines, referred to as the absorption spectrum of a star. Normally the line we refer to in the red shift is in the bright red region of the spectrum. This is also referred to as the Doppler shift. We will briefly discuss absorption spectra in Chapter 3.

would be coming to earth from far out in space (early in time) and would be coming uniformly from all directions. That is, some of that early radiation is still being absorbed by the matter far out in space and reemitted toward earth (and in all other directions). However, because most of the radiation would have already escaped and the matter or bodies are very far apart, the radiation we receive would be very, very weak.

Furthermore, because the bodies (farther out in space) are moving close to the speed of light, the radiation would be greatly red shifted. Thus, that original intense, high energy should be observable as weak radio waves or look like radiation from a cold body near the absolute zero of temperature.[6] Penzias and Wilson observed that space did appear as a cold body near absolute zero radiating radio waves. These observations are consistent with the Big Bang theory. Thus, although the evidence for the Big Bang theory is somewhat meager, it seems sufficient for this theory to favor it over other theories.

We must emphasize that the material presented in this and the previous discussion is brief and sketchy but presents key details of two hypotheses proposed to explain the observations that relate to the origin of the world about us. Since man has been on earth for such a brief time, we assume that man's ideas will develop further as the body of scientific knowledge increases and our technology advances.

The Earth and Its Evolution

The formation of our solar system and, in particular, planet earth, is thought to have happened about 4.5 billion years ago. After much of the gaseous materials (lighter elements such as hydrogen and helium) on earth were boiled away by the sun's heat, mainly heavier materials and elements, for example, silicates and metals, remained. There would also have been trace quantities of heavier radioactive elements, such as uranium.[7] Because the radioactive elements were newly formed and in their early stages of decay, they probably supplied a major source of heat as they decayed (Chapter 10). Also, the space surrounding earth must have been heavily laden with meteorites and other debris that were attracted toward the earth due to the earth's gravitational attraction. The kinetic energy (energy of motion) of the debris was given up as heat energy as it fell to the earth. These two sources of heat are thought to have been the major contributors to the melting of the earth. Eventually, when most of the debris had fallen

[6]The absolute zero of the metric temperature scale is $-273°C$. For reference, room temperature would be $22°C$ or about $72°F$.

[7]In the process of radioactive decay, elements emit particles or electromagnetic waves in order to obtain a more stable structure. The radioactive emissions carry energy, much of which will be dissipated in the form of heat during collision with other matter.

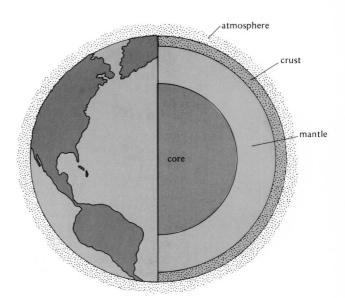

Figure 2-17. Structure of the earth. The earth is divided into three major layers: the core, the mantle, and the crust. Most of the atmosphere lies in an envelope about 10 mi thick.

to earth, the earth's atmosphere was left free from dust and debris. The earth was then able to radiate more heat into space, and the earth gradually cooled. When the outer layer was sufficiently cool, the **crust** formed. We believe that the earth at this stage of evolution was the beginning of the earth as we know it today.

The basic structure of the earth is as shown in Figure 2-17. The thickness of the crust varies from about 3 mi under the ocean basins to about 25 mi under mountain ranges. Oxygen and silicon compose 75% (by weight) of the earth's crust with smaller quantities of elements like aluminum, iron, calcium, sodium, potassium, and magnesium (see Table 2-2).

Table 2-2. Main Elements in the Crust

Element	Weight (%)
oxygen	46.4
silicon	28.2
aluminum	8.2
iron	5.6
calcium	4.2
sodium	2.4
magnesium	2.3
potassium	2.1
Total	99.4

More specifically, the crust is made up of rocks, which are composed mainly of silicates. Familiar examples of silicates are quartz, mica, feldspar, and common sand. All other elements exist only in trace quantities, making up less than 1% of the crust. The **mantle** is divided into two parts: upper, 600 mi thick, and lower, 1200 mi thick. It, too, is composed of silicates but with different metallic composition than the crust. As the **core** is approached, the amount of iron and nickel in the silicates increases substantially. Large quantities of free iron and nickel probably comprise the core (2200 mi thick). The core is thought to be molten while the mantle seems to have properties of a thick or viscous fluid. This means that the cool and solid crust is literally floating on semifluid rock. Such information on the earth's structure is obtained from earthquake data.

One would think that all of the heavier silicates (more metal content) and elements (such as uranium), would have sunk to the core while the earth was molten, thus leaving only the lighter silicates at the surface. However, we find some heavier materials in the crust as we know it today. Hence, the molten earth must have been a turbulent, dynamic "cauldron" with much mixing occuring. The mixing, however, was not uniform because the composition of the earth's crust is anything but homogeneous.[8]

One might think that the "upwelling" of materials from within the earth due to volcanic action might be responsible for the appearance of at least some of the heavier elements and materials. However, present thought on this subject is that the material in the flow of "lava" originates in a layer just beneath the crust, that is, in the upper layer of the mantle. This layer has only a slightly different composition of silicates from the crust.

The atmosphere of young earth also was not as we know it today. We might suppose that the solar wind would have blown it away as it did much of the dust and gas in interplanetary space. However, the earth is sufficiently massive so that its gravitational attraction probably held most atmospheric materials in its vicinity. Even though our atmosphere extends out to around 400 mi, most atmospheric material is in a layer about 10 mi thick (Figure 2-18), which is called the **troposphere.**

We think that the original atmosphere was composed chiefly of carbon dioxide, methane, ammonia, water vapor, and perhaps small quantities of hydrogen and helium. Free oxygen in the atmosphere did not appear until about 2 billion years ago, when the first oxygen generating, photosynthetic cells appeared. This is shown on a geologic time scale in Figure 2-19. After sufficient free oxygen had been generated (about 800 million years ago) diatomic oxygen (O_2) and ozone (O_3) formed. These gases in the atmosphere filtered out the ultraviolet rays from the sun which were harmful to cells

---

[8] It should be emphasized that any discussion of earth as it may have existed even a hundred thousand years ago is merely "conjecture" and should not be taken as known fact.

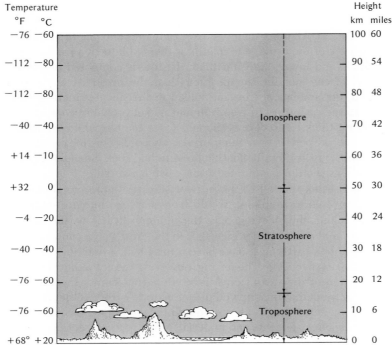

Temperature

°F	°C		

Height

km miles

Figure 2-18. Layers of the atmosphere.

and organisms on earth. However, the oxygen did not approach its present level of about 21% content until about 200 million years ago. The increase in oxygen content was accompanied by the evolution of higher order plants and animals: plankton (free floating water organisms), land plants, insects, reptiles, and mammals, in that order. Grasses and a diversity of grazing mammals were abundant around 20 million years ago. Man's ancestors may have appeared as early as 11 million years ago. The various stages of the earth's evolution are summarized in Figure 2-19.

Formation of the earth's crust was no doubt accompanied by cracking, wrinkling, and shifting of surface layers. The earth's surface features are still changing and will forever be changing (albeit slowly) according to a new modern theory called **plate tectonics.** The plate tectonic theory is based on the division of the earth's crust into plates floating on the semifluid mantle. Currents are believed to exist in the earth's fluid interior that cause these plates to drift. When one plate breaks or pulls away from another, a ridge-rift-valley structure forms, such as is characteristic of the ocean floor. When two plates collide, a mountain range (like the Andes Mountains) is formed, as shown in Figure 2-20. The sliding of plates relative to one

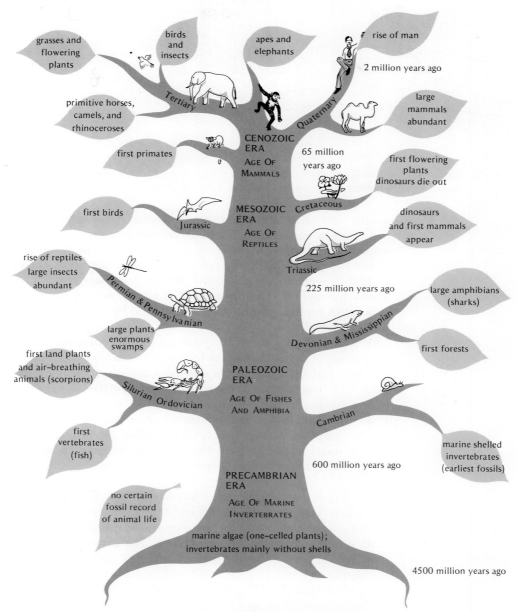

Figure 2-19. Geologic time scale (millions of years before the present). The earth was formed about 4.5 billion years ago.

49

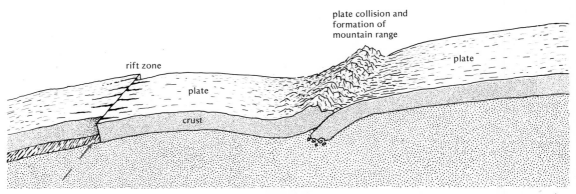

Figure 2-20. Collision of plates (plate tectonics). Plates breaking away form the valley-rift system observed on the ocean floor, whereas the collision of plates form mountain ranges.

another can cause large faults such as the San Andreas fault in California and result in earthquakes (**diastrophism**) which occur frequently in California, Central America, and along the western coastline of South America. When two plates collide and one slides up over another, a great amount of heat is generated, which melts rocks and perhaps results in volcanic action (**volcanism**). Thus, many of the earth's surface features and the concept of continental drift can be explained by the theory of plate tectonics.

Volcanism began about 10^9 years ago. From our observations of the sculpturing of the earth's surface by volcanism and diastrophism, we would conclude that the earth has an enormous amount of energy in it (Figure 2-21). Unfortunately, we do not clearly understand the source of this energy or how this energy manifests itself in the processes that we have discussed. If we were able to understand, man might be able to use this energy for his own purposes. Although energy from within the earth tends to develop surface features, energy from the sun tends to erase them (with some exceptions) through erosion (Figure 2-22) by winds and falling or moving water.

Finally, in considering man's interaction with his environment, it is convenient to discuss the earth's constituents in very broad terms. The earth's solid matter (crust plus upper mantle) is called the **lithosphere.** The earth's waters are called the **hydrosphere.** The **atmosphere** designates the thin layer of gases above the surface. A recent addition to these terms is the **biosphere.** We include here the **biotic** (biological) community, including man, which is confined largely in or near the earth's surface, that is, a layer 10 ft down and 50 ft up.

The study of the interactions of man, animal, and plant with the environment is called **ecology.** The four spheres are collectively referred to

Figure 2-21. A volcano erupting—volcanism. According to the new theory of plate tectonics, volcanic action results from the heat (and melting of plate materials) produced by friction when plates slide over one another. [*Courtesy of U. S. Geological Survey, photograph by K. Segerstrom.*]

as the **ecosphere,** (Figure 2-23). In addition to those changes wrought by nature, man has been contributing to changes in landscape (Figure 2-24), particularly in the past 50 years in a manner not to be considered insignificant! This finally brings the small portion of the earth's crust and atmosphere that we inhabit to the appearance that it has today.

Figure 2-22. Erosion. Winds and water tend to destroy the surface features of the earth. The vegetation in this region (Ducktown, Tennessee) was destroyed by fumes (SO_2) from a nearby copper smelter, see Chapter 14. [*Courtesy of U. S. Geological Survey, photograph by A. Keith.*]

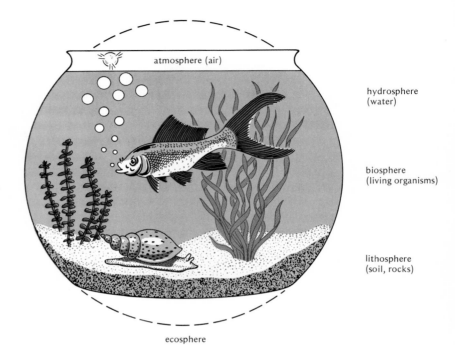

Figure 2-23. The ecosphere. The four interacting spheres making up the ecosphere are the atmosphere, hydrosphere, lithosphere, and biosphere.

atmosphere (air)

hydrosphere (water)

biosphere (living organisms)

lithosphere (soil, rocks)

ecosphere

Figure 2-24. Man changing the earth's surface features. Berkeley Pit, open-pit copper mine, Butte, Montana. [*Courtesy of The Anaconda Company.*]

The Future of Earth   We have set forth some ideas for the creation of earth, and we have traced an early history and an evolutionary sequence for our planet. What can we say about its future? Unlike the ancient Assyrians who based their predictions about the future on myth and fantasy, we, as scientists and members of a technological society, want to base our predictions on verifiable scientific fact. Unless we can observe other planets and planetary systems similar to our own, we can say little directly about our own. As most of us realize, a child learns by observing others. Unfortunately, not even through our largest telescopes are we able to observe the reflected light from planets that are trillions of miles or even farther away. However, since we can observe billions of stars in various stages of evolution, we can say something about the future of our star, the sun. After our discussion of stellar evolution, we will follow with a hypothetical problem that proposes a mass evacuation of planet earth as a possible but unsatisfactory solution to the problems of overpopulation, resource depletion, and pollution on earth.

In considering the future of the sun, let us return to a theory of formation of our solar system. The theory proposed that the sun and the planets were formed from the same nebula simultaneously. The most reliable dating methods suggest that this happened 4.5 billion years ago.[9] Hence, our sun also is about 4.5 billion years old. The present thought is that stars similar to our sun have a life cycle of 10 billion years which makes our sun a middle-aged star. Further information about the evolutionary sequence of a "typical" star must be obtained from our observations of other similar star types.

In viewing stars, we are reminded that light from the closest stars takes only a few years to reach us and the information is representative of immediate history. As we look farther out to stars that are millions of light-years away, we observe the star as it was millions of years ago. Finally, as we look to our limits of the universe, about 10 billion lt-yr away, we then have information about the ancient history of stars and galaxies, perhaps to the extent of "seeing the beginning of time!" Thus, as we view stars out in space, we observe them in the various stages of their life sequences.[10]

One of the most helpful summaries of various star types in their various stages of evolution is shown in Figure 2-25, the Hertzsprung-Russell diagram. The diagram relates the luminosity (brightness) of stars to the color of the star. A star's color is indicative of the temperature of its outer layer.

---

[9] The most reliable dating method is the observation of radioactive elements in the earth's crust, for example, the uranium-lead decay series. This will be discussed in Chapter 10.

[10] The reader should note that stars in various stages of their evolution may be found roughly at the same distance from earth. However, the light from those 10 billion lt-yr away would appear as they were 10 billion years ago, the "beginning" of the universe.

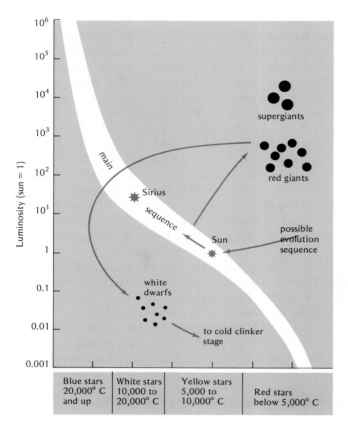

Figure 2-25. The Hertzsprung-Russell diagram. This diagram shows the relationship between color-temperature and luminosity of stars. The red giants are extremely bright, but their light is red, not white, because their surface temperature is relatively low. The white dwarfs are small white stars, some with diameters not much greater than that of the earth. The sun is moderate in mass and in luminosity.

Most stars fall in the main sequence, as does our sun. Also shown is a proposed life sequence for our sun. In the sun's early history, it was a cool red star. As gravitational contraction continued, the sun heated up to become the smaller, hotter yellow star that it is now.

As the sun converts its hydrogen fuel to helium, it may become hotter. In about 5 billion years when most of the hydrogen has been converted to helium, the main energy supply will again come from gravitational contraction of the core. The outer layers of the sun will expand and the sun will become a red giant, perhaps engulfing earth. The sun will then convert helium to carbon in its nuclear power plant and contract again to the main sequence as a much hotter star. Eventually, it is thought that the older sun will shrink to a size about that of the earth—the white dwarf stage—and remain in that stage for a long period of time before becoming a "cold clinker" in space. However, the later stages are hardly of practical interest to us since life on earth will have ceased long before that point in time.

Although our digression on the life sequence of the sun is interesting, it is only of academic interest because the time scales are so long as compared with a man's life span or even with the span equivalent to man's recorded history. Earth's future, astronomically speaking, is reasonably certain for a very long period of time. Thus, we will concern ourselves with problems of immediate consequence, such as those that will affect us in the next few decades. We will use scientific fact and available data to try and see into our immediate future. Most of the following chapters in our text are dedicated to this purpose. Our predictions will be for a finite planet with finite resources populated with energy-hungry inhabitants, you and me.

To conclude our discussion on the future of earth, let us propose a hypothetical problem. Historically, as man has become more mobile in our society, there has been a tendency for him to pack his bags and move away from those places where he has fouled the environment or when he has depleted readily available resources. Thus, let us assume that man has fouled the earth's environment (or has depleted earth's resources) and feels the necessity to move away from earth to another planet. Referring back to our basic astronomical observations, we find that earth and its sun are very isolated from other stars that might have planetary systems similar to our own. It also is clear that even if we were able to travel at the speed of light it would take four years to get to the next nearest star or to one of its planets (unknown to us) that might be inhabitable by man. But, Einstein's theory of special relativity precludes our traveling at or near the speed of light, not to mention our lack of technology. Hence, a travel time estimate might be more like hundreds of years to the nearest stars and hundreds of thousands of years to stars farther away.

Putting aside the problem of travel time let us attempt the evacuation based on assumptions that we believe are reasonable. Let us assume that a typical passenger rocket ship would carry 100 people (noting that such a ship has not yet been built) at a cost of 100×10^6 (100 million) dollars, and that such a ship leaves each week. How much would it cost and how long would it take to evacuate the 3.5×10^9 (3.5 billion) people of earth (as of 1968)? We find that we need 35×10^6 rocket ships to do the evacuation. At a cost of $\$100 \times 10^6$ per ship, the total cost of the program would be $\$3500 \times 10^{12}$ (3500 trillion), which is just 3500 times the gross national product of the United States in 1971. If one ship leaves each week, we easily see that a little less than 10^6 (million) years would be required for the evacuation; see Example 2-4.

More frequent rocket ship departures, such as daily departures, would

---

EXAMPLE 2-4

(a) How much would it cost to evacuate 3.5×10^9 people from earth assuming a rocket ship could carry 100 people at a cost of $\$100 \times 10^6$?

$$3.5 \times 10^9 \text{ people} \times \frac{1 \text{ ship}}{100 \text{ people}} = 3.5 \times 10^7 \text{ ships}$$

$$3.5 \times 10^7 \text{ ships} \times \frac{\$100 \times 10^6}{\text{ship}} = \$3500 \times 10^{12}$$

Our gross national product in 1971 was about $1000 billion or 10^{12} dollars.

(b) How long would it take to perform an evacuation of earth if one rocket ship leaves every week?

$$3.5 \times 10^7 \text{ ships} \times \frac{1 \text{ week}}{\text{ship}} \times \frac{1 \text{ year}}{52 \text{ weeks}} = 6.7 \times 10^5 \text{ years or (670,000 years)}$$

---

require about 10^5 years for the evacuation of 3.5×10^9 people, which still will not do the job (Figure 2-26). Furthermore, we have assumed a static population for earth, that is zero population growth. However, the population growth for the world is not zero and is not likely to be in the near or somewhat distant future. This introduces an interesting and more realistic extension to our hypothetical problem. If the population is growing at the rate of 2% per year, what effect would this have on our hypothetical evacuation?

Using 3.5×10^9 people, we get an annual increase in population of 7×10^7 people. In our hypothetical problem, we were removing 100 people per week or only 5200 people per year, considerably less than the 7×10^7 people, the world's population increases in a year. Hence, we would have to evacuate people about 14 thousand (1.4×10^4) times faster than proposed in Example 2-4 just to keep the earth's population constant at 3.5×10^9 people. Actually, a rocket ship must leave earth every 0.72 sec with 100 people aboard for the earth's population to be held constant. This still leaves the original number of 3.5×10^9 persons stranded.

From this example we may conclude that the prospect of many people leaving our spaceship earth is really is most unlikely. Therefore, any predictions that we make for the future of our technological society will have to be brought "down to earth."

Conclusion

After having made a few astronomical observations, a few astronomical speculations, and a brief survey of earth's history, we now ask the question about what this had to do with our planet and its environmental problems?

Figure 2-26. Evacuation of earth.

Among the significant points in this discussion is the fact that a great amount of both scientific and technological advancement preceded man's present awareness of the true isolation and finiteness of the tiny traveler in space called Earth.

When James Lovell in 1968 looked back and described earth as a grand oasis in space, many people, probably for the first time, realized the smallness of earth and began to worry about the finiteness of its life support system—the ecosphere. Specifically, many began to ask questions about the extent of earth's mineral resources, about the extent of our energy resources, and about the importance of overpopulation. With this in mind, Rep. Morris K. Udall (*Arizona Magazine,* 1969) said,

> . . . Scientists call a spaceship a closed system, meaning that everything needed for a trip must be carried on board and nothing can be thrown away. On long voyages everything—even human wastes—must be recycled and reused. The earth is a closed system, too. It is our spaceship, and it has everything on board

that we will ever have—all the air, water, metal, soil, and fuel. But, it takes on more passengers all the time. At some point, this must stop.

Somehow, I can't help but believe that if all mankind could see the earth as Captain Lovell saw it from the far side of the moon last Christmas Eve, we would change our attitudes and our policies. We would again realize that here, on our grand oasis in the great vastness of space, it is man's relation to his environment that will determine our survival and our happiness.

Suddenly, in the late twentieth century, after centuries of scientific and technological achievement, man has suddenly struck a discord in the "music of the spheres." Will he now be able to establish a harmonious relationship with his environment? The answer probably lies in the complex considerations of energy and energy flow in the ecosphere. Where does it come from? How much is there? How do we use it? Once used, where does it go? Thus, we begin our discussion of energy in Chapter 3 with our major source—the sun.

Questions

1. Why did we consider astronomy during its early history a technology?

2. What significant change in philosophy marked a new era of curiosity and learning?

3. Describe the rotational and revolutional motions of earth.

4. Why does the sun rise in the east and set in the west? (*Hint:* CCW motion is defined as movement from west to east.)

5. Why is it difficult to observe planets, particularly those outside our solar system?

6. What is an astronomical unit (AU)?

7. (a) List the nine planets according to distance from the sun. (b) List their physical characteristics.

8. What features separate the planets into two distinct groups, terrestrial and jovian?

9. Describe the unit of light-years in terms of astronomical units and miles.

10. What do we believe is the significance of the swirling motion typical of spiral galaxies?

11. In 22.4 liters (about 20 quarts) of hydrogen gas, there are 6×10^{23} atoms. Compare this with our estimate of the number of stars in our "visible" universe. Does this stir your imagination?

12. Set forth the major features of the protoplanet hypothesis.

13. Compare earth in its earliest history with earth as we know it today.

14. Discuss the observations that favor the protoplanet hypothesis.

15. Set forth the essential feature of the Big Bang theory.

16. Discuss the observations that favor the Big Bang theory.

17. What is the Hubble red shift?

18. Sketch the proposed earth's structure and describe the various layers.

19. Describe the proposed formation of oxygen in the earth's atmosphere and the time scale involved.

20. Describe the major features of plate tectonics and how they explain continental drift.

21. Describe the four spheres which make up the ecosphere.

22. As we look to the "outer limits" of the universe, we may be seeing the "beginning" of time. Explain.

23. Describe the life sequence of a typical star such as our sun.

24. State why you may or may not think that evacuation of earth is possible.

25. From the material in Chapter 2, suggest a reasonable time for a group of people leaving earth to find and reach another planet with hopes of colonizing. With a knowledge of the longevity of our present "social institution," do you think that such a trip is possible, assuming the technology is available?

Numerical Exercises

1. Find the number of seconds and hours in 1 year or 365 days.

2. Using the fact that the circumference of a circle is $2\pi r$ (where $\pi = 3.14$ and $r =$ radius) and that the average earth-sun distance is 93 million mi, find the distance traveled by the earth around the sun in 1 year. Do not use powers of ten notation.

3. Do problem 2 using powers of ten notation. How would you compare the ease of manipulation of the numbers?

4. If it takes 365 days for the earth to make one complete revolution, divide your answer to problem 2 by the time in hours to determine the average speed (in miles per hour) of earth as it moves about the sun.

5. (a) Express 6,000,000,000,000 miles (6 trillion) in powers of ten.
   (b) Express 0.0000005 m (meters) (read as five tenmillionths).
   (c) Express 2×10^8 (population of United States in 1970) in decimal form.
   (d) Express 2×10^{-5} in decimal form.

6. If the plane of the ecliptic is 7.5×10^9 mi across (diameter $= 2 \times$ radius), what is its circumference? Note this is the approximate distance that Pluto travels in one complete revolution above the sun.

7. (a) From Table 2-1, determine the circumference of Mercury's orbit and Jupiter's orbit. How do these orbits compare?
   (b) Compare the distance with their respective periods of revolution.
   (c) What does this tell you about their speed as they travel about the sun? (*Hint:* you may have to divide the orbital distance (circumference) by the period of revolution to make the inference.)

8. Using the planet-sun distances given in Table 2-1 for each of the nine planets, calculate the planet-sun distances in miles. Be sure to use powers of ten notation.

9. (a) You have determined that a sample of matter removed from the moon's surface has a mass of 64 g

(grams) and a volume of 32 cm³ (cubic centimeters). What is the density of the matter?

(b) What kind of material do you think that it may be?

10. Show that Jupiter's diameter is 11 times that of earth.

11. (a) A light-year is 6×10^{12} mi. If Polaris, the North Star, is 32 lt-yr away from us, determine its distance in miles.

(b) Determine the Earth-Polaris distance in astronomical units.

12. If the Milky Way galaxy is 100,000 lt-yr across and about 10,000 lt-yr thick, what is its diameter and thickness in miles?

13. Determine how far from the center of the Milky Way that our solar system is located. (*Hint:* see section entitled Our Galaxy.)

14. Show that "our" revolutional speed about the center of the Milky Way is 6.48×10^5 mi/hr. (*Hint:* Using 180 mi/sec, multiply by the number of seconds in an hour.)

15. (a) If the Great Spiral in constellation Andromeda is 2×10^6 lt-yr away, how far is it in miles?

(b) If we wanted to travel there and our fastest rocket ship could travel at half the speed of light (93,000 mi/sec), how long would the trip take?

(c) Realistically, we might expect a future spaceship to travel 10^{-4} of the speed of light. How long would the trip take then?

16. If there are 100×10^9 galaxies and 77% of the galaxies are spirals, how many spirals are there?

17. (a) The outer limits (visible limits) of the universe have been set at 10 billion lt-yr. Show that this distance is 6×10^{22} miles.

(b) How long has light been traveling to reach us from a star or galaxy 10×10^9 lt-yr away?

18. (a) If the earth's crust is on the average 10 mi thick, and the volume of a spherical shell is $4/3\pi$ $(r^3_{outer} - r^3_{inner})$ find the volume in cubic miles, in cubic feet, in cubic meters, and in cubic centimeters.

(b) If the average density of the crust is 2.5 g/cm³ (grams/cubic centimeter), what is the total mass of the crust?

(c) From Table 2-2, what are the actual masses of each of the various constituents of the crust?

19. (a) How much would it cost to evacuate 2×10^8 people of the United States, assuming each rocket ship could carry 200 people at a cost of $100 \times 10^6 each? How does this compare with the gross national product?

(b) How long would it take to perform the evacuation if a ship left each week? Each day?

20. Assuming a 2% annual population growth for the United States, how many additional ships would be necessary to evacuate the added people each year? What would be the added cost and the time required? What are the implications of your calculation?

Energy
from
the Sun

The sun has long been recognized as the earth's major source of energy, but in reality, the sun is truly the earth's only significant source of energy. Without the sun, life on earth would soon cease. The oceans would freeze; the nitrogen and oxygen in our atmosphere would liquefy while the carbon dioxide would freeze. Only small quantities of argon, hydrogen, and helium along with trace quantities of even rarer gases might remain. All of the components necessary for sustaining life as we know it would become unavailable, and such life would cease.

The sun's energy comes to earth in a variety of forms, for example, light, heat, radio waves, and ultraviolet rays. Some of these energy forms are necessary for sustaining life on earth, but a few are harmful to life on earth. Fortunately, our ecosystem has evolved in such a way that the harmful energy forms are largely prevented from reaching earth, with only the vital forms of energy reaching the planet's surface. The geographical distribution of energy depends on the time of year, the latitude, atmospheric conditions, and on the circulation of the atmosphere and the oceans. Eventually, all the energy that arrives from the sun must be reradiated back into space or the average surface temperature of the earth would increase. This complex process of reception, circulation, and reradiation has been called the **energy cycle.**

Because the sun is so important for our survival, we begin by discussing the sun's physical properties and the nature of the energy it emits. A discussion of "our energy budget" follows, and finally, some of the pertinent details of energy flow through our ecosphere is discussed.

The Sun

As the sun disappears over the western horizon, perhaps on a summer evening, it appears beautiful and restful. Most of us have wondered about what the sun is, how large it is, and where it is. Even though the sun appears quiet and restful, it is, in fact, a seething inferno that generates its own energy. The sun is a star.

As stars go, the sun is rather average. Even though the sun is an average star, it occupies a position of central importance to us, it is our nearest star and our life giver. We consider it to be a middle-aged star about half way through its life sequence of about 10^{10} years (10 billion years). The interior temperature may be as high as 20 million °C whereas the exterior temperature is about 6000°C or about 12,000°F. About 70 chemical elements have been identified in the sun with hydrogen and helium by far the most abundant. The earth–sun distance is about 9.3×10^7 mi (93×10^6 or 93 million mi) or 1.5×10^8 km (kilometers), which is a mere short distance as compared to the next nearest star, Proxima Centauri, which is 2.4×10^{13} miles away.

Physical Characteristics

The sun is enormous in size as compared to earth and the other planets (see Figure 2-4). Its diameter of 864,000 mi (1.39×10^6 km) is 108 times the 8000 mi (1.3×10^4 km) diameter of earth. The sun's volume is about 10^6 times that of earth, which simply means we could drop a million earths inside the volume occupied by the sun. It is nearly spherical and rotates slowly, with a 27-day time for one rotation (period of rotation). The mass of the sun has been determined by its gravitational effect [1] on the earth, that attractive force between any two bodies which effectively keeps the earth from flying off into space. The sun's mass is calculated to be about 2×10^{33} g (grams) or about 300,000 times the mass of earth.

If we divide any mass by the volume occupied by that mass, we obtain the quantity known as density, which has been calculated for the sun to be 1.4 g/cm³ (grams per cubic centimeter). This is about one fourth the 5.5 g/cm³ density of the earth. For reference, liquid water has a density of 1.0 g/cm³. The sun is broadly characterized as a large, intensely hot sphere of gas.

The Structure of the Sun

The sun is made up of various layers, each characterized by a particular function and by the way that these layers transmit energy from the core. The layers are illustrated in Figure 3-1. The surface that we observe on

Figure 3-1. Structure of the sun. The energy that is emitted from the photosphere is generated in the core via the fusion of hydrogen atoms into helium atoms.

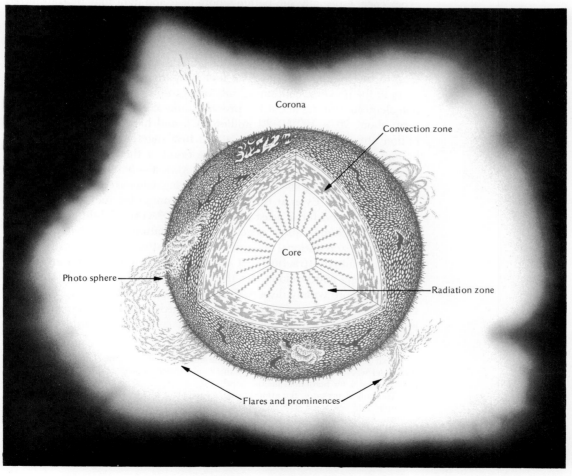

an early summer evening (except during eclipses) is called the **photosphere.** The photosphere is not to be confused with the corona, which we observe during eclipses and which extends out beyond the photosphere at times over 100,000 mi. The intense light from the photosphere normally blocks any view of the corona. The photosphere is a well defined layer with a temperature of about 6000°C.

The next layer below the photosphere is the **convective envelope,**[1] which is characterized by the violent movement of hot gases outward, thus transferring the energy from deep in the sun's interior to the photosphere. Beneath the convective layer lies the **radiation zone,** which is again characterized by the way in which energy is transferred; energy is transferred from the core through this layer by way of radiation. As we know from experience, especially sun-bathing, both heat and light may be transferred by radiation which requires no medium (such as air) for the transfer.

The very hot core of the sun produces the energy that is then transferred through the various layers, ultimately to be emitted into space. Interestingly, we believe that the process of generating energy is similar to that used in the hydrogen bomb. The fuel is hydrogen, which is the simplest atom in the universe containing only one proton in the atom's nucleus and one electron orbiting about the proton. Under conditions of high temperature, about 20×10^6°C (20 million °C), and high pressure, 10^9 atm (1 billion atmospheres),[2] hydrogen atoms fuse together ultimately forming helium, the next simplest element. The result of the fusion process is the conversion of some mass into energy. Almost 4.5 million metric tons[3] [4.5×10^9 kg (kilograms)] of mass are being converted per second into 4×10^{26} joules.[4] Even at this high rate of consuming fuel, we expect the sun to be around for at least another 5×10^9 years (5 billion years).

The high temperature necessary for initiating the fusion process is provided by gravitational collapse of stellar matter, which is mostly hydrogen. The mutual attraction that matter exhibits (gravitational attraction) tends to bring matter closer together. By virtue of this mutually attractive force, the more widely dispersed matter has potential energy. An analogy is a body held at some position above the earth's surface. From our experience, we know that as the body falls its energy of position—potential

---

[1] Heat transfer by convection involves the flow of material. Some familiar examples to the reader would be, transfer of heat by the winds, the movement of air currents in a room across a heater for household heating, and so on.

[2] The unit of pressure, the **atmosphere,** is equivalent to 14.7 lb/in.[2], the pressure of our atmosphere at sea level. The concept of pressure of the atmosphere, familiar to most of us, involves taking the weight of a column of air and dividing the weight by the cross-sectional area of the column, that is 1 in.[2] (square inch).

[3] 1 metric ton is equivalent to 1000 kg of mass or 2204 lb weight.

[4] The reader will have to accept this unit of energy until Chapter 4. For comparison, a toaster uses electrical energy at the rate of about 1×10^3 joules/sec (watts).

Figure 3-2. The sun's corona. The corona is the sun's atmosphere and is at a phenomenal temperature of 10^6 °C. It may be observed only during full eclipses of the sun by the moon. The corona reaches out hundreds of thousands of miles. [Courtesy of Hale Observatories.]

energy—is converted into energy of motion—kinetic energy (Chapter 4). Thus, as the sun's matter falls together, potential energy is transformed into kinetic energy of the atoms (in particular, hydrogen atoms). The average kinetic energy of gas atoms or molecules is a measure of the temperature. It is this energy that is ultimately transformed into the heat necessary to "trigger" the nuclear fusion reaction.

The **corona** is the sun's atmosphere. As mentioned earlier, we observe the corona only during times of total eclipse of the sun by the moon (Figure 3-2). The outer edges of the corona at times may even reach earth. An atmospheric layer of lesser importance is the **chromosphere**, which lies between the photosphere and the corona.

Energy from the Sun

The energy emitted from the sun is in the form of high speed particles (for example, electrons and protons) and electromagnetic waves. Radio, heat (infrared), light, ultraviolet and x rays are all examples of **electromagnetic waves.** Some of these forms of energy (mainly light, heat, and some radio waves) reach the surface of the earth, but most of the ultraviolet, all of the x rays, and the high speed particles are filtered out or stopped by the earth's atmosphere.

We refer to the electromagnetic waves emitted by the sun as the sun's **energy spectrum.** When the light or visible portion of the spectrum is passed through a prism, it is separated into its component colors: red, orange, yellow, green, blue and violet (see Figure 3-3). Tiny droplets of water vapor in the atmosphere can also act as prisms and separate the sun's light into colors forming a rainbow. We also call this spectrum a **continuous spectrum** because it includes all hues from red to violet. Actually, the entire energy spectrum of the sun is continuous, beginning with radio waves and ending with x rays. However, the sun does not emit the same amounts of all kinds of electromagnetic radiation.

Although a full understanding of the true nature of electromagnetic waves needs considerable background in electric and magnetic behavior, the concept of wave motion can be understood in reference to water waves (Figure 3-4). Most everyone has either tossed a rock into a quiet pond or watched a leaf fall into water or perhaps has watched waves roll into the beach from the open ocean. We observe that a disturbance, a ripple, travels away from the point of contact with water. The wave continues to travel until it reaches the shore and, as the wave dies away, its energy is dissipated, usually in the form of heat (but sometimes in the form of destruction). The most characteristic feature of the wave is a crest-trough-crest structure as

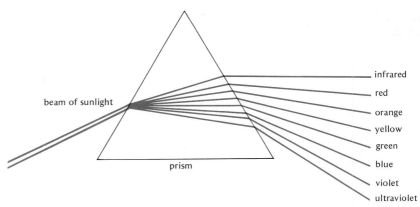

beam of sunlight

prism

infrared
red
orange
yellow
green
blue
violet
ultraviolet

Figure 3-3. Separation of the visible spectrum into its component colors (dispersion). As light passes through a prism, it is refracted or bent. However, the degree of bending varies with color (wavelength). Thus, each color is bent differently, which separates light into its components.

Figure 3-4. Water waves. A rock, a leaf, and so on, dropped into a quiet pond causes a disturbance—waves that travel outward in a circular pattern. The waves have a crest-trough-crest structure and carry energy. As the wave passes, the water is displaced (up and down) in a direction perpendicular to the direction of motion.

shown in Figure 3-5. Another, but much less obvious, feature is that waves carry energy, which becomes evident when we observe the destruction and beach erosion along the shore line after a storm. Perhaps you have observed a surfer on a surfboard traveling in the crest of a wave. The energy for his motion comes from somewhere—the wave!

A Question for the Reader: Where did the surfer's wave get its energy?

By reference to Figure 3-5, you can see that waves are quantitatively characterized by the crest-to-crest distance, the number of crests passing a point (or you) per second, and the speed of the wave. The crest-to-crest distance is the **wavelength** (λ, lambda) and is measured in feet, meters, and so on. The number of crests per second passing a point is the **frequency** (or pitch) of the wave. The passing of one wavelength is known as a cycle, so, the frequency is measured in cycles per second (cps or Hz for Hertz). These wave properties are summarized by the relation

$$\text{wavelength} \times \text{frequency} = \text{speed}$$

or $$\lambda \times f = v \qquad (3\text{-}1)$$

that is

$$\text{meters} \times \frac{\text{cycles}}{\text{second}} = \frac{\text{meters}}{\text{second}}$$

Although in general the speed of waves v depends on the type of wave, the speed of an electromagnetic wave is constant at 186,000 mi/sec or 3×10^8 m/sec and equation (3-1) is written as

$$\lambda \times f = c$$

This is the speed of light (in a vacuum). It should be noted from equation (3-1) that a wave of higher frequency has a shorter wavelength, and vice versa. Again, referring to Figure 3-5, we call your attention to the features of the waves that we utilize for our discussions of electromagnetic waves: (1) that they have a creat-trough-crest structure, that is described by

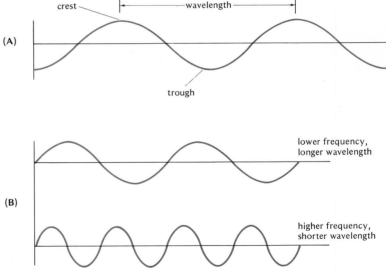

Figure 3-5. Crest-trough-crest structure of a wave. The wave structure shown here is for a transverse wave, which means that the medium is displaced in a direction perpendicular to the direction of wave motion. The waves are characterized by a wavelength, a frequency, and a speed of travel: $\lambda f = v$. High frequency waves have a short wavelength; low frequency waves have a long wavelength. The speed, in general, depends on the type of wave. For example, sound waves travel at 1100 ft/sec in air at sea level. All electromagnetic waves, including light waves, travel at 186,000 mi/sec in free space, and we usually write the speed of light as c instead of v.

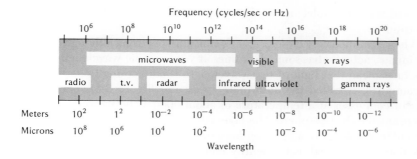

Figure 3-6. The electromagnetic spectrum. The visible portion is only a small part of the total spectrum. In general, the emission of electromagnetic waves by a hot body is not uniform over the entire spectrum.

equation (3-1) and (2) that they carry energy [1].[5] Furthermore, electromagnetic waves differ from waves in water in that they travel without the displacement of matter; this is evident by the fact that light can travel through empty space or a vacuum.

The entire electromagnetic spectrum is shown in Figure 3-6. The wavelength is given both in meters and in micro-meters or microns (μ); $1 \mu = 10^{-6}$ m. A particular feature to note is that the visible portion of the spectrum represents a very small part of the total energy spectrum. The sun's energy spectrum is similar to that of an "ideal emitter," radiating energy at a temperature of 6000°C. The greatest amount of energy is emitted in the visible region with a peak at about a wavelength of 0.6 μ. About one half of the total sun's energy is given off in the visible region. From Figure 3-7 we see that a wavelength of 0.6 μ corresponds to yellow light. This is the reason that the sun is a yellow star, that is, the temperature of the photosphere is 6000°C. Cooler bodies have peaks in their emission spectra in the visible red or infrared region, whereas hotter bodies have

[5]Often wave phenomena include reflection, refraction, diffraction, and interference. These are excluded as being beyond the scope of this text.

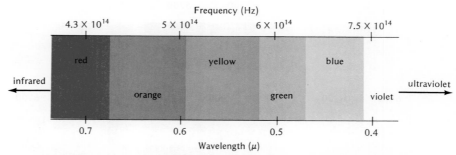

Figure 3-7. The visible electromagnetic spectrum. The visible spectrum begins with red light at 0.74 μ and ends with violet at about 0.4 μ.

peaks in the blue-violet or the ultraviolet region of the electromagnetic spectrum.

When examined very closely, a star's emission spectrum, although largely continuous, actually has some dark lines at very specific wavelengths. The dark lines mean that light of that wavelength is not coming to earth. This characteristic is called the **absorption spectrum** (Figure 3-8). There is a layer of cool gases at a temperature of about 5000°C between the photosphere and the chromosphere. This layer contains elements that absorb these particular wavelengths of light coming from the photosphere. Study of these dark lines has enabled us to identify constituents of the sun.

Figure 3-8. The sun's spectrum showing absorption lines. When observed closely, the continuous spectrum of the sun has superimposed on it a series of dark lines—the absorption spectrum. Atoms between the photosphere and the chromosphere, absorb very definite wavelengths. About 70 known elements have been identified by such absorption lines. [Courtesy Hale Observatories.]

Solar Spectrum made with the 13–foot Spectroheliograph

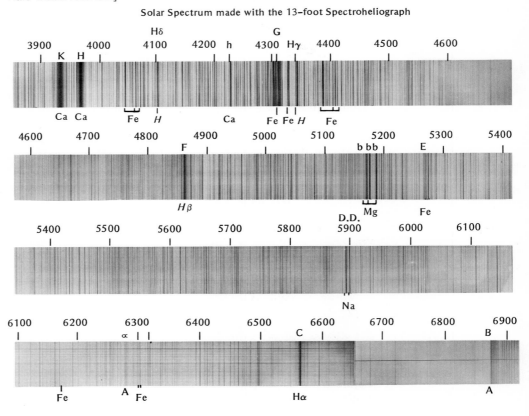

EXAMPLE 3-1

The Hα line has $\lambda = 0.656\,\mu$. What is its wavelength in meters and what is its frequency?

$$0.656\,\mu \times 10^{-6}\,\frac{m}{\mu} = 0.656 \times 10^{-6}\,m$$

$$f = \frac{c}{\lambda} = \frac{3 \times 10^8\,m/sec}{0.656 \times 10^{-6}\,m} \cong 5 \times 10^{14}\,\frac{1}{sec} = 5 \times 10^{14}\,Hz$$

Note: $1/sec \equiv cps \equiv Hz$

One of the lines absorbed by the hydrogen gas in the cool layer is the $0.656\,\mu$ (red) line and is called the Hα (H alpha) line. Using 3×10^8 m/sec for the speed of light, we obtain a frequency of 5×10^{14} Hz (Example 3-1) for this Hα line. When we discussed the Hubble red shift in Chapter 2, it was the shift in the Hα line that was used to determine the recessional speeds of stars.

A frequency of interest to you is that of FM radio broadcasting at a frequency of 100 MHz (megaHertz) (100×10^6 Hz) with a wavelength of 3 m. See Example 3-2.

Comparison of the respective frequencies and wavelengths in these examples shows that a smaller frequency wave has a longer wavelength.

From our study of the sun thus far, we may summarize some of the more important features. The sun gives off all types of electromagnetic waves (except gamma) but the emission is not completely uniform over the entire spectrum. Of the energy emitted by the sun, one half is in the visible region (albeit not uniformly) of the spectrum, $0.7\,\mu$ to $0.4\,\mu$, with a peak emission at $0.6\,\mu$ (Figure 3-9). However, it is important to realize that there are substantial emissions of energy in the other regions—radio, heat, and so on. Fortunately, our atmosphere acts as a window (with a screen). It allows mainly visible energy to penetrate and screens out most of the harmful short wavelength radiation—ultraviolet and x rays (see Figures 3-11 and 3-12). The ozone layer of the atmosphere is instrumental in this screening action. If substantial quantities of these screened-out forms

---

EXAMPLE 3-2

What is the wavelength of an FM radio station broadcasting at 100 MHz (100×10^6 Hz)?

$$\lambda = \frac{c}{f} = \frac{3 \times 10^8\,m/sec}{100 \times 10^6\,1/sec} = 3\,m$$

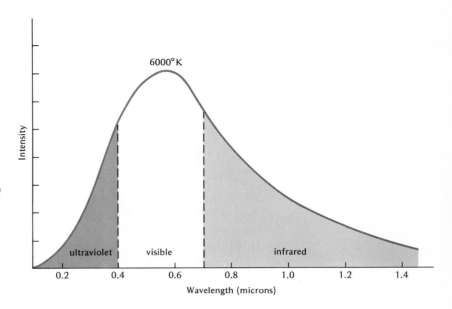

Figure 3-9. The spectral distribution of the sun. The sun emits energy like an "ideal emitter," radiating at about 6000 °C. As can be seen, the emission is not uniform over the entire spectrum. About half of the sun's energy is emitted in the visible region; the rest is distributed throughout the other regions: infrared, radio, ultraviolet, and x-rays.

of energy were to penetrate the atmosphere, they would be extremely harmful to the biosphere. Nonetheless, sufficient ultraviolet rays penetrate the ozone layer to give the "sun worshippers" beautiful suntans (and burns)!

Our Energy Budget

Until we had satellites and various space craft outside the earth's atmosphere, it was difficult to determine how much of the sun's energy was intercepted by the earth due to the screening effect of the earth's atmosphere. In addition, weather (cloud layers) and other atmospheric conditions cause the degree of this screening to change. We are interested in the amount of energy received by earth because small permanent changes, 1% or less, might cause drastic changes in climate. Most of you are familiar with the moderating effect on temperatures of clouds on a cloudy day. With the advent of the space age, accurate measurements of the sun's energy received by earth have become possible.

The Solar Constant

The energy being intercepted by earth is given as the amount of energy crossing the earth's diametric plane as shown in Figure 3-10. The diametric plane can be envisioned by taking a slice through the earth exactly perpendicular to the sun's rays. As a result, the diametric plane is a circular plane

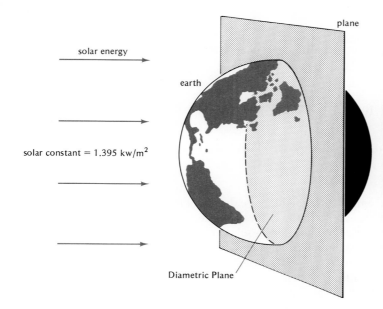

Figure 3-10. The earth's diametric plane. If a plane were passing through the center of the earth, that is, along a diameter and perpendicular to the sun's rays, there is formed a large circular plane, which we call the diametric plane. The area of the plane is 1.275×10^{14} m².

passing through the center of the earth and perpendicular to the direction of the sun's rays. The rate at which energy crosses or meets the diametric plane is 1.395 kw/m² (kilowatts per square meter) which is called the **solar constant.** This means that the rate of energy flow across each square meter of plane is 1.395 kw (or 1395 joules/sec). As mentioned earlier, this rate of energy flow is about the same rate at which electrical energy is used by a large electric toaster in your kitchen. If such a toaster worked continually for 1 hr, the total electrical energy consumed would be 1.395 kwh (kilowatt-hour).

If the area of the earth's diametric plane is 1.275×10^{14} m² (square meters), then the total energy flow into earth is 1.779×10^{14} kw which corresponds to a yearly reception of 1.558×10^{18} kwh (about the energy that 10^{14} or 100 trillion toasters would consume in 1 year while toasting 24 hr/day). See Example 3-3.

The solar constant has been found to be constant, at least during the short period of time (since 1969) that it has been accurately measured from outside the earth's atmosphere. Our concern about changes of the solar constant stems from possible changes in climate that might result. Interestingly, we do know that solar flares and other solar activity, recorded for over 100 years, seem to have had little effect on the earth's climate. It has been estimated that a change of a few per cent in the solar constant, and hence the solar input to earth, would cause the average temperature at the surface of the earth to change by several degrees. Seeming small

---

EXAMPLE 3-3

Given a solar constant of 1.395 kw/m², determine the energy flow across the earth's diametric plane with an area of 1.275×10^{14} m², and then determine the total energy received by earth for the period of 1 year.

$$1.395 \frac{\text{kw}}{\text{m}^2} \times 1.275 \times 10^{14} \text{ m}^2 = 1.779 \times 10^{14} \text{ kw (the energy flow)}$$

$$1.779 \times 10^{14} \text{ kw} \times \frac{24 \text{ hr}}{\text{day}} \times 365 \text{ days} = 1.558 \times 10^{18} \text{ kwh (total energy received in 1 year)}$$

---

changes in temperature might cause drastic climatic changes. In fact, many theories of climate variations, for example, ages of glaciation, have been based on this idea of variation in solar constant. However, at this time we say that the solar input is constant.

Energy Accounting

The earth's "energy balance sheet" is shown in Table 3-1. Due to the complexity of the details involved in the reception, circulation, and reradiation of solar energy, it is difficult to define precisely the entries to the balance sheet. However, our accounting gives us general information about the relative importance of the entries with regard to possible energy resources. Our unit for measuring quantites of energy is the kilowatt-hour (kwh).[6]

The total energy intercepted by the earth over the period of 1 year is 1.558×10^{18} kwh. Of that, about 30% $(0.467 \times 10^{18}$ kwh) is immediately reflected back into space. The energy (30%) that is reflected immediately is known as the earth's **albedo**. It is not available for use on earth as an energy resource at all. The albedo is composed mainly of short wavelength radiation, blue light, ultraviolet light, and x rays. Photographs from space show earth as a blue marbled sphere and attest to the effectiveness of the atmosphere, clouds, and earth for reflecting the shorter wavelength electromagnetic radiation. From space, the earth's atmosphere appears as a blue haze, blue being among the shorter wavelengths in the visible spectrum. The remaining 70% $(1.091 \times 10^{18}$ kwh) enters the earth's atmosphere and flows throughout the ecosphere.

The 70% of intercepted energy is further subdivided into two major parts: about 50% $(0.779 \times 10^{18}$ kwh) light and about 20% $(0.312 \times 10^{18}$ kwh) heat (infrared) and radio waves. Because the earth's

[6]The kilowatt-hour unit will be discussed in detail in Chapter 4. We consumers of electricity pay for electric energy in kilowatt-hours. If you were to burn 3400 wooden kitchen matches completely, you would obtain about 1 kwh of heat energy.

Table 3-1. Energy Balance Sheet for Earth*

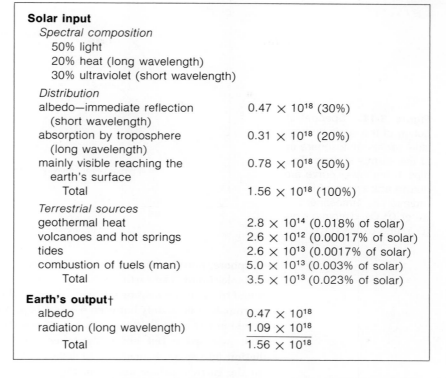

Solar input	
Spectral composition	
50% light	
20% heat (long wavelength)	
30% ultraviolet (short wavelength)	
Distribution	
albedo—immediate reflection (short wavelength)	0.47×10^{18} (30%)
absorption by troposphere (long wavelength)	0.31×10^{18} (20%)
mainly visible reaching the earth's surface	0.78×10^{18} (50%)
Total	1.56×10^{18} (100%)
Terrestrial sources	
geothermal heat	2.8×10^{14} (0.018% of solar)
volcanoes and hot springs	2.6×10^{12} (0.00017% of solar)
tides	2.6×10^{13} (0.0017% of solar)
combustion of fuels (man)	5.0×10^{13} (0.003% of solar)
Total	3.5×10^{13} (0.023% of solar)
Earth's output†	
albedo	0.47×10^{18}
radiation (long wavelength)	1.09×10^{18}
Total	1.56×10^{18}

*All quantities here are approximate.

†The equality of the earth's output with the solar input is known as the "radiation balance."

atmosphere is transparent to most of the visible spectrum, the 50% as light (that is, λ between 0.7μ and 0.4μ) reaches the surface of the earth (Figure 3-11). The 20% as heat and radio waves (λ greater than 0.7μ) is absorbed by the lower troposphere and provides a major source of energy for, first, atmospheric circulation and, in turn, oceanic circulation.

You will recall that the troposphere is the atmospheric layer nearest the earth's surface and contains most of the air including large quantities of water vapor. The water vapor and carbon dioxide are very effective absorbers of long wavelength radiation and are thus responsible for the aforementioned absorption of larger wavelength radiation. Some of the heat absorbed by the troposphere is emitted toward earth and some is transferred into the upper atmospheric layers, thus providing additional heat at the earth's surface and for circulation in the upper atmosphere. The heat from the troposphere and the visible energy (about 50% of the sun's energy input) reaching the earth's surface are together called the **insolation.**

Because no drastic warming trend at the surface has been observed, that is the earth's climate has been roughly the same for many years, the earth must be in **radiation balance.** More specifically, the earth, including its

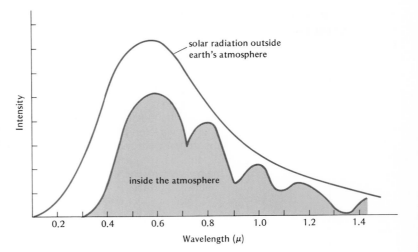

Figure 3-11. Spectral distribution of the sun's energy outside earth's atmosphere and at the earth's surface. The dips in the lower curve are due to absorption of incoming radiation by atmospheric water (H_2O), diatomic oxygen (O_2), and ozone (O_3).

atmosphere, must radiate an amount of energy back into space just equal to the solar input. The earth's radiation as measured by man-made satellites is found to be in the heat or infrared region of the electromagnetic energy spectrum, that is, mainly between $8\ \mu$ and $14\ \mu$. Again, the atmosphere acts as a window (Figure 3-12), allowing heat energy between $8\ \mu$ and $14\ \mu$ to escape into space but absorbing other wavelengths. The earth's heat radiation into space is made up of heat emitted into space both directly from the earth's surface and from the various atmospheric layers. Thus, the earth's radiation derives from a complex combination of absorption, transfer, and reradiation of the solar input.

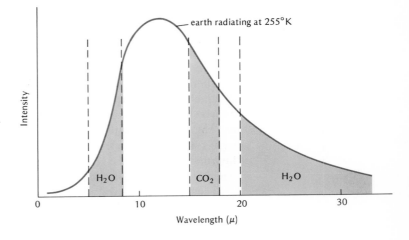

Figure 3-12. Spectral distribution of earth's emission inside the atmosphere. The earth behaves like an ideal emitter, radiating at a temperature of $-18°C$ ($255°K$) with a peak emission at $12\ \mu$. The colored areas show the portion of the earth's energy that is absorbed mainly by the atmospheric water and carbon dioxide.

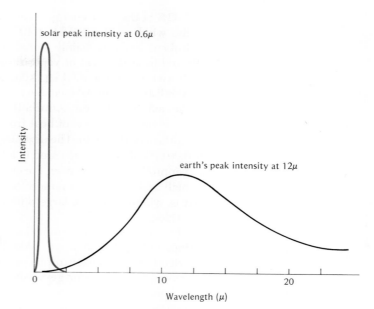

Intensity

solar peak intensity at 0.6μ

earth's peak intensity at 12μ

0   10   20

Wavelength (μ)

Figure 3-13. Comparison of earth's emission with the sun's emission. The earth's spectral or wavelength distribution, with peak at 12 μ, is shown compared with the sun's, with peak at 0.6 μ.

Figure 3-13 compares the distribution of electromagnetic energy from the earth with that of the sun. As mentioned previously, the sun radiates energy like a perfect emitter at about 6000°C. A body at that temperature has a peak radiation intensity at about 0.6 μ. The earth's emission has peak intensity at about 12 μ (Figure 3-11) which is indicative of a perfect emitter whose temperature is 255° degrees absolute or 255°K (degrees Kelvin)[7] which is about 0°F or −18°C. However, the earth's average surface temperature is not 0°F, but is closer to 40°F. The apparent discrepancy is due to the earth not being truly a perfect emitter of radiation.

M. K. Hubbert [2] provides us with an even more detailed breakdown of the solar input (the 70%). By summing both the light energy converted into heat which takes place at the earth's surface and the heat in the atmospheric layers, 47% (7.32×10^{17} kwh) of the total sun's input is directly lost from earth to outer space as heat. The remaining 23% (3.58×10^{17} kwh) is channeled into winds, oceanic circulation, photosynthesis, and the hydrologic or water cycle (the warming, evaporation and precipitation of water). Most of the 23% goes into running the hydrologic cycle. The energy going into photosynthesis is only about 3.3×10^{14} kwh

[7]The temperature scale here is the absolute or Kelvin scale. The absolute or Kelvin scale references temperatures to an absolute zero which the Fahrenheit and Celsius or centigrade scales do not. For example, the ice point for water is 0°C or 273°K. Temperature scales will be discussed more fully in Chapter 7.

or 0.02% of the sun's energy received by earth, whereas the energy stored in the winds is about 3.24×10^{15} kwh (0.2%). The heat going into the hydrologic cycle is mainly stored in the form of latent heat, sometimes referred to as the **heat of vaporization.** It takes 540 kcal (kilocalories)[8] or 0.628 kwh to evaporate 1 kg (kilogram) (2.2 lb) of water.

Also listed on our balance sheet are those small energy sources not related to the sun. The largest contribution (2.8×10^{14} kwh or 0.018% of solar input) is the conduction of heat from the earth's molten core, out through the surface of the earth. The next largest is the tidal source (2.6×10^{13} kwh or 0.0017% of solar input) used in a few, very specific locales for electrical power production. However, the energy available from harnessing the tides is small and will remain important only where the tides are readily used, that is, where there is a large difference in water level between high and low tides.

The final heat source from the earth itself is due to volcanoes and hot springs (2.6×10^{12} kwh or 0.00017% of solar input). Because volcanic activity is often a catastrophic event, it is of little practical use to us. However, hot springs or geysers are now being used, but, again, they are important mainly as a local resource and may be of great significance in certain areas, for example, in the geothermal fields of California (Figure 3-14). Although still in the early stages of development, nuclear energy (fission) may also become an important energy resource. It has been estimated that its potential is probably equal to our total fossil fuel resources.

After having done our energy accounting, the major sources of energy

---

[8] The unit of kilocalorie (kcal) that we have used here is the calorie that the nutritionist uses to measure heat values for food. For example, a Danish pastry will yield about 500 kcal of heat upon combustion and presumably the same amount of heat on digestion.

Figure 3-14. Geothermal fields in California. Heat from within the earth turns water into steam which then escapes through fissures in the crust. Geothermal heat may be an important energy source in localized areas. [Courtesy of J. R. Jackson, Exxon Co.]

on earth become clear. For example, direct utilization of the sun's light and heat could provide a continuous large source of energy if the use is economically and technologically feasible. Such direct use of the sun's energy is as yet to be shown. While energy from the water cycle is potentially large, its utilization is mainly through falling water which is limited in supply and is highly localized to particular regions on earth, that is, not in the Sahara desert. Smaller amounts of energy are available from the winds and from the burning of our fossil fuels, which are now being consumed at an ever increasing rate. With the advent of the steam engine, the use of windmills decreased rapidly. However, due to our increasing energy needs, windmills may again dot the landscape. Our discussion of energy resources and their use particularly for the generation of electricity will continue throughout the book.

Geographical Distribution of Energy from the Sun

The preceding discussions have given us an overview of our energy budget. The fact is that the reception of the sun's energy is not uniform over the surface of the earth. The tropical zones, which are the regions between 23.5° North latitude (Tropic of Cancer) and 23.5° South latitude (Tropic of Capricorn), receive a substantial portion of the incoming energy (Figure 3-15). The temperate zones, those regions between the latitudes of 23.5° and 66.5° latitude, receive lesser amounts. The polar zones, the regions from 66.5° to 90° latitude, receive the least amount of energy. Interestingly, the albedo of the polar regions is considerably greater than the tropical regions, which means that little of the sun's energy falling on the area is absorbed.

Thus, the uneven distribution of energy received from the sun should, in principle, cause the equatorial or tropical zones to become unbearably hot whereas the polar regions should become unbearably cold. However, heat energy is circulated to the polar regions by the winds and ocean currents, producing a moderating temperature effect on both regions.

The energy received by various latitudes also depends on the time of day and the time of year, that is, the **seasons.** The daily variation of energy received by a region is rather obvious. As the earth rotates counterclockwise on its axis, that is, west to east, the sun appears to rise in the east and set in the west. The time between rising and setting on the average is 12 hr. Most of the energy is received around midday because at that time the sun is the most directly overhead or at the highest altitude above the horizon (again depending on latitude). Interestingly, the sun is never overhead anywhere in the continental United States since all of the land area lies above 23.5° North latitude.

Furthermore, the degree to which the sun is overhead at noonday depends on the time of year or season (Figure 3-16). During the summer

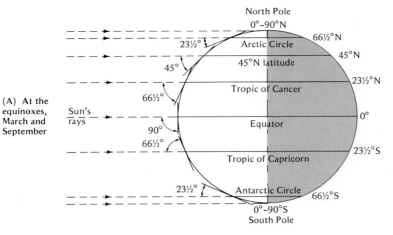

(A) At the equinoxes, March and September

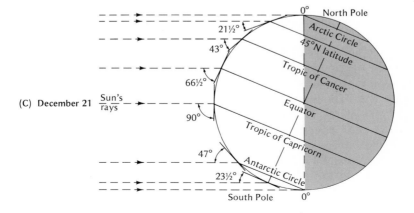

(B) June 21

(C) December 21

Figure 3-15. Energy received by various latitudinal zones on earth. The earth is shown in various positions shown also in Figure 3-17. **(A)** The sun at noon is overhead at the equator. **(B)** Sun overhead at the Tropic of Cancer. **(C)** Sun overhead at the Tropic of Capricorn. The tropic zone between Cancer and Capricorn is relatively "seasonless." Most of the energy received by the earth is in the tropic zones.

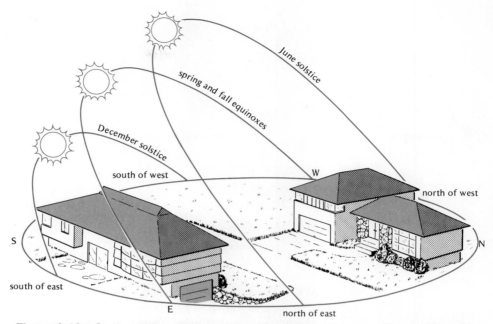

Figure 3-16. Sun's position at noon for various times of the year. The position of the sun at noon along with points of "sunrise" and "sunset" are shown at different seasons. The June and December solstices occur on the twenty-first day and is the time when the sun appears to "rest" in the same position at noonday before it begins its trip further south at noonday as the autumn season approaches and vice versa for the December solstice.

months, the sun is the most directly overhead, thus the regions experiencing summer receive the most energy during the longer daylight hours (while the nights are shorter). Figure 3-17, which is a diagram of the earth in various positions about the sun, exhibits the three main causes of seasons: (1) the tilt of the earth's axis, (2) the parallelism of the axis, and (3) the 365.25 day revolutional period of the earth about the sun.

The earth's spin axis is tilted at an angle of 23.5° relative to the ecliptic plane. This angle of tilt is maintained in the same direction, or is pointed at the North Star throughout the earth's 365.25 day trip around the sun. This is called the **parallelism of the axis.** On December 21, the North Pole of the axis is pointed away from the sun so that the sun is overhead at the Tropic of Capricorn, 23.5° latitude South, (in the Southern hemisphere).[9] The degree of heating is greatest for the Tropic of Capricorn at this time.

[9]For reference, the Tropic of Cancer passes just north of Havana, Cuba and the Tropic of Capricorn passes just south of Rio de Janeiro.

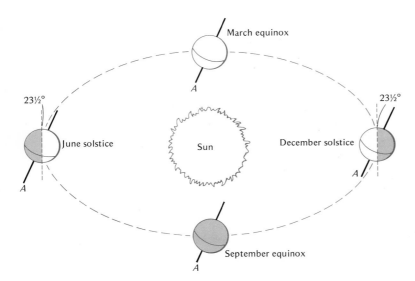

Figure 3-17. Seasons. The earth's axis maintains a constant angle of inclination (23.5°) to a line perpendicular to the plane of the ecliptic and is called "parallelism of the axis." In June, the tilt of the axis is such that the sun is more directly overhead in the northern hemisphere, thus giving the summer season. In December, this same situation is true for the southern hemisphere. During the March and September equinoxes, the sun is directly overhead at the equator at noon, signifying the change in seasons.

When the sun is more directly overhead, a unit of sunshine heats a smaller area as shown in Figure 3-18. When the sun appears to be lower on the horizon, with a more glancing incidence of its rays, the same unit of the sun's energy must heat a larger area, also shown in Figure 3-18. Conversely, on June 21 the sun is directly overhead at the Tropic of Cancer[9] (in the Northern hemisphere). The sun is directly overhead at the equator on March 21, the vernal equinox and also on September 23, the autumnal equinox. These seasonal variations of the sun's angle also determine the length of the day (and night) at the poles. For example, March 21 is essentially dawn at the North Pole, June 21 is midday, and September 23 is nightfall. Of course the six months of night run from September to March.

Hence, the energy received by the earth at various latitudes depends on how directly the sun is overhead, which in turn depends on the time of year or the season. It is interesting to note that during a 24-hr day in the summer season, more energy is received at the pole than at the equator, even at the low incidence of the sun's rays. This is due, of course, to the 24 hr of light during the "day" at the pole. However, due to the high albedo of the polar regions, the heating effect of this increased amount of energy is still substantially smaller.

Finally, the energy received from the sun depends on the atmospheric conditions. Although a cloud layer is somewhat opaque to visible radiation, it is even more opaque to heat. This, as we mentioned earlier, is due to the effectiveness of water vapor for absorbing heat energy (longer wavelength radiation). The light energy that gets through the clouds is absorbed by the earth and is then reemitted in the form of heat. The clouds reflect

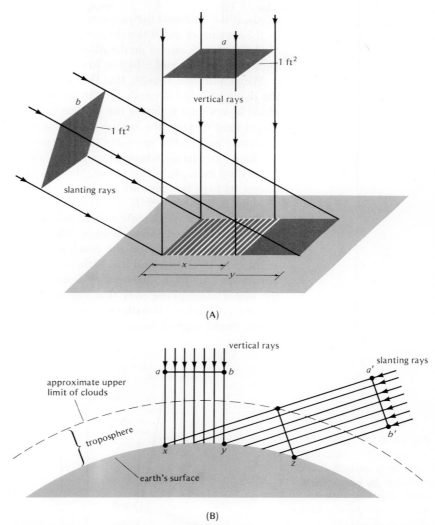

Figure 3-18. Vertical rays of the sun versus slanting rays. In both **(A)** and **(B)** the two bands of rays contain the same amount of energy. The energy of the vertical rays is spread over a smaller area on striking earth than is the energy of the slanting rays, so that the heat per unit area is greater. Because the sun is more vertical during the summer months, the summers are warmer in the temperate and polar zones. The regions in the vicinity of the equator, the tropical zones, are relatively seasonless.

this heat back toward the earth with the heat then being circulated about by the moving air and water (Figure 3-19). The overall effect on the weather of clouds then is one of moderation. Put simply, a cloudy day has a lower high temperature and a higher low temperature than a clear day.

We should note, though, that the moderating effect of the clouds during the daylight hours depends greatly on the height of the cloud layer; low cloud layers tend to lower the surface temperatures, whereas high cloud layers tend to raise the surface temperatures. If we had no atmosphere or more importantly, *no* water in the atmosphere, we would have larger (perhaps very large) temperature differences between day and night. An interesting example is the planet Mercury, which has little or no atmosphere; its high temperature of the day is about 640°F, whereas the low at night is about −190°F.

Dust or other particulate matter may also have a substantial effect on the energy received and on both weather and climate. For example, after the volcano, Krakatoa, on the Sundra Strait between Java and Sumatra erupted in 1883, England had little or no summer season for three years, apparently due to the particulate matter from the eruption circulated in the upper atmosphere over that region. This brings to mind one of our concerns (Chapter 14) with burning large quantities of coal which results in the discharging of large quantities of fly ash into the atmosphere.

Figure 3-19. Effect of atmospheric conditions on energy reception and emission by earth. Dust, water vapor (and clouds), and carbon dioxide keep some light energy ($\lambda = 0.6\ \mu$) from reaching earth while very effectively trapping the heat that would normally be radiated into space from earth. This heat-trapping effect is similar to the heat trapping by the glass in a greenhouse, and thus is called, "the greenhouse effect."

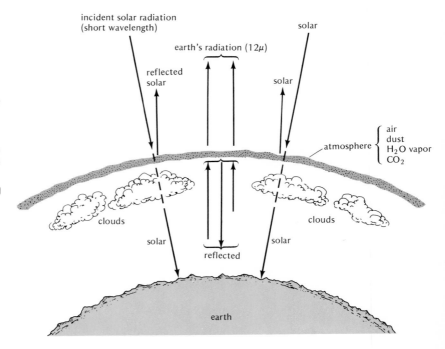

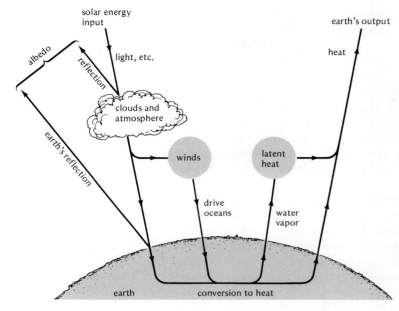

Figure 3-20. The earth's energy cycle. The diagram of the energy cycle displays the gross characteristics of reception, circulation, and reradiation of energy by earth as given on the balance sheet. The energy input is not cycled and reused as is matter, such as carbon, but is eventually lost for further use by reradiation into space. However, this action is typical of heat engines, and the ecosystem has been appropriately called an atmospheric heat engine.

Energy Flow in the Ecosphere

An overview of the gross features of the energy budget are shown in Figure 3-20. Energy enters the ecosphere, is used, and then dissipated in the form of heat. The heat is radiated back into space and thus lost for any further use by earth inhabitants. For this reason, heat is sometimes referred to as "degraded energy," even though in the process of degrading energy (for example, burning coal, wood, or oil to obtain heat) we obtain the "work" necessary to run our society on spaceship earth. Of necessity, the amount of energy that comes to earth must eventually be lost by earth, otherwise, the earth would experience an overall warming trend.

From our energy budget we saw that some of the energy goes into heating up the atmosphere and the hydrosphere, causing atmospheric currents (winds) and oceanic currents. A major portion of energy goes into evaporating water, which we called **latent heat.** The energy stored in these processes is then circulated globally, representing the major flow of energy throughout the ecosphere. Further, energy storage in plants by way of **photosynthesis** provides the food necessary for life on earth and is responsible for the minute amount of the sun's energy that is stored in our fossil fuels (coal, oil, and natural gas) which is equivalent to only seven days of sunshine.

The reception, circulation, and reradiation of solar energy has often been referred to as the **energy cycle.** However, this is not a cycle in the biological sense, it is a thermodynamic (heat flow) cycle. A biological cycle refers to the continual cycling of matter, as in the carbon dioxide cycle, whereas a thermodynamic cycle refers to use of energy to drive an engine whereupon heat is discarded as waste to complete the engine's "operating cycle" (Chapter 7). The features of the earth's energy flow do seem to fit this engine analogy in a broad sense.

The importance of this analogy is that it allows us to characterize earth and the societies living on it as complex machines that degrade energy in the process of obtaining the "work" necessary to keep the machine running. By such a model, we are able to project energy needs and to analyze the problems resulting from such energy consumption, including pollution. This model will be discussed in detail in Chapters 7, 8, and 13.

Atmospheric and Oceanic Circulation

Atmospheric and oceanic circulation together produce the major features of **climate.** As we saw earlier, there is a great inequality of energy received in various locations over earth. During the winter months for the northern latitudes and, in particular, the polar regions, the temperature would be considerably lower if it were not for the heat transported by winds and oceanic currents such as the Gulf Stream. For example, Iceland is just below the Arctic Circle (66.5° latitude North), but, due to the Gulf Stream, has an average winter temperature rather like the temperature in the New York–New Jersey region.

A simple model, proposed by George Hadley (sixteenth century) for air circulation, is shown in Figure 3-21). When air is heated, it expands, becomes less dense and rises, whereas cool air is more dense and tends to fall. The cool air from the polar regions flows toward the equator, while

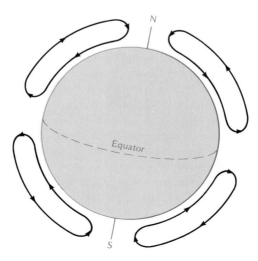

Figure 3-21. Air circulation. The cell-like pattern of circulation was proposed by George Hadley, a sixteenth century meteorologist. The cells represent the rising and moving toward the poles of warm air in the tropic zones and the flow of cold air down from the polar regions to replace it.

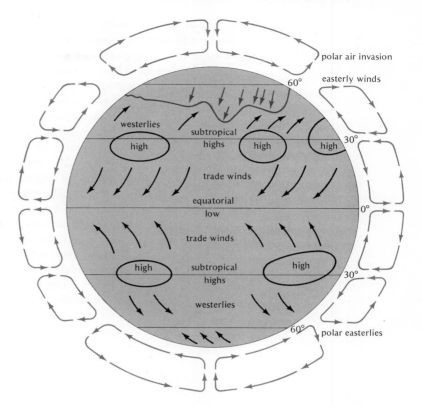

Figure 3-22. Prevailing winds. The rotating earth underneath the air mass creates the prevailing winds as shown. Even the earth's rotation combined with multiple cells are inadequate to explain completely the winds and weather.

the warm air rises and moves northward creating a cellular flow pattern called "Hadley's cells. In reality, each of the larger cells consist of a series of smaller cells. The effect of the earth's rotation is to create prevailing wind directions, the polar easterlies, the temperate westerlies, and the tropical tradewinds as shown in Figure 3-22.

Although these circulations are important for climate, can they be used in the production of electrical power? Most of us recall that in the past windmills have played an important role. The windmill has declined in importance in the United States since the 1930s, but perhaps the future holds a new and even more significant role for windmills as a way of harnessing wind energy (Chapter 15).

The Hydrologic Cycle

A large amount of the sun's energy is stored in the evaporation of water. Absorption of solar energy raises the average temperature of water. Roughly speaking, a higher temperature means that the water molecules have higher energy. The energy can be thought of as giving larger vibrational and translational motions to the molecules (Chapter 5). Some of the water molecules will have sufficient energy to break the forces between molecules and escape the surface of the liquid; thus some water evaporates.

Recall the amount of heat required to evaporate 1 kg (kilogram) of water is 540 kcal (kilocalories) and this stored energy is called latent heat. In Figure 3-23, we see that the moving masses of air, convection currents, sweep the water vapor from the air immediately above the oceans well up into the atmosphere. In moving up from the surface through the troposphere, the temperature decreases at a rate of about 4°F/1000 ft (or 7°C/km) of elevation (Figure 3-24). Since air at lower temperatures can hold less water vapor than air at higher temperatures, water begins to condense to form clouds. Eventually, the water falls (precipitates) in the form of rain or snow. When the water condenses from the air to form clouds or drops, the latent heat is now the heat of condensation and is given up to the surrounding clouds and atmosphere.

The combination of both evaporation and precipitation with both atmospheric and oceanic circulation provides us with our weather. In contrast with climate, weather refers to atmospheric conditions only over a short time period. Sometimes the precipitation is local, but more often

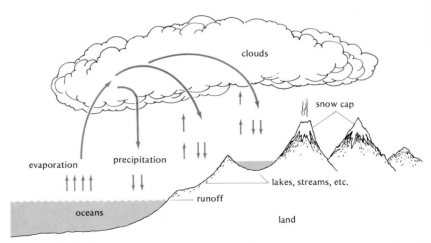

Figure 3-23. The hydrologic cycle. The water cycle is a model proposed to explain the evaporation, circulation, and precipitation of water. Since three quarters of the earth's surface is covered by water and all but 2% of the water on earth is found in the ocean, most of the evaporation and precipitation takes place there. About 23% of the total precipitation is over the land, with about 7% of the total or 30% (30% of 23% = 7%) of the actual precipitation over the land providing our runoff. The mean annual rainfall is about 0.86 m over the surface of the earth.

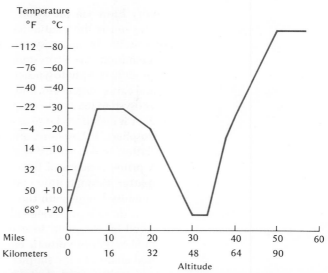

Figure 3-24. Variation of temperatures with altitude. The various atmospheric layers are shown with a temperature profile.

the moisture is carried many miles away by the convection currents (winds) from the site of the evaporation. The transit time for the moisture varies from hours to weeks, with an average of about 10 days. When the precipitation occurs, we have another transformation taking place, the energy by virtue of the water droplet's height (potential energy) is transformed into energy of motion (kinetic energy). The droplet's kinetic energy is given up in the form of heat to the atmosphere as it falls and finally to the land or water below when it hits the surface.

Some of the precipitation received by the land makes up stream flow or runoff. From our experience, we know that runoff flows downhill, ultimately back to the oceans from whence it came. We also know that for centuries, our society has used the **runoff** in various ways, to run grist mills, to move goods, to generate electrical power, to cool industrial facilities, and, in general, to furnish energy used to perform tasks for man. Water is still an important resource and will continue to be. Our simplified view of the water cycle[10] [3] provides us with the essentials for our later discussions of hydroelectric power and the need for water to cool fossil-fueled and nuclear-fueled power plants.

---

[10] Our discussion of the water cycle was simplified to the extent of neglecting the process of transpiration—evaporation from plants and the formation of ground water.

Energy Flow via the Food Chain—Ecology

The sun is the continuing source of energy for our ecological system (ecosystem). By an ecosystem we mean organisms and their environment. By definition, the ecosystem must include organisms in three categories: (1) producers—photosynthetic organisms (plants), (2) consumers—plant or animal eating organisms, and (3) decomposers—decay producing organisms. An ecosystem may be very spatially limited, as in organisms in a tiny spring pool, or it may be as large as earth and its interacting spheres—**atmosphere, hydrosphere, lithosphere,** and **biosphere.** For this discussion, we will mean the latter, which we have already termed the **ecosphere.**

A prime concern of the ecologist is the flow of energy and the cycling of matter through this complex system. Energy-flow through a **biotic** community begins with the storage of energy in plants by photosynthesis. Carbon dioxide and water along with the sun's energy (visible spectrum) are used by the plant to manufacture plant sugar (chemical energy) and plant fiber. Photosynthetic energy then flows through the ecosystem by way of **food chains.** In the simplest picture, we can speak of a single food chain, but, in reality, food chains interact to form a complex **food web.**

A food chain represents a series of steps of eating and being eaten. A simple food chain, the plant-mouse-weasel chain, is shown in Figure 3-25. The plants are the first step or level and are called the **producers.** The mouse, a **herbivore,** eats plants and converts the photosynthetic energy into animal tissue. The weasel, a **carnivore,** is a flesh eater. One can then imagine a larger animal, a second level carnivore, feeding on the weasel. Finally,

Figure 3-25. Plant-mouse-weasel food chain. The food chain represents the way in which energy flows through the biosphere. Because a mouse uses much less energy for "locomotion," about 80% of the plant food assimilated goes into flesh, which can be passed on to the next level of consumers in the chain. The weasel is much less efficient, about 20%, and so would pass considerably less on to a next level if there were one. Finally, the matter is returned to earth by the organisms promoting decay after the weasel dies and is thus made ready for further use by plants and animals.

solar energy

plants and grass   mouse   weasel

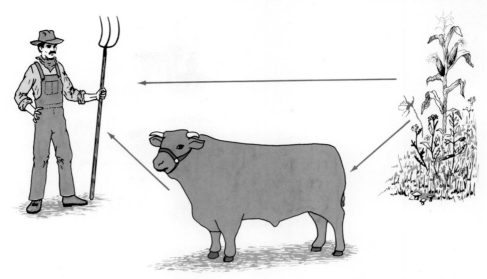

Figure 3-26. Man's food chain. Man's chain is more complex since he is an omni-vore—a plant and animal consumer. In man's chain, we see the beginning of a com-plex food web.

the food chain is terminated with the death and subsequent decomposition effected by microorganisms of decay. In man's food chain (Figure 3-26), the grasses are eaten by cattle, which are in turn eaten by man. However, man is **omnivorous** because he eats both plant and animal tissue.

When it comes to storing energy, photosynthesis is not a very efficient process for converting the sun's energy to plant matter. Only about 2% of the energy incident upon a leaf or a whole plant is ultimately converted to food energy. So, we can say that the efficiency of the plant for converting energy is 2%. This is one of the reasons why so little of the sun's energy (0.02%) is stored in plants by photosynthesis. Furthermore, each succeeding level of consumer passes on less energy. That is, a larger animal expends a larger portion of its energy intake in body maintenance. Less goes into building tissue which is stored energy (food) for the next consumer level. This is shown pictorally by the **trophic pyramid** (food pyramid) in Figure 3-27.

The trophic pyramid is simply a way of showing that smaller amounts of food are available for each succeeding consumer level of the food chain. The mass of each higher level on the pyramid is smaller than the preceding one. Only about 10% of the stored energy of any one step is passed on to each following step of the trophic pyramid. Thus, 1000 lb of forage grass would yield 100 lb of steer which, in turn, would yield 10 lb of human flesh.

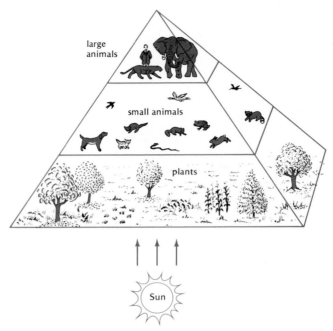

Figure 3-27. Trophic pyramid. The food pyramid represents the food for the next levels of consumers. An average figure would be about 10%, that is, about 10% of the mass of each level is available for supporting the following level.

An interesting consequence arises as a result of our study of the trophic pyramid. As man increases his numbers, he will have to subsist on the food from lower pyramid levels because the shorter the food chain the more energy there is to go around. Ultimately, we can visualize an overpopulated world living on **phytoplankton** (algae). Anyone for a phytoburger?[11]

Termination of the food chain is followed by rapid decomposition of the cells and tissues into carbon dioxide and water, accompanied by the consumption of oxygen. Heat is released and radiated into the environment, ultimately to be a part of the earth's heat energy which is reradiated into space. The components of decay were initially the components of photosynthesis. Carbon dioxide and water were photosynthesized into food while oxygen was liberated. This complex cycling of matter is called the **carbon cycle.** Contained within the carbon cycle is the **oxygen cycle.** The carbon and oxygen cycles are shown in Figure 3-28.

[11]One can further imagine the forming of phytoplankton in tanks using our own body wastes, diluted to optimal levels, for nutrients—fertilizer.

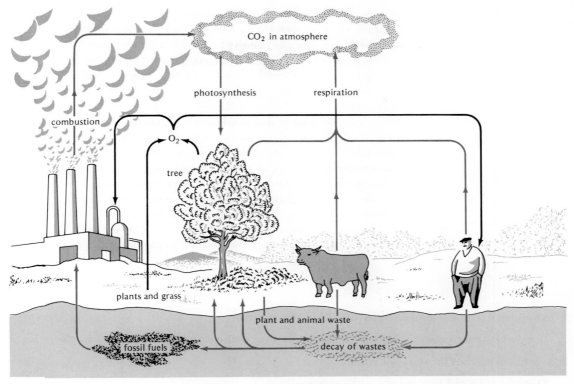

Figure 3-28. The carbon and oxygen cycles. A schematic diagram shows the various steps of taking in carbon dioxide (CO_2), water (H_2O), and sunlight to produce plant fiber and sugar while producing oxygen. The plant food is in turn used by animals. The animals respire, giving up carbon dioxide, and ultimately decay, giving up more carbon dioxide and water while using oxygen. The carbon dioxide is then available for further use. The oxygen cycle can be viewed as a part of the carbon cycle as shown. The very small amount of photosynthetic energy stored as fossil fuels over a period of 600,000 years is only about equivalent to 7 days of the sun's energy intercepted by earth.

It should be noted that other elements "cycle" in nature, for example, nitrogen and phosphorus. However, these cycles have been omitted here both due to their complexity and due to the fact that they are not necessary for our discussions. It also is to be noted that only a very small amount of the sun's energy over a period of 600,000 years has been stored in partially decayed plants—fossil fuels.

Conclusion

As we have seen from our considerations of energy flow in the ecosphere, matter cycles but energy does not. There is a continual cycling of matter in natural cycles such as in the carbon, oxygen, and water cycles. However, there must be a continuing supply of energy to power these cycles and to maintain life. Our major source of this energy is the sun. Even the sum total of earthly supplies of energy, that is, stored fossil fuels, nuclear fuels, and geothermal energy, is small when compared with energy from the sun. For comparison, total combustion of earth's known fossil fuels would provide an amount of energy equivalent to only seven days of sunshine (see Chapter 6).

Without the sun, our planet would immediately freeze, becoming a dark, cold chunk of rock in space. In the author's opinion, too little emphasis has been made on direct utilization of the sun's energy as a source of power. New nonpolluting sources of energy for man's use must be explored. In the following chapter, we will present a more quantitative description of energy, developed around electrical power generation using falling water —the hydrologic cycle. Other forms of energy and their use will be discussed throughout the remainder of the book. The prospects for direct utilization of the sun's energy have all been placed in the final chapter.

Questions

1. What characteristic does the sun have which makes it a star?

2. (a) By what process does the sun obtain its energy? (b) What conditions are necessary for this process?

3. In what forms does the sun emit energy?

4. What is a wave and how are waves characterized?

5. From your experience, how do we know that waves carry energy?

6. What is meant by the visible spectrum?

7. (a) One half of the sun's energy spectrum is in what part of the electromagnetic spectrum? (b) At what particular wavelength is the greatest amount of the sun's energy emitted?

8. Explain the statement, "The earth's atmosphere acts as a window that screens out the harmful forms of solar radiation."

9. Explain the significance of the solar constant and why it is important for us on earth for it to remain constant.

10. What might be the effect on earth of a decreased solar constant? An increased solar constant?

11. Describe the contributors to the earth's energy budget including the major energy "stores."

12. (a) What constitutes the earth's albedo? (b) What constitutes insolation?

13. Explain the significance of the term radiation balance with regard to the earth.

14. (a) At about what wavelength(s) does the earth radiate energy into space? (b) Suggest various ways in which the earth converts light into this radiated energy.

15. (a) Besides the sun and fossil fuels, what other sources of energy are available to man on earth? (b) Compare the potential of these fuels with that of the sun.

16. "A small change in albedo could cause a drastic change in climate." Explain this statement.

17. (a) What three major factors cause seasons on earth? (b) What would happen if the tilt of the earth's axis were increased? Decreased? Zero?

18. The earth is about 2 million mi closer to the sun in December than in June. Why isn't December the warmest time of the year in the northern hemisphere?

19. Explain how cloud and dust layers moderate the daily temperature difference at the earth's surface.

20. Explain the concept of the energy cycle.

21. What part of the hydrologic cycle provides earth with a potential energy resource?

22. Define what is meant by the following terms:
    (a) ecosystem   (d) food pyramid
    (b) food chain   (e) latent heat
    (c) food web

23. By what process is energy introduced into the food chain?

24. Explain how energy flows through the ecosphere.

Numerical Exercises

1. Show that the diameter of the sun is 1.39×10^6 km by converting the sun's diameter of 864,000 mi into kilometers.

2. If the volume of a sphere is given by $\frac{4}{3}\pi r^3$, where r = radius and π = 3.14, show that the sun's volume is 10^6 times the earth's.

3. The density of a body is the mass divided by the volume occupied by the mass ($\rho = m/v$). Show that the density of the sun is about 1.4 g/cm^3 (using the volume as determined in problem 2).

4. Using the volume for earth as found in problem 2 and the average density of earth as 5.5 g/cm^3, find the mass of earth in grams. Compare the earth's mass to the sun's mass.

5. Einstein proposed that energy and mass were equivalent, the equivalence given by $E = mc^2$, where E is energy in joules, m is mass in kilograms, and c is the speed of light, 3×10^8 m/sec. Show that the mass of 4.5×10^9 kg being converted into energy every second by the sun is 4×10^{26} joules.

6. If all of the sun's mass, 2×10^{30} kg, were to be converted into energy at the rate of 4.5×10^9 kg/sec, how long would the complete conversion of the sun's mass to energy take? How does this compare with the projected lifetime of 5×10^9 years we quoted? What is the reason for the discrepancy?

7. AM radio uses a frequency of 550 kHz (kilo-Hertz) for carrying "sound" information. What is the wavelength of the wave? Compare your finding with the FM wave in Example 3-2.

8. Express a wavelength of 10^{-7} m in microns. Where does this wave lie in the electromagnetic spectrum?

9. Does red light with $\lambda = 0.7\,\mu$ have a higher or lower frequency as compared to blue light with $\lambda = 0.4\,\mu$?

10. Of particular interest in the field of radio astronomy is energy coming from "empty" outer space with a wavelength of 21 cm or 0.21 m. In a radio astronomer's jargon, this is the 21-cm line of atomic hydrogen. What is the frequency of this electromagnetic wave and where does it lie in the spectrum (see Figure 3-6)?

11. The sodium lines in the sun's absorption spectrum occur at about $0.589\,\mu$. What is the wavelength expressed in meters?

12. The land area of the continental United States, excluding Alaska, is 3.02×10^6 mi^2.
    (a) Convert this to square meters.
    (b) Assuming that this land area is perpendicular to the sun's rays, determine the rate of energy reception by the United States.
    (c) How much energy is this for 1 year (in kilowatt-hours and using a 12-hr day)?
    (d) What per cent of the total energy received by earth is part c?

    It should be noted that this is a grossly simplified problem, but it is an *estimation* of the energy received by the United States. The variables are time of day, time of year, and latitude. A simple plane projection to correct for latitude, gives a corrected land area perpendicular to the sun's rays as 2.4×10^6 mi^2.

13. Show that the yearly sun's energy going into photosynthetic energy is 0.02% of the solar input.

14. What per cent of the earth's energy output is not directly related to the sun's input of energy? (Use Table 3-1.)

15. If the insolation were to decrease by 1% *or* the albedo were to increase by 1%, what would happen to the average surface temperature of the earth? What would happen if the change were 10%? (*Note:* It has been suggested that a 1% change would cause a 2°F change in surface temperature.)

16. (a) If the troposphere is about 10 km in thickness, what is the temperature at the outer region assuming a 7°C decrease every kilometer? What fraction of a mile is a kilometer? (3.28 ft = 1 m) Compare your answer with Figure 3-24.
    (b) What would be the temperature of the troposphere at 5 km in altitude? (*Note:* Use 20°C as the reference temperature at the earth's surface.)

17. If plants are 2% efficient at storing sunlight and the amount of energy stored in plants for 1 year is 3.5×10^{14} kwh, what must have been the total sunlight falling on vegetation?

18. (a) If the mouse, in the plant-mouse-weasel food chain, is 2% efficient in converting plant energy into flesh and if the weasel is only 2% efficient, what is the combined efficiency of mouse-weasel for providing food for higher level consumers?
    (b) Assuming 100 lb (45.5 kg) of forage is consumed by mice, how much flesh in pounds and in kilograms would be available for the next consumer level after the weasel? Do you think that a carnivore preying on the weasel could exist?

    (It should be mentioned that the mice would be expected to consume only about 0.5% of the net grass production. Furthermore, the weasel could be expected to catch only about 30% of the mice. So, in reality, the problem is even more complex than we have presented it here.)

References

[1] For a clear, but more detailed discussion of determining the mass of the earth and the sun and wave phenomena the reader is referred to V. H. Booth and M. C. Bloom: *Physical Science*. The Macmillan Company, New York, 1972.

[2] M. K. Hubbert: The energy resources of the earth. *Sci. Amer.*, **224**(3):60 (Sept. 1971); *Energy Resources: A report to the Committee on Natural Resources, National Academy of Sciences—National Research Council Publication 1000-D, Washington, D. C., 1962.

[3] For a more complete discussion of the water cycle, refer to B. Skinner: *Earth Resources*. Prentice-Hall, Inc., Englewood Cliffs, N. J., 1968.

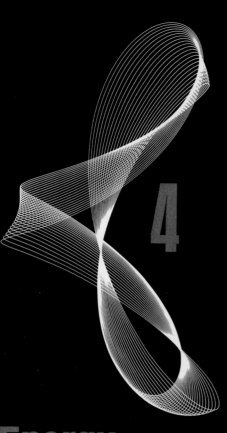

Energy, Work and Power

Up to this point, the term, energy, has deliberately been used only in a descriptive way. For all of us an understanding of the concept of energy is both useful and necessary before attempting to handle specific quantities of energy. As we view the physical world, we find that energy is one of the most fundamental and important concepts in science. Energy is essential to our everyday experience. From the time we turn off the electric alarm clock in the morning to the time we jump into our automobiles, powered by gasoline engines, until we sit down to the evening meal, the use of energy in various forms is a central feature of our daily activity. Without it, the activity of our mobile, technological society would grind to a halt. Put simply, our society is based on a continuing supply of inexpensive, accessible energy.

To discuss energy and its use in society, we need a common basis, usually known as units, for measuring quantities of energy. The introductory sections of Chapter 4 give a qualitative view of work and energy with a summary of the commonly used units. The later sections present work, energy, and power quantitatively via a discussion of water power, while noting that energy from falling water has played a significant role in the development of our industrial society. In our quantitative treatment, we include motion and Newton's laws. We conclude with a discussion of the equivalence of work and heat via the Joule experiment because nearly all the energy used in our society ends up as heat energy.

The Physical Nature of Work and Energy

In any descriptive treatise it is important to make plausible to the reader basic principles that at first may seem obscure but which, after some thought, really are matters of common sense. You may realize by now that energy in its various forms is probably one of the most familiar concepts to you, not one of the least familiar as you may have first imagined. Perhaps in your early morning stupor, you have touched a hot coffee pot and have become painfully aware of heat. Or, perhaps on a long drive you have left your sunglasses at home and have become painfully aware of the excessive light glare on a bright, sunny day. On the other hand, you have entered your warm, comfortable house, heated by burning fuel oil and abundantly furnished with electrical appliances. All of these experiences involve some form of energy.

From early childhood, you have been accustomed to seeing things fall. But, have you ever considered that the falling action is the result of a force being applied to the falling object? Or that, in the falling action, energy is being changed from one form to another? Furthermore, the moving object is capable of doing something for us that may or may not be useful. That something useful we usually refer to as **work.** For example, if a glass is sitting on the ground and you drop a rock onto it, the result is a smashed glass. If you desire to have the glass smashed, that is useful. It is the energy of motion of the rock that is transferred to the glass in such a way that the glass ends up in bits and pieces. Put another way, energy was expended in breaking the glass. Work was then performed.

Work and Energy

What then is work? Some of you may picture work as an activity that you must perform in order to earn money. And, in a sense, this is close to the scientific definition—you have expended force (your effort at the job) and you have accomplished something (earned money). Scientists use a rather simple definition of work, one which is associated with the most elementary form of labor—a lift, a push, or a pull that moves something.

Figure 4-1 illustrates the concept of work. Central to the concept is a force (an action) that displaces bodies or objects. Formally, **work** is defined as *the force applied to an object multiplied by the distance the object moves along the line of action* (equation (4-1)).

$$\text{work} = \text{force} \times \text{distance} \tag{4-1}$$

Conversely, no work is done if the object to which the force is being applied does not move, for example a force applied by a man to the wall of a building.

Now that we have defined work, we can say something about the units. The force unit in the English system is the **pound (lb),** whereas, in the metric system it is the **newton (nt).** The equivalence between newtons and pounds

Work is:

a lift

a push

a pull

Figure 4-1. Work is a lift, a push, or a pull that moves something.

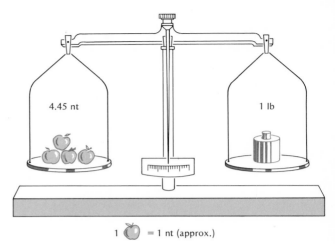

1 🍎 = 1 nt (approx.)

Figure 4-2. The equivalence between newtons (nt) and pounds (lb).

is shown in Figure 4-2. The distance or length units in the two systems are the **foot (ft)** and the **meter (m)**, respectively (1.00 m = 3.28 ft). Thus, the work done in applying a force of 4.45 nt (1 lb) to an object while moving it through a distance of 1 m is 4.45 nt × 1 m or 3.28 ft × lb. The unit of work is then the **newton × meter (nt-m)** in the metric system and the **foot × pound (ft-lb)** in the english system. The metric unit more commonly used is called the **joule (J)** (with the "ou" pronounced as in mouse).

$$1 \text{ J} = 1 \text{ nt-m} \tag{4-2}$$

Intimately related to this concept of work is power. **Power** is defined as the *rate of doing work,* or *work done per unit of time.*

$$\text{power} = \frac{\text{work}}{\text{time}} \tag{4-3}$$

Another definition of power is energy flow, that is, energy produced or consumed divided by time. The units of power are foot × pounds per second (ft-lb/sec) or joules per second (J/sec). Again, more commonly, the joule per second is called a **watt.**

$$1 \text{ watt} = 1 \frac{\text{J}}{\text{sec}} \tag{4-4}$$

Equation (4-3) simply says that a machine (or person) that can do twice the work of another machine in the same amount of time is twice as powerful. Some commonly used work and power units are summarized in Table 4-1, while selected relationships among these units are given in Table 4-2.

Table 4-1. Some Common Work, Energy, and Power Units

	Metric	*English*
work—energy unit	joule (J) kilowatt-hour (kwh) kilocalories (kcal)	foot-pound (ft-lb) British thermal unit (Btu)
power unit	watt (w) kilowatt (kw)	ft-lb/sec horsepower (hp)

Table 4-2. Relationships between Selected Units

1 kilowatt-hour (kwh) = 3.6×10^6 joules (J)
1 kilocalorie (kcal) = 4185 J
1 British thermal unit (Btu) = 778 ft-lb
                                     = 0.252 kcal
1 kwh = 860 kcal
1 kwh = 3412 Btu
1 Btu = 1 wooden kitchen match
1 horsepower (hp) = 550 ft-lb/sec = 746 watts (W)

An example of the use of work, energy, and power units appropriate for our emphasis is as follows: A 100-w light bulb consumes electrical energy at the rate of 100 J/sec. If we leave the bulb on for 1 hr the amount of energy consumed is 100 w-hr (100 w × 1 hr), or 0.1 kwh. A 200-w light bulb consumes energy at the rate of 200 J/sec or at twice the rate of the 100-w bulb. If we leave the 200-w bulb on for 1 hr the amount of energy consumed is 200 w-hr or 0.2 kwh.

From the work concept alone, a concept (albeit simplistic) of energy easily follows. A worker (Figure 4-3) carrying a bucket of water up a hill must do work, that is, apply a force that is equal to the weight of water

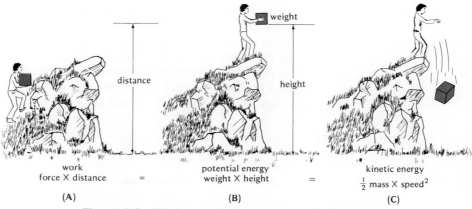

(A)   work
force × distance

(B)   potential energy
weight × height

(C)   kinetic energy
$\frac{1}{2}$ mass × speed^2

Figure 4-3. Work, potential energy, and kinetic energy.

and bucket to the bucket and carry it to some height. Note that the vertical rise, or height, is the appropriate distance here because the lifting action is vertical. By virtue of its position, the water (and bucket) have the capacity to do work. We call this *energy of position* or **potential energy** (PE). Potential energy is the weight times the height as in equation 4-5.

$$PE = weight \times height \qquad (4\text{-}5)$$

Furthermore, it is easily seen from equation (4-5) that the units of potential energy are foot $\times$ pound in the English system or in the metric system newton $\times$ meter (joule)—the same as the work unit. It further follows that if we define the potential energy as zero at the bottom of the hill, then the change in potential energy is equivalent to the work done in carrying the bucket up the hill. If the water is poured from the bucket, you know from your experience that it will travel faster and faster the farther the water falls. At the point just before it strikes the ground or a wheel, the water no longer has energy of position. Now it only has energy by virtue of its motion. We call the *energy of motion*, **kinetic energy**. Equation (4-6) shows how kinetic energy (KE) depends on speed.

$$KE = 1/2 \times mass \times (speed)^2 \qquad (4\text{-}6)$$

In contrast to potential energy, the kinetic energy depends on the mass of the falling object, not the weight as for potential energy in equation (4-5). The distinction between mass and weight is discussed later in this chapter. Also, equation (4-6) has the appropriate energy unit, that is, joule, and so on. When the water strikes a wheel, it applies a force to the wheel, moving the wheel through a distance; that is, the water does work on the wheel.

Thus, a simple mechanical view of energy suggests that **energy** is the *capacity to do work*. However, we should reemphasize that this is a simple, working definition. As we study energy and its use by society, we will find situations where this definition is not completely appropriate. Also illustrated in this example is the concept of energy transformation. In Figure 4-3, we observe the transformation from work to potential energy, potential energy to kinetic energy, and kinetic energy back to work again. In our simple example, *no energy is destroyed (or created), it is merely transformed between various types*. This statement is known as the **law of conservation of energy**. In later chapters, we will return to these principles in greater detail.

Other Forms of Energy

In our discussion of work and energy, we have alluded to various other types of energy. For example, when we refer to the energy of a moving object, as for a wheel, we often refer to the wheel as having mechanical energy (albeit kinetic energy). We call the energy available in oil and other substances, chemical energy. The chemical energy in oil may be considered as the potential and kinetic energies associated with molecular bonds. When the bonds are changed or broken in the burning process, the heat released comes from a change in the bond energy and can be used, for example, to heat water for steam–electric power generation.

We use electrical energy for our lights and the many appliances in our homes and businesses. The flow of electrical energy involves the flow of electrical charge down a wire, which most of you know is called a **conductor.** The flowing electrical charges have kinetic energy that can be converted into various forms of work and energy, for example, heat, light, chemical energy of storage batteries, pumping water, and so on. Although the details of electrical conduction are not necessary for our discussion, a brief treatment is provided in Appendix C.

As we have seen in Chapter 3, the sun is our main source of energy in the forms of light and heat. Light energy is necessary for the photosynthetic process that ultimately supports our life via chemical energy stored in plants. The heat part of the solar spectrum plays a big role in the circulation of air which sweeps the evaporated water (water vapor) up into clouds, later to be precipitated at high elevations. The potential energy of the water at higher elevations can be converted into work. A detailed discussion of energy from precipitated water—runoff—follows in the next section. These are but a few examples of the role of energy in our lives.

In conclusion, one of the more important points to remember is that work and energy are quantities, whereas power is the rate of doing work or the rate of consuming (producing) energy. Furthermore, the units we use to measure these quantities and rates form the basis for our discussion of what energy is and how we use it.

Water Power— A Quantitative Approach to Energy

Energy available from falling water is both familiar and particularly convenient for studying energy units along with certain quantitative aspects of energy. Water power has played a significant role in providing energy for the development of our industrial society. One of the regions of the United States most dependent on water power is the southeastern sector which is serviced by the Tennessee Valley Authority, T.V.A. Nearly all the available stream flow or runoff in that region has been developed.

Other regions of the country, for example, the northeastern sector, have large population densities and a high demand for electric power but little potential for additional development of water power. Even though water power will continue to be an important source of electrical energy in these and other parts of the country (and of the world), it is definitely a regional source of energy and cannot be expanded sufficiently to meet all the burgeoning demands. In the following section, we discuss the availability of water power for the United States and then begin the quantitative aspects of water power.

Availability of Water Power

The quantity of runoff (Chapter 3) is of interest to us for two major reasons: one, its potential for hydroelectric power production, and two, its potential for cooling fossil-fuel-fired electric power generating stations. The latter use will be discussed in Chapters 8 and 13. As discussed in chapter 3, the principal source of runoff is rainfall or precipitation.

Figure 4-4 summarizes the distribution of the annual precipitation in the continental United States. Each year an average of about 1.5×10^{15} gal (gallons) of water fall as rain. About 70% (1.05×10^{15} gal) of the rainfall goes into evaporation, transpiration, and absorption by the soil. Only the remaining 30% (0.45×10^{15} gal) of the rainfall goes into the runoff.[1] Only about 100×10^{12} gal (22% of 0.45×10^{15} gal) of this runoff is presently used. Of this, steam–electric power (30%) and agriculture (40%) consume about 70% of the 100×10^{12} gal withdrawn yearly. The balance (30%) goes mainly into municipal and industrial use.

Projections by the Water Resources Council show that steam–electric power will be by far the largest user of runoff by 2000 A.D., withdrawing at least 40% of the available runoff. The other users will withdraw another 27% making the projected need 67% [1] of the total runoff. The projection of the Water Resources Council of 67% assumes that our demand and the

---

[1]This amount of water corresponds to enough water to fill about 3.3 billion Olympic size swimming pools!

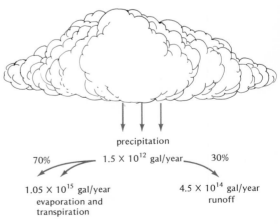

Figure 4-4. Distribution of annual precipitation in the United States. Only about 30% of the annual precipitation is available for our use. The remaining 70% is lost through evaporation and absorption.

precipitation

1.5×10^{12} gal/year

70%

30%

1.05×10^{15} gal/year
evaporation and
transpiration

4.5×10^{14} gal/year
runoff

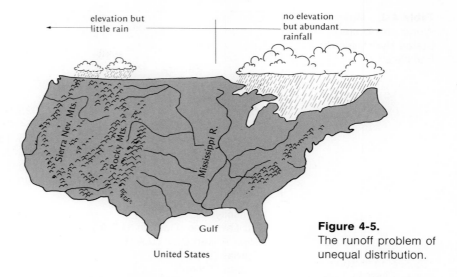

elevation but little rain

no elevation but abundant rainfall

Sierra Nev. Mts.

Rocky Mts.

Mississippi R.

Gulf

United States

Figure 4-5.
The runoff problem of unequal distribution.

subsequent growth of the electrical power industry will taper off somewhat in the 1980s. Other projections have suggested that we might even require 80 to 100% of the runoff for the various users, particularly for electrical power generation, if the rapid growth continues to the year 2000.

The availability of runoff for direct power production depends on many factors, but two in particular are of note: (1) the amount of rainfall in a particular area and (2) the topography of the area (Figure 4-5). The section of the United States which is the more mountainous (and has a higher land elevation above sea level) is the western part of the country. However, the land west of the Mississippi River receives only 35% of the annual precipitation of the 48 states. Furthermore, most of that 35% drops on the Pacific northwest (western Washington and western Oregon). The typical normal rainfall for most western states is only about 15 in. (yearly). The land east of the Mississippi river gets 65% of the total annual precipitation, but this area is considerably less mountainous. The consequence is that the runoff has less vertical distance to fall as it flows toward the sea. As you recall, the vertical distance the water travels is a measure of its potential energy.

Another important factor is the suitability of a particular region for damming the stream flow. Hydroelectric plants are locatable only where a sufficiently large amount of water, say in a river, can be dammed with a sufficiently high dam so that the water falls through a sufficient height to turn a water turbine. Obviously, too, a region with a high population density is not a very suitable land area to be used for water storage.

In the United States, east of the Mississippi River, most of the hydroe-

Table 4-3. Major World and United States Hydroelectric Facilities*

Facility	Present Power/ Total Power Capacity (10^3 kw)	Date Completed
United States		
Grand Coulee (Wash.)	2025/9771	1941
Robert Moses (Niagara)	1954/2050	1961
Hoover (Nev.)	1345/1345	1936
Fontana (N.C.-T.V.A.)	225/225	1944
Watauga (Tenn.-T.V.A.)	50/50	1948
Wilson (Ala.-T.V.A.)	630/630	1962
World		
Krasnayarsk (U.S.S.R.)	5000/6000	1968
St. Lawrence (Canada/U.S.A.)	1880/1880	1958
High Aswan Dam (U.A.R.)	1750/2100	1967
Furnas (Brazil)	900/1100	1963
Reban (Turkey)	620/1240	—
Kariba (Zambia)	600/1500	1959

*SOURCE: 1972 *World Almanac,* Cleveland Press; *T.V.A. Today,* 1972, T.V.A. Publications.

Figure 4-6. Norris Dam, completed in 1936, was the first dam built by T.V.A. Located on the Clinch River in east Tennessee, it is 265 ft high and 1860 ft long. The power installation consists of two 50 Mw units. [Courtesy of the Tennessee Valley Authority.]

lectric power potential had been developed as long ago as 1950. For example, by that time the Tennessee Valley Authority (T.V.A.) in the southeastern sector of the country had completed most of the present 27 major dams on the Tennessee River and its tributary rivers. T.V.A. stretches from Knoxville, Tennessee, to the Ohio River and furnishes power to seven states. In 1950, T.V.A. hydroelectric plants generated 14 billion kw of electricity [2], which has slightly increased to 16 billion kw in 1971. Since 1950, numerous coal-fired electric power plants have been built in the region, simply because the few remaining locations having hydroelectric potential were insufficient by themselves to meet the area's demand.

Table 4-3 lists some of the major United States and world hydroelectric facilities and their power capacity. One of the T.V.A. hydroelectric facilities is shown in Figure 4-6.

As we mentioned in Chapter 1, man has used water power for centuries to do tasks. However, the major development of water power did not begin until the late 1800s when it quickly grew to provide about 4% of the total power requirements of that time. Wood and coal provided the other 96%. Interestingly, in 1970 hydroelectricity still provided only about 4% of the United States' total power requirements, or 17% of the electric power requirements. Although hydroelectric power has played a significant role in the development of various regions, the increase in demand (steam–electric power included) in the United States, particularly since World War II, has been mainly met by the use of coal, oil, and natural gas. Because in the United States water resources are either mostly developed (or will not be developed because of environmental impact) now, the percentage of the total will continue to decrease, although, in various locales, water power will still remain a very important energy source. A discussion of the hydroelectric power world-wide potential for the future appears in Chapter 15, What Next? Let us now turn to a quantitative study of water power.

A Study of Motion

We live in a world of quantities. Weights and measures are constantly used by everyone. Our daily lives are permeated with numbers, facts, and figures. Usually attached to these quantities is a descriptive word or words. Without such descriptions, numbers in our practical physical world are meaningless. These words that we attach are called units (or sometimes dimensions).

The concept of units cannot be overstressed because we pay for all our consumption in terms of units; for example, we pay for our consumption of electrical energy in units of kilowatt-hours. Also, we continually are faced with distances, as measured in feet, meters, miles, and kilometers or time, as measured in seconds, days, weeks, or years. We further, consciously or unconsciously, estimate the speed we must travel to cover a given distance in a given time. Speed is measured (for example) in units of miles per hour or kilometers per hour. Thus, whether or not we are outwardly aware of our daily use of units, the concept is quite common in our everyday experience. Our approach to building our "unit reserve" while studying the concepts of work and energy will be a traditional one, beginning with speed.

The most basic concept in considering the motion is speed. It probably is the most familiar concept of motion to most of you. **Speed** simply is *the ratio of the distance traveled to the time required for covering the distance.* However, most of you are aware that your speed varies as you drive along. On a trip from New York City to Washington, D.C., that is about 200 mi, we might "average" 50 mi/hr (miles per hour). Thus, we will define an average speed.

$$\text{average speed } (\bar{v}) = \frac{\text{total distance traveled}}{\text{total time elapsed}} = \frac{d}{t} \tag{4-7}$$

We also note that the 200-mi or 322-km (kilometers) trip would require about 4 hr. Using equation (4-7), we determine an average speed of 50 mi/hr or 80.5 km/hr (kilometers per hour) (see Example 4-1).

Speed is a rate of covering distance. If the rate is constant, that is, if we travel at constant speed, we cover the same distance in equal time intervals. In 1 hr we travel 80.5 km; in 2 hr, 161 km; in 3 hr, 241.5 km; and 322 km in 4 hr. But, if our speed varies on the trip, our instantaneous speed may be higher or lower than the average. One way to determine our average speed is to record the distance traveled and the time for a number of distance intervals throughout the trip. Then an average speed could be determined for each interval. This gives a better profile of the trip, but ultimately our average speed for the whole trip is obtained simply by adding the distances traveled during the intervals to get the total distance and then dividing by the total time, equation (4-7).

EXAMPLE 4-1

The distance for our trip from New York City to Washington, D.C., as read on our car's odometer, was 200 mi or 322 km, and the time necessary for the trip was 4 hr. What was our average speed in miles per hour and in kilometers per hour?

$$\bar{v} = \frac{d}{t} = \frac{200 \text{ mi}}{4 \text{ hr}} = 50 \text{ mi/hr}$$

$$\bar{v} = \frac{d}{t} = \frac{322 \text{ km}}{4 \text{ hr}} = 80.5 \text{ km/hr}$$

In our practical physical world, we often find the need to change units, for example, from miles per hour to feet per second, or from kilometers per hour to meters per second, or conversions between the English system and the S.I. units.[2] In Example 4-2, we see that 30 mi/hr = 44 ft/sec, and 36 km/hr is the same as 10 m/sec. Example 4-3 shows the conversion between feet per second and meters per second.

In Chapter 2 we mentioned that a man (or woman) standing on the equator is moving in space along with the earth on which he is standing at a speed of 1054 mi/hr. Example 4-4 shows the determination of this unusual number.

[2] The S.I. system is essentially the metric system, meter-kilogram-second system.

EXAMPLE 4-2

(a) If you are moving at 30 mi/hr, calculate your speed in feet per second.
(b) If you are moving at 36 km/hr, calculate your speed in meters per second.

$$30\,\frac{mi}{hr} \times 5280\,\frac{ft}{mi} \times \frac{1\,hr}{3600\,sec} = 44\,\frac{ft}{sec}$$

or

$$30\,\frac{mi}{hr} = 44\,\frac{ft}{sec}$$

$$36\,\frac{km}{hr} \times \frac{10^3\,m}{km} \times \frac{1\,hr}{3600\,sec} = \frac{10\,m}{sec}$$

or

$$36\,\frac{km}{hr} = 10\,\frac{m}{sec}$$

EXAMPLE 4-3

Show that 73.3 ft/sec is equal to 22.3 m/sec.

$$73.3\,\frac{ft}{sec} \times \frac{1\,m}{3.28\,ft} = 22.3\,m/sec$$

EXAMPLE 4-4

A man standing on the equator travels in space a distance of 25,300 mi in 24 hr. What is his average speed?

$$\bar{v} = \frac{d}{t} = \frac{25,300\,mi}{24\,hr} = 1054\,mi/hr$$

Most of you are aware that increasing motion means to accelerate. For example, as you travel along in your automobile, increasing your speed means pushing further on your accelerator pedal. Conversely, decreasing your motion, decelerating, means to let up on the pedal. To a scientist, both speeding or slowing down are changes in motion and are called **accelerations**. Decelerations are just "negative" accelerations. You may also change your motion by changing your direction as in going around a corner. Thus, a change in motion is a change in speed and/or a change in direction (Figure 4-7). We will not treat changes in direction here. **Acceleration** is defined as *the change in speed divided by the time necessary for the change.*

$$\text{acceleration} = \frac{\text{change in speed}}{\text{time}}$$

or
$$a = v/t \qquad (4\text{-}8)$$

The change in speed is simply v, the final speed, if we take the initial speed as equal to zero. The definition in equation (4-8) is really the average acceleration as equation (4-7) was for the average speed.[3] The units for acceleration are miles per hour squared (mi/hr^2), kilometers per hour squared (km/hr^2), feet per second squared (ft/sec^2), meters per second squared (m/sec^2), and so on. Usually acceleration is given in feet per second squared or meters per second squared. The additional time unit in the denominator is obtained from the fact that speed, which is the rate of covering distance, distance per time, is divided by time.

Thus, acceleration has units of distance per time per time which is

[3] Traditionally, in physical science, acceleration is usually taken to be constant, in which case the average is the same as the instantaneous value.

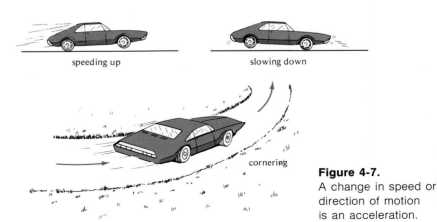

speeding up   slowing down

cornering

Figure 4-7.
A change in speed or
direction of motion
is an acceleration.

EXAMPLE 4-5

If the speed of your automobile changes from 0 to 60 mi/hr in 10 sec, what is your acceleration?

First

$$60\frac{mi}{hr} \times \frac{44 \text{ ft/sec}}{30 \text{ mi/hr}} = 88\frac{ft}{sec} = 26.8\frac{m}{sec}$$

$$a = \frac{v}{t} = \frac{88 \text{ ft/sec}}{10 \text{ sec}} = 8.8\frac{ft}{sec^2}$$

Or, in S.I. units,

$$a = \frac{v}{t} = \frac{26.8 \text{ m/sec}}{10 \text{ sec}} = 2.68\frac{m}{sec^2}$$

---

distance per time^2: feet per second squared, meters per second squared, and so on. Hence, acceleration is the rate of change of speed. A simple application of equation (4-8) is shown in Example 4-5. If your speed changes from 0 to 60 mi/hr (88 ft/sec) in 10 sec, your acceleration is 8.8 ft/sec^2. This is equivalent to 2.68 m/sec^2 in metric or S.I. units.

We have mentioned previously that bodies in space generally fall toward each other unless prevented from doing so by a swirling motion, as is observed in our solar system or in galaxies in outer space. Thus, a body released near the surface of the earth will fall freely toward the earth (Figure 4-8). Its speed will increase while falling, that is, a freely falling

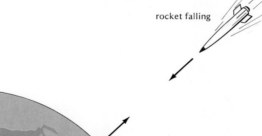

rocket falling

earth

Figure 4-8. Bodies fall toward each other due to mutually attractive forces. A body falls toward earth and earth falls toward the body. Any body (at the surface of earth) attracts earth with the same force that earth attracts the body.

body will accelerate. In reality, the earth falls towards the aforementioned falling body, too. However, the earth is so massive that no such motion of earth is detectable.

We also mentioned that this falling toward each other is due to the mutual attraction that bodies have for each other, the pull (force) of **gravity** (see Figure 4-8). Observations (Figure 4-9) show that the speed of such a body falling near the earth's surface increases by a nominal 32 ft/sec (or 9.8 m/sec) for every second of fall, ($v = at$). Thus, the **acceleration of gravity** ($a = g = v/t$) is 32 ft/sec² or 9.8 m/sec². Since the acceleration of gravity is used so often, scientists substitute g for a so that $v = gt$ for free-fall problems.

The concept of the acceleration of gravity is important for our discussion of the distinction between mass and weight. Acceleration is also intricately related to the change in potential energy to kinetic energy during free-fall of a body, via Newton's second law (following). It should be noted that all bodies (ignoring air resistance) accelerate in free fall at the value given for the acceleration of gravity. We have seen that a change in speed of a body means that the body has experienced an acceleration. However, with zero acceleration, that is, no force applied to a body, the motion of

Figure 4-9. Galileo dropping a rock from the leaning tower of pisa. Shown is the speed and acceleration for several times during fall.

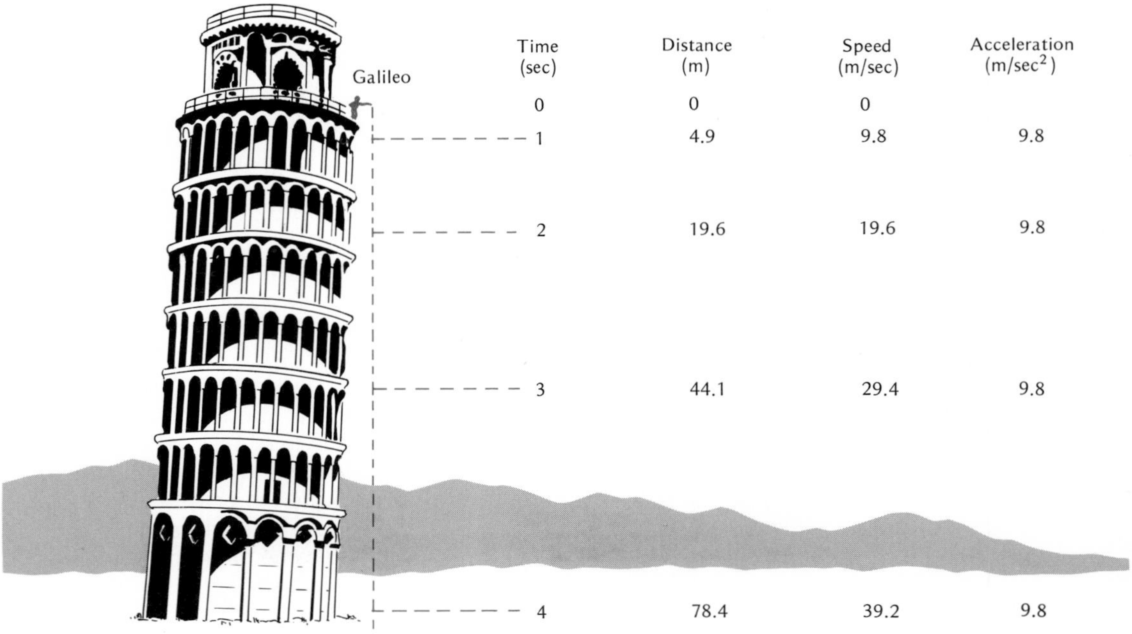

Time (sec)	Distance (m)	Speed (m/sec)	Acceleration (m/sec²)
0	0	0	
1	4.9	9.8	9.8
2	19.6	19.6	9.8
3	44.1	29.4	9.8
4	78.4	39.2	9.8

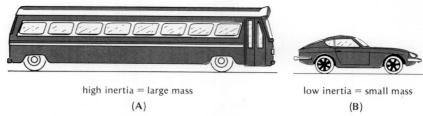

high inertia = large mass   low inertia = small mass
(A)   (B)

Figure 4-10. An illustration of Newton's first law of motion—the law of inertia. Inertia is a measure of a body's resistance to a change in motion. Mass is a measure of inertia. Neither vehicle will move unless acted upon by a force—a push.

a body will not change. This observation is embodied in Sir Isacc Newton's first law of motion (*Principia,* 1700):

> Every body continues in its state of rest, or of uniform motion in a straight line [constant speed and direction],[4] unless the state is compelled to change by an external force [which will produce an acceleration].

That property of a body which resists a change in motion is its **inertia,** (Figure 4-10). Thus, Newton's first law is sometimes referred to as the **law of inertia.** A measure of a body's inertia is its mass. Hence, we now have a physical definition of mass; **mass** is a *measure of a body's inertia.* Another useful definition is that **mass** is *the quantity of matter in a body.* (This is the definition used widely by chemists—Chapter 5.) The most widely used units for designating quantities of mass are the gram or kilogram. The mass or inertia of a body is easily illustrated by a simple example as shown in Figure 4-10. Shown are a sports car and a bus. Which do you think has the greater resistance to a push or to a change in motion, that is, which has the greater inertia?

Newton's second law of motion immediately follows the observations set forth by the first law and from the above example. To produce an acceleration, that is, a change in motion of a body, a "net" force must be applied (Figure 4-11). For a given force, the size of the resulting acceleration

[4]The comments in square brackets, which are made for clarity, are the authors. The statement of the three laws here are paraphrased from *Principia* as necessary for clarity.

Figure 4-11. Newton's second law—a body accelerates due to a force applied to it. The magnitude of the acceleration depends on the net force applied and the inertia-mass of the body.

force $\text{acceleration} = \dfrac{\text{force}}{\text{mass of car}}$

depends on the mass of the body; that is, for the same force (push) the resulting acceleration for the sports car will be greater than for the bus. Thus, Newton's second law says

> The acceleration of a body is directly proportional to the net force acting upon a body and inversely proportional to the mass of a body.

Newton's second law expressed as an equation is

$$\text{acceleration} = \frac{\text{force}}{\text{mass}} \tag{4-9}$$

or
$$a = \frac{F}{m} \tag{4-9a}$$

or
$$F = ma \tag{4-9b}$$

Since in this relationship (4-9b) a mass in kilograms is multiplied by an acceleration which has the units, meters per second squared, we encounter on the right hand side of this expression the units (kilograms × meters/seconds²). The equation requires force on the other side of the equation to have the same units.

The inconvenience of writing all these units is simplified by defining in the S.I. system **1 newton** (1 nt) as the force necessary to cause a 1 kg body to accelerate by 1 m/sec². In the English system of units, the force is measured in pounds. Our emphasis is on the S.I. system,[5] but we occasionally make the comparison with English units to help develop familiarity. We also use the English system where general use unfortunately necessitates a knowledge of dual units.

An illustration of the use of equation (4-9a) is shown in Example 4-6. The man in Figure 4-11 is able to apply a force of 890 nt (200 lb) to his sports car of 10^3 kg (2200 lb) mass. Using equation (4-9a), we find that he obtains an acceleration of 0.89 m/sec². That is, when this amount of force is continually applied to the car, every second the speed of the car increases by 0.89 m/sec ($v = at$). Although our concern is for falling water, Newton's laws apply to the motions of all bodies.

[5] The International System (S.I.) of units is summarized in Appendix A.

EXAMPLE 4-6

A man applies a force of 890 nt or 200 lb to his sports car, which has a mass of 10^3 kg or 2200 lb. What is the acceleration of the automobile?

$$a = \frac{F}{m} = \frac{890 \text{ nt}}{10^3 \text{ kg}} = 0.89 \frac{\text{kg-m/sec}^2}{\text{kg}} = 0.89 \frac{\text{m}}{\text{sec}^2}$$

Newton's laws of motion include a third law, which is really a statement about the nature of forces.

For every action force, there is always an equal but opposite reaction force.

This law has many consequences. For example, the man trying to push his sports car on a slick pavement would be in trouble (Figure 4-11). If

Figure 4-12. Action and reaction forces—Newton's third law. Any time there is a force applied to an object—action—there is always a reaction force applied by the object to the source of the force (a man, for example). However, a body's motion changes—acceleration—by virtue of the action (force applied to it), not by the reaction (force applied by it).

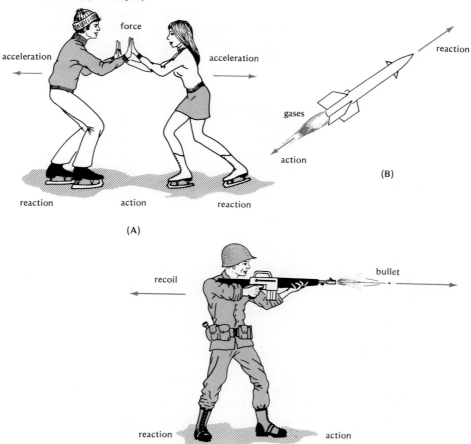

he applies a force of 890 nt to the automobile, the automobile likewise applies an 890-nt force to the man, but in the opposite direction. The significance of the slick road is that there is no frictional force to keep the man from sliding. Thus, both accelerate in opposite directions, each accelerating with a value depending on their respective masses (equation 4-9a). The reaction force is responsible for such things as rocket propulsion, the "kick" one receives when firing a shotgun, and many more phenomena (Figure 4-12).

The distinction between mass and weight follows from Newton's second law. An astronaut standing on the moon weighs about one sixth of what he weighs on earth. However, his mass is the same on the moon as on earth. That is, mass is invariant (except when a body is traveling close to the speed of light). *Weight is the force of gravity on a body* (Figure 4-13). The force of gravity, or the acceleration due to gravity, on the moon is one sixth the force of gravity on the earth. Weight on earth, in agreement with Newton's second law, is expressed as

$$\text{weight} = \text{mass} \times \text{acceleration of gravity}$$

or
$$w = m \times g \tag{4-10}$$

where g again has been substituted for a and has the value of 9.8 m/sec^2 or 32 ft/sec^2. According to equation (4-10), the astronaut weighs one sixth of his earth weight on the moon because the moon's g is one sixth of the earth's or about 1.6 m/sec^2. A woman weighing 100 lb on earth and 16.7 lb on the moon has a constant mass of 45.4 kg.

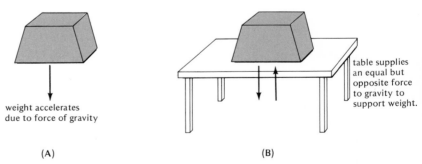

weight accelerates
due to force of gravity

table supplies an equal but opposite force to gravity to support weight.

(A)

(B)

Figure 4-13. Weight is the force of gravity on a body. By Newton's second law ($a = F/m$), a body accelerates because of a force applied to it. An unsupported body will accelerate toward the earth at 9.8 m/sec^2 or 32 ft/sec^2. The value of the force of gravity on a body which is weight is just the acceleration multiplied by the mass of the body. To support a body, a force that is equal and opposite to the force of gravity (weight) must be applied, for example, by a platform, a scale, or the earth's crust.

Table 4-4. Motion Units*

Quantity	Symbols	Metric, S.I. (abbreviations)	English
distance (length)	*l, d, h, x* etc.	centimeters (cm) meters (m) kilometers (km)	inches (in.) feet (ft) miles (mi)
time	*t*	seconds (sec, s) hours (hr)	seconds hours
mass	*m*	kilograms (kg) grams (g)	slugs
force	*F*	newtons (nt)	pounds (lb)
Rate speed	*v, c*	$\dfrac{\text{meters}}{\text{second}} \left(\dfrac{\text{m}}{\text{sec}}\right)$	$\dfrac{\text{feet}}{\text{second}} \left(\dfrac{\text{ft}}{\text{sec}}\right)$
		$\dfrac{\text{kilometers}}{\text{hour}} \left(\dfrac{\text{km}}{\text{hr}}\right)$	$\dfrac{\text{miles}}{\text{hour}} \left(\dfrac{\text{mi}}{\text{hr}}\right)$
acceleration	*a, g*	$\dfrac{\text{meters}}{\text{second}^2} \left(\dfrac{\text{m}}{\text{sec}^2}\right)$	$\dfrac{\text{feet}}{\text{second}^2} \left(\dfrac{\text{ft}}{\text{sec}^2}\right)$

*The units included here are those that we use commonly. A summary of units and conversions may be found in the Appendix A.

This comparison of units of measure also brings to focus our problem with a dual system of units. Other examples will be given as numerical exercises at the end of the chapter. You will note that now most goods purchased in a grocery store have both the English weight in pounds and the metric (S.I.) mass given. For example, a 1-lb can of beans also has a mass of 454 g (grams) or 0.454 kg. Note that some producers will use grams and some kilograms. Also, note the confusing weight-mass equivalence, 1 lb = 0.454 kg which is specific only at the earth's surface. What is worse, in Europe you would buy a kilo (kg) of potatoes, whereas, in the United States you would buy the equivalent 2.2 lb of potatoes. We are still caught in a dual system of units, so "grin and bear it!" Someday we may be entirely metric, but this change will be slow despite legislation.

A summary of the units we have developed in our discussion of motion is shown in Table 4-4.

Work and Energy

The conversion of work into potential energy usually involves lifting something. To lift means to apply a force equal to a body's weight. Doing work means lifting through a distance; work then is

$$\text{force} \times \text{distance} = \text{weight} \times \text{distance}$$

But, weight times distance (lifted) is what we earlier called potential energy, equation (4-5). Now that we know that weight is the mass of a body times the acceleration of gravity, we have the more commonly used equation (4-11) for potential energy (PE).

$$\text{PE} = m \times g \times h \quad \text{or} \quad \text{PE} = mgh \tag{4-11}$$

The conversion of work directly into kinetic energy involves a push or a pull; a push (or a pull) is a force that produces an acceleration. Thus,

$$\text{work} = \text{mass} \times \text{acceleration} \times \text{distance moved}$$
$$w = mad$$

We know from previous equations that $a = v/t$ and $\bar{v} = d/t$ where $a = g$ (so that $g = v/t$) if we are talking about free-fall and $\bar{v}$ is the average speed. but, the average speed $\bar{v}$ is the sum of the initial speed (which is zero) and the final speed divided by two, in the true sense of an average. So

$$\bar{v} = v_{\text{final}}/2 \quad \text{or} \quad d = v_{\text{final}}t/2$$

Combining these two equations with the definition of work, we obtain the following form for the kinetic energy (KE).

$$\text{KE} = \tfrac{1}{2}mv^2 \tag{4-12}$$

In reality, potential energy is converted into kinetic energy by gravity doing work. That is, the earth pulls (applies a force) on a body (the body's weight) and moves it through a distance (as the body falls). The details of these energy conversions are shown in Example 4-7.

EXAMPLE 4-7

(a) Show that the work done in lifting a body of mass m through a height h is the potential energy of the body. (*Note:* $F = \text{weight} = mg$.)

$$\text{work} = F \times h = mgh = \text{PE}_{\text{top}}$$

(b) Show that work done in pushing a body through a distance d is the kinetic energy of the body. Note from Equation 4-8, $a = v/t$ and $d = vt/2$, and from equation (4-9b), $F = ma$.

$$\text{work} = F \times d = ma \times d = m\frac{v}{t} \times \frac{vt}{2} = \frac{1}{2}mv^2 = \text{KE}$$

Thus, the work done on the body to move it through a distance d increases its speed to v and manifests itself as kinetic energy.

(c) Show that as a body falls, potential energy is converted into kinetic energy. Note that $g = v/t$ (page 114) and $d = h = vt/2$.

$$\text{PE}_{\text{top}} = mgh = m\frac{v}{t}\frac{vt}{2} = \frac{1}{2}mv^2 = \text{KE}_{\text{bottom}}$$

The significance of these examples is that they show in a quantitative way that energy is conserved; that is, energy is transformed without being destroyed. It also illustrates the relationship of simple motion to the concepts of energy and work. Let us now apply our motion principles to a specific quantitative energy calculation, energy from falling water.

As explained in our qualitative treatment of the hydrologic cycle, solar energy goes both into evaporation of water and into winds. The winds carry water vapor to higher elevations. For convenience, we take the higher elevations to mean the height of the waterfall being studied. Also, for simplicity, we consider the energy conversions taking place for a small quantity of water, a volume of 1 m^3 (cubic meter), and later generalize the problem to include stream-flow.

Because the density of water is 1000 kg/m^3 or (1 g/cm^3), 1 m^3 of water has a mass of 1000 kg. To determine the quantity of solar energy stored as potential energy of the 1 m^3 cube at the head of the waterfall, we only need to determine how much work a man must do to transfer the "cube" to the top of the waterfall, that is, to a height h (see Figure 4-14). With this simplification, our example can be handled easily with the use of basic mechanics. In our previous qualitative description, we have seen that the

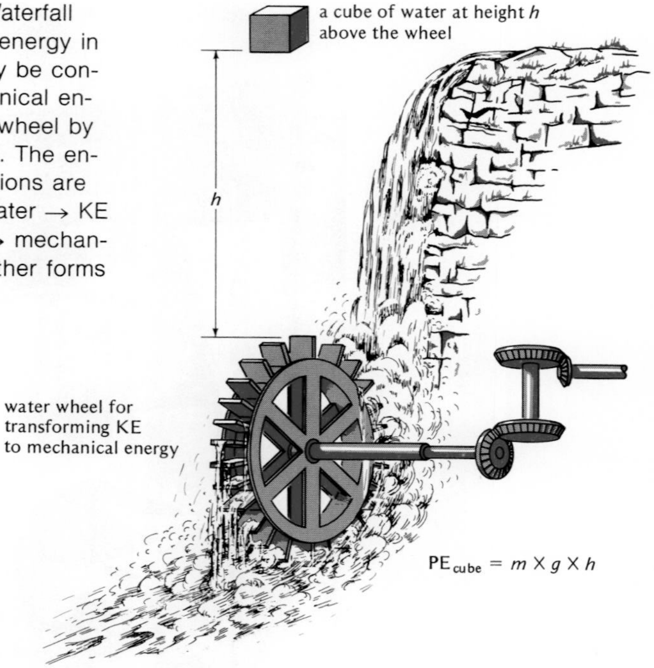

Figure 4-14. Waterfall and wheel. The energy in falling water may be converted to mechanical energy of a water wheel by doing work on it. The energy transformations are solar → PE of water → KE of water(work) → mechanical energy → other forms of energy.

a cube of water at height h above the wheel

h

water wheel for transforming KE to mechanical energy

$PE_{cube} = m \times g \times h$

work in transporting the cube of water to the top was transformed into potential energy of the cube. Furthermore, as the cube falls, the potential energy is completely transformed into kinetic energy. This kinetic energy then is capable of producing work, as in driving a water turbine which in turn may drive an electric generator. Figures 4-15 and 4-16 show a water turbine and turbine-generator facility.

For Example 4-8, we have chosen to use the height appropriate for Niagara Falls (Figure 4-16), 193 ft or about 60 m. That is, we say that the falls have a hydrostatic head of 60 m. The flow taken for the turbine at Niagara is actually taken from the Niagara River about 2 mi upstream and channeled under the town through large ducts to the turbine. So, our problem is necessarily a gross simplification.

First, let us determine the amount of work that we would have to do in order to transport the cube of water having a mass of 10^3 kg up through a height of 60 m. Then, what is the potential energy of the cube?

From Example 4-8, we see that the work done by solar energy in transporting the cube through a height of 60 m is 5.88×10^5 nt-m. For convenience, the work unit of 1 nt $\times$ 1 m (1 nt-m) is defined as 1 J (joule).

Figure 4-15. A water turbine being installed in one of the T.V.A.'s facilities. [Courtesy of the Tennessee Valley Authority.]

(A)

Figure 4-16.
(A) Niagara Falls.
(B) Niagara Falls power project. The water for power production is actually withdrawn about 2 mi upstream from the falls. The amount of water withdrawn depends on the time of year and flow. The facility is presently being operated at about the full capacity of 2000 Mw. [Courtesy of the Power Authority of the State of New York.]

(B)

EXAMPLE 4-8

(a) Find the work necessary to transfer a quantity of water $1 \, m^3$ or $10^3 \, kg$, through the height h of 60 m. (*Note:* $g = 9.8 \, m/sec^2$.)

$$\text{work} = F \times h = mgh = 1000 \, kg \times 9.8 \, \frac{m}{sec^2} \times 60 \, m = 5.88 \times 10^5 \, nt\text{-}m$$

The unit of work, newtons $\times$ meters (nt-m) is defined as 1 J (joule). Thus,

$$\text{work} = 5.88 \times 10^5 \, J$$

(b) What is the potential energy of the cube at 60 m? We showed in Example 4-7 that work done in lifting is equivalent to potential energy. Therefore

$$\text{work} = PE_{\text{top}} = mgh = 5.88 \times 10^5 \, J$$

EXAMPLE 4-9

The cube of water flows over the edge of the falls which are 60 m in height. First find the speed at the bottom, and then determine the kinetic energy of the cube just before splashing down.

From Equations 4-8 and 4-9 we have seen that $v = gt$ (or $t = v/g$) and $h = vt/2$. Substituting $t = v/g$ into the expression for h, we obtain

$$h = \frac{vt}{2} = \frac{v}{2} \times \frac{v}{g} = \frac{v^2}{2g}$$

Multiplying through by $2g$ yields

$$v^2 = 2 \, gh \qquad \text{or} \qquad v = \sqrt{2gh}$$

Now, $h = 60$ m and $g = 9.8 \, m/sec^2$, so

$$v^2 = 2 \times 60 \, m \times 9.8 \frac{m}{sec^2} = 1176 \frac{m^2}{sec^2}$$

For v we take the square root, and obtain $v = 34.3$ m/sec.

To obtain the kinetic energy

$$KE = 1/2 \, mv^2$$
$$= 1/2 \times 1000 \, kg \times 1176 \frac{m^2}{sec^2}$$
$$= 5.88 \times 10^5 \, kg \frac{m}{sec^2} \times m = 5.88 \times 10^5 \, nt\text{-}m$$
$$= 5.88 \times 10^5 \, J$$

Historically, the newton-meter and the equivalent joule are units of work. These are to be shown in their equivalence to heat units (kilocalories) in the Joule experiment discussed at the end of the chapter.

We also have seen that the potential energy at the top is just the work done in transporting the cube. In Example 4-9, we determine the kinetic energy that the cube of water has, just before it "splashes down." We find that, at the splash point at the bottom of the waterfall, the cube has kinetic energy of 5.88×10^5 J, the same as the potential energy on the top. Thus, in transforming the energy, we observe that none is destroyed or created. The energy has only been transformed, consistent with the law of conservation of energy.

The kinetic energy of 5.88×10^5 J is the energy of 1 m³ or 10^3 kg of water with a "head" of 60 m. Ideally, all of the 5.88×10^5 J of energy could be transformed into electrical energy, although practically, it is not possible to convert all of the available kinetic energy into "useful" work.

As a final example of our applications of mechanics, we extend Example 4-9 to include rate of water fall through the turbine and an estimate of the "potential" power output of a hypothetical, hydroelectric power plant operating according to our assumptions. In Example 4-10, we assume that

---

EXAMPLE 4-10

Assume that the river feeding the Niagara Falls is 200 m across, 6 m deep and flowing at a speed of 3 m/sec. First, what is the volume of flow going over the falls?

$$\frac{\text{vol}}{\text{sec}} = \text{width} \times \text{depth} \times \text{speed} = 200 \text{ m} \times 6 \text{ m} \times 3 \frac{\text{m}}{\text{sec}}$$

$$= 3600 \frac{\text{m}^3}{\text{sec}}$$

Since there are no water sources or water sinks to provide or capture flow on the way down, all the water crossing the edge must appear at the bottom.

If 1 m³ of water dropping over the falls provides 5.88×10^5 J of energy, the 3600 m³/sec crossing over the edge would ideally provide 3600 times this amount per second. Thus, the rate at which energy is available for turbine operation is

$$3600 \frac{\text{m}^3}{\text{sec}} \times 5.88 \times 10^5 \frac{\text{J}}{\text{m}^3} = 2117 \times 10^6 \frac{\text{J}}{\text{sec}}$$

But, the rate of obtaining energy is power, so

$$P = 2117 \times 10^6 \frac{\text{J}}{\text{sec}} = 2117 \times 10^6 \text{ w (watts)} = 2117 \text{ Mw (megawatts)}$$

the Niagara River in the vicinity of the falls in 200 m across, 6 m deep and flows at a speed of 3 m/sec. Using these dimensions, we calculate that 3600 m³/sec of water flows over the edge of the falls, and hence 3600 m³/sec of water is potentially available for driving the turbines. Because the energy available from 1 m³ with a "head" of 60 m is 5.88×10^5 J, 2.12×10^9 J/sec are available from a flow of 3600 m³/sec (see Example 4-10).

Our result of 2117×10^6 J/sec is the rate at which this waterfall delivers energy. The definition of power is the rate of obtaining energy, in analogy with the rate of doing work. The unit of power, joule per second, is called a watt. Power plant ratings are usually given in kilowatts (10^3 watts = kw) or megawatts (10^6 watts = Mw). Thus, our hypothetical plant could operate at 2117 Mw, which is equivalent to two large (1000 Mw each) coal- or oil-fired electrical generating facilities. For comparison, the Niagara Falls generating station is presently generating at 1950 Mw and has a rated capacity of 2054 Mw. Hence, the Niagara facility is capable of generating essentially at its total capacity. We want to reemphasize that the example is designed to illustrate principles and not to approximate the actual operating situation at the Niagara facility.

In 1 hr the "quantity" of energy delivered is 2.1×10^6 kwh (kilowatt-hours) (the numbers are rounded for convenience). In 2 hr the energy delivered is 4.2×10^6 kwh, and so on. When we pay our electric bills, we pay for quantities of energy, not power. Power is the rate at which the plant delivers energy to our homes. For another example, consider an appliance that has a power rating of 1 kw (a toaster). The toaster delivers energy at the rate of 1 kw, which is 1000 J/sec. If the toaster is used for 1 hr a quantity equaling 1 kwh ($= 3.6 \times 10^6$ J) of energy has been consumed. Refer to Tables 4-1 and 4-2 for a summary of energy and power units and some conversions. For other conversions, the reader is referred to Appendix A.

Even though hydroelectric power represents a small and actually decreasing percentage of the electric power generated in the United States, its problem analysis is particularly straight forward and provides a good foundation both in energy units and in the concept of energy transformations necessary for our later discussions of power production using fossil and nuclear fuels. We have developed our story of energy, work, and power using S.I. or metric units because power plants the world over use these units, namely, the power units of watt, kilowatt, and megawatt. In the following sections, we also refer to the heat energy units, the **British thermal unit (Btu)** and the **kilocalorie (kcal)**. These units and the appropriate conversions were included in Table 4-1 for completeness.

The Equivalence of Work and Heat— Friction

Most of our previous discussions have been on energy transformations not outwardly involving heat. However, we have alluded to the fact that the desired energy transformations are not absolutely complete. From time to time such energy loss shows up in the surroundings as heat, usually arising from work done against friction. We will not attempt a mathematical treatment of frictional forces and work against frictional forces here; however, the concept is shown in Figure 4-17. A man is shown attempting to push a heavy box across a floor, an action that you surely have experienced. Even though he is not attempting to lift the box against gravitational forces, he must apply a force to get the box to move. One of the forces that opposes his motion or force is the force of friction. The magnitude of the frictional force depends on the weight of the box and the roughness of the floor. After sliding the box, if you feel the bottom, you will find that it is warm.

An automobile moving through the fluid, air, must do work against the friction of air molecules, often referred to as air resistance. The result in each action described above is the evolution of heat. This also is evidenced from our space probe when the spacecraft reenters the earth's atmosphere. A very high temperature heat shield is necessary to keep the craft from burning.

A question for the Reader: You have increased the speed of your automobile up to 30 mi/hr (with $KE = \frac{1}{2}mv^2$) at which time you put the gear shift into neutral and take your foot off the throttle. You gradually come to a stop without applying the brakes. Where has your energy gone?

In Chapter 3, we saw that 47% of our energy budget was converted directly into heat and reradiated into space (and lost). Many are now

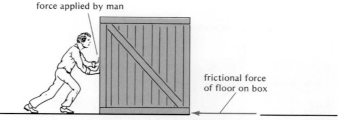

force applied by man

frictional force of floor on box

Figure 4-17. Concept of frictional force. Frictional forces oppose the motion of objects. In order to move any body, a force sufficient to overcome the frictional force must be applied to an object, and, in so moving, will have done work against friction. After moving the box, we would note a feeling of warmth on the bottom because the work done against friction is dissipated as heat.

questioning the potential use of this large energy resource in producing work. In the next two chapters, we will discuss the burning of coal and oil to obtain heat for doing work. Even though the last two examples involve the use of heat to produce work, historically one of the most cited experiments was the opposite conversion, namely, of work into heat. Hence, we will begin here, particularly because most of the rest of our story deals with using heat to produce work.

The experiment showing the equivalence between work and heat was performed by James P. Joule in 1845. In Figure 4-18, we show a diagram of Joule's experiment. It consists of a weight connected via a pulley to a paddle wheel in a container of water. Work is done to lift the weight through a distance giving the weight potential energy. The weight is allowed to fall, which turns the paddle wheel; work has been transformed into mechanical energy. The paddle wheel must turn against the opposing frictional forces of the water molecules. Upon performing his experiment, Joule observed that the temperature of the water increased, indicating the transformation of work or of the mechanical energy of the paddles into heat. The work-heat equivalence which he determined and which now has been determined more precisely as 4185 J equaling 1 kcal (kilocalorie) of heat. The equivalence is usually referred to as the **mechanical equivalent of heat.** A kilocalorie (kcal) is the amount of heat energy will raise the temperature of 1 kg of water 1°C. A similar experiment in English units would show that 778 ft-lb (foot-pounds) of work would produce 1 Btu

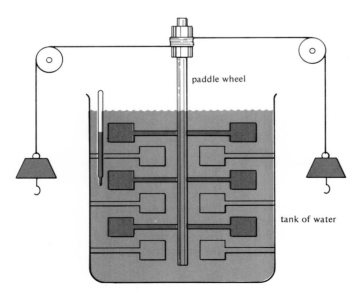

paddle wheel

tank of water

Figure 4-18. The Joule experiment. The equivalence of work and heat were shown by James P. Joule, an Englishman, in 1840. Weights could be shifted from the hook on one side to that on the other to turn the paddles in the tank. The potential energy of the weights when elevated could be calculated, and the rise in temperature of the known mass of water measured.

(British thermal unit) of heat. A British thermal unit is the amount of heat required to raise 1 lb of water 1°F.

Thus, although Joule's experiment clearly showed the equivalence between work, mechanical energy, and heat, it also showed that heat was a form of energy. Furthermore, it emphasizes the importance of friction as a way of dissipating energy in the form of heat.

Conclusion

At this point, we have focussed our attention on one of the two principal ways in which man has harnessed or utilized his environment to make his existence on earth more comfortable. By the study of falling or moving water, we have introduced the central concept of mechanics. The principles of mechanics, for example, Newton's laws of motion, govern motion. Motion involves energy. The transformations between various forms of energy are fundamental to an understanding of how man uses his environment, as in the use of moving water to do work or to generate electricity.

That energy of motion is transformable to heat is conceptually important, both as another illustration of an energy transformation and as an introduction to the opposite transformation of heat energy into energies of motion. The utilization of heat energy is the second principal way in which man has utilized his environment. This topic is the theme covered in Chapters 5 through 9 and incorporates most of the ideas and terminology of this chapter on energy, work, and power.

Questions

1. In what ways may water be considered as the driver of nature?

2. Is "runoff" water an inexhaustible resource?

3. Discuss the geographical limitations to the availability of runoff.

4. Why has (T.V.A.) water power in the southeastern sector of the country not been able to keep pace with the region's need for power?

5. Describe how the potential energy of water may be converted into work.

6. Distinguish between work and power. Give examples of each.

7. Is work the same as energy? Is energy the same as work?

8. After a child's tower of blocks has fallen, where has the potential energy that they had while stacked gone? Has it been destroyed?

9. Discuss what we mean by energy conversion. Give everyday examples of energy being converted.

10. What is the nature of the difficulty in giving a one sentence, all inclusive definition for energy?

11. Describe the importance of the law of conservation of energy.

12. Define the following:
    - (a) speed
    - (b) acceleration
    - (c) force
    - (d) mass
    - (e) work
    - (f) potential energy
    - (g) kinetic energy
    - (h) mechanical energy
    - (i) heat energy

13. Discuss the pertinence of (a), (b), (c), and (d) in question 12 to energy considerations.

14. (a) Distinguish between average speed and constant speed. (b) Distinguish between average acceleration and constant acceleration.

15. Describe what we mean by gravity.

16. Objects on the moon weigh about one sixth what they weigh on earth. Why is that?

17. (a) How are mass and inertia related? (b) Which has the greatest inertia, a loaded truck or an empty truck?

18. State Newton's three laws of motion and give examples of how each is applied in everyday experiences.

19. Which is greater, the attraction of the earth for 1 kg of lead or the attraction of 1 kg of lead for the earth?

20. Newton's third law says that action and reaction forces are equal and opposite. From this law, one might conclude that in all situations involving forces, the forces must balance giving a zero net force. Thus, there could never be a change in motion. What is the fallacy in the conclusion?

21. By what effects may forces of different magnitudes be compared?

22. Distinguish between mass and weight. What are the units of each (S.I.)?

23. (a) Discuss energy units and power units. (b) Discuss, "power is the rate of doing work; power is the rate of energy production or consumption."

24. Give five everyday examples of the scientist's definition of work.

25. Which of the following cases has the greatest kinetic energy on impact:
    - (a) 1000 nt object falling 1 m
    - (b) 100 nt body falling 10 m
    - (c) 10 nt body falling 100 m

26. Describe the energy transformations taking place in the following physical situations:
    - (a) The pendulum of your grandfather's clock.
    - (b) An automobile rolls down a hill and strikes another one at the bottom, thus moving the second automobile.
    - (c) A swiftly moving stream turning a paddle wheel.
    - (d) Water falling down a cliff running a turbine in an electrical power generating plant.
    - (e) A pile driver.
    - (f) An elevator moving upward before and after the cable breaks.
    - (g) A physical science student pulling his sled up the hill and coasting down.
    - (h) A windmill (begin with considering the sun's energy).

Numerical Exercises

1. The American ship, Savannah, was the first ship (1818) to use steam in crossing the ocean. The 3600-mi trip from Savannah to Liverpool took 29 days and 4 hr. What was the ship's average speed in miles per hour? Convert 3600 mi to kilometers and find the average speed in kilometers per hour.

2. Today, the same trip, from Savannah to Liverpool, can be made by plane in about 6 hr. What is the plane's average speed in miles per hour and in kilometers per hour.

3. In problems 1 and 2, convert the average speed from miles per hour and kilometers per hour to feet per second and meters per second.

4. In 1934, the Union Pacific (train) made the trip from Cheyenne to Omaha in 6 hr at an average speed of 84 mi/hr. How far did the train travel?

5. In July, 1958, the world's first nuclear submarine, the U.S.S. Nautilus, traveled submerged via the North Pole from Pearl Harbor to Iceland while averaging 8 km/hr, for the distance of 2940 km. How long did the trip take?

6. If a spaceship begins its long trek to a new sun, Alpha Centauri, which is 24×10^{12} mi away and if the ship is able to maintain a speed of 24,000 mi/hr, how long will the trip take in hours? In days? In years?

7. A Porsche 914 accelerates from 0 to 60 mi/hr (88 ft/sec) in 4 sec. Find the acceleration in feet per square second and in meters per square second. What per cent of the acceleration of gravity is this?

8. A truck slows to a stop from 60 mi/hr (88 ft/sec) in 20 sec. What is the deceleration (negative acceleration) of the truck?

9. A stone is dropped from a 200-ft high cliff. After 1 sec, its speed is clocked as 32 ft/sec (9.8 m/sec); after 2 sec, 64 ft/sec (19.6 m/sec); after 3 sec, 96 ft/sec (29.4 m/sec); and so on, until it hits. What is its acceleration during each interval and for the entire 3-sec interval (both English and metric)? How does this compare with the acceleration of gravity?

10. A sports car of mass 10^3 kg (2200 lb) accelerates from 0 to 26.8 m/sec (60 mi/hr) in 4 sec. What force must the engine provide for this acceleration? Note that according to Newton's third law the automobile will accelerate due to a force (a reaction force) applied to it. Where must this force come from?

11. A 154-lb man has a mass of 70 kg. What is his weight in newtons?

12. In Europe, you would like to buy a 2-lb bag of potatoes. How many kilos of potatoes would you ask for?

13. An automobile starts from rest and accelerates uniformly at 10 m/sec² for 15 min.
    (a) What will the speed of the automobile be in 15 sec?
    (b) How far will it have traveled?

14. A horse is able to move a cart having a mass of 100 kg by pulling with a force of 445 nt (100 lb). He moves the cart for 100 m. How much work has he done?

15. The 100-kg cart in problem 16 is traveling at a speed of 8 m/sec (about 15 mi/hr). What is the kinetic energy of the cart?

16. A small engine is able to do 7470 J of work in 10 sec. How powerful is it (in watts)? Find its power rating in horsepower.

17. A rock, with a mass 10^3 kg (2200 lb), is poised on the edge of a cliff 30 m above a road. What is the potential energy of the rock relative to the road below? When the rock falls and hits the road, where does its potential energy go?

18. A waterfall has a hydrostatic head of 80 m. What is the potential energy of 10^3 kg of the water?

19. If the river above the falls in problem 18 is 50 m wide, 10 m deep, and flows at a speed of 10 m/sec,
    (a) what is the volume of flow in meters per second?
    (b) what is the mass flowing over the edge in kilograms per second?

20. What is the potential power rating of the falls in problem 19? (Neglect any energy losses in falling, and so on.)

21. Find the British thermal unit equivalent to 10^4 ft-lb of work (778 ft-lb = 1 Btu).

22. Convert 12,000 Btu to kilowatt-hours (3412 Btu = 1 kwh).

23. Convert 1000 kwh of energy to kilocalories (1 kwh = 860 kcal).

24. What is the kinetic energy of each cubic meter of water in a stream that is moving with a speed of 10 m/sec?

25. Water, with a head of 40 m, is supplied to a turbine at the rate of 10^6 kg/sec. Assuming 100% efficiency at converting the kinetic energy of the water to mechanical energy, what is the power output of the turbine in watts?

26. A 418.5-nt weight (about 90 lb) at a height of 10 m is connected via a pulley to a paddle wheel which is submerged in a container of water.
    (a) As the weight falls to the gound, how much total energy is available for heating the water?
    (b) If there is 2 kg of water in the container, what is its rise in temperature, in degrees Celsius?

References

[1] Water Resources Council, 1968, National Water Assessment.
[2] From *T.V.A. Today*, T.V.A. Publications, 1972.

Fossil Fuels Are Molecules

5

The use of moving or falling water to perform work for man was widespread in pre-Industrial Revolution society, The economic impact on early societies due to variability of water flows (runoff) from season to season and from drought year to very wet year, was diminished by the development of a practical steam engine (by James Watt) mainly for pumping water. The steam engine was used to pump water from below the water wheel of the mill up to the reservoir so that the water might rerun over the water wheel. The steam engine was also used to pump out water that seeped into underground coal mines. By removing this water, large additional quantities of coal could be obtained from such mines.

Operation of the steam engine was based on heat obtained from combustion reactions. Although wood initially served as a source of fuel, "black rocks" or coal were found to be far more efficient as a source of heat energy to power the steam engines. The development

133

and widespread use of the steam engine characterized the period known as the Industrial Revolution.

This chapter and the next four chapters (through Chapter 9) focus on the use of heat energy to provide power for man's use. However, before undertaking a study of the transformation of heat energy into electrical energy, let us examine some sources of heat energy. These sources are coal, petroleum, and natural gas, which together are known as the fossil fuels. The fossil fuels are substances made up of molecules which, in turn, are made up of atoms. Thus, the chapter begins with a discussion of matter and some of the changes that matter undergoes. The nature of the atom and then the composition of molecules is then followed by a discussion of how scientists count very large numbers of these very small molecules. Comparisons of different kinds of atoms and some of the molecules they form lead to a general discussion of the periodic law and the periodic table, both of which are central to the science of chemistry. In conclusion, the chapter then describes briefly some of the molecules that are found in the fossil fuels.

Matter and Change

Matter

Chemistry is the study of all matter, its composition, its properties, and the changes between various forms that matter undergoes. Matter is anything that occupies space and has mass. Everything that we see or feel is matter, as well as many things we cannot see or feel. Such things as wood, food, coal, soil, or machinery are examples of matter we can see and feel. Air or an odor in the air is an example of matter we cannot see, yet we know that the air or an odor is all around us.

Matter can be grouped into two categories: pure substances and mixtures. **Pure substances** are materials that cannot be separated into two or more other forms of matter by physical means. **Mixtures** are materials that consist of two or more pure substances.

Let us illustrate. Pick up a sample of earth from most anywhere and examine it closely. Even with your naked eye you can distinguish many different ingredients, such as bits of wood, ash, soot, clay, broken glass, translucent grains of silica or sand, and many other things. Some of these ingredients are natural, some are added by man. If you examine the sample with a magnifying glass, additional components can be seen and identified, such as pollen grains, flakes of mica, and crystals of various minerals. In fact, to the practiced observer, the sample of earth tells the history of the spot from which it was taken, just as the artifacts found in diggings reveal to the archeologist the history of a civilization.

The sample of earth is a mixture. Some of the components of mixtures may themselves be pure substances. For example, the soot is probably pure carbon and cannot be broken or separated into other substances. Figure 5-1 is a schematic representation of a mixture of soot and sand. Silicon dioxide (sand) is a pure substance that cannot be easily separated into its components, silicon and oxygen. Thus, the main distinction between mixtures and pure substances is that a mixture is easily separated into its components whereas pure substances are not.

On a larger scale, a mixture of aluminum cans and tinplated steel (iron) cans can be separated physically by a magnet. Iron is attracted to the magnet whereas aluminum is not. Not all mixtures can be so conveniently identified and separated mechanically. A mixture of sand and salt (useful on icy spots in winter) is difficult to separate mechanically even though microscopic examination shows that the individual grains or crystals of sand and salt do look different. However, we all know that salt dissolves readily in water whereas sand does not. We also know that if we let the water evaporate completely from a salt solution, a white solid (salt) will be left behind. Evaporation of sea water, using energy from the sun, is used in many places in the world to obtain salt (sodium chloride). So, if a mixture of sand and salt is stirred with water, the water dissolves the salt, leaving the sand undissolved. By carefully pouring off the water solution, followed by evaporation of the water, we recover the salt free from sand.

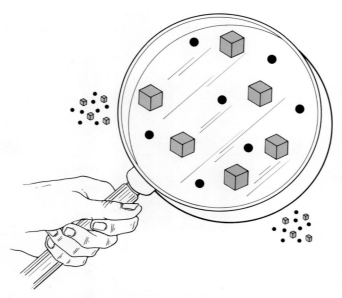

Figure 5-1. Salt-sand mixture under a magnifying glass.

Separately, salt and sand are each pure substances. It also follows from our example that a pure substance is matter that is chemically uniform and homogeneous. No matter what physical process we use in trying to separate the pure substance into other materials, we find that we still have the original material.

The States of Matter and Physical Change

Most matter, as we perceive it, exists in one or more states or phases. The common states of matter are **gas, liquid,** and **solid.** We are all aware of liquid water, solid ice, and water vapor. When liquid water is cooled to its freezing point, the liquid changes to ice or the solid state. When liquid water boils, it changes from a liquid to a gas, vapor, which because it has no color is invisible to the eye. Please note that the cloud we *see* from a hot tea kettle is actually made up of tiny droplets of liquid water. However, right at the spout of a rapidly-boiling kettle, no droplets are visible. Here, just above the tea kettle, one has gaseous water or water vapor. As the water vapor expands, it cools and condenses to form water droplets (Figure 5-2). These are examples of matter changing from one state to another.

Figure 5-2. Water boiling in a tea kettle. Note the cloud of water droplets forming at a distance from the spout. The water vapor, which cannot be seen at the spout, cools and condenses to form the cloud or steam.

Likewise, a piece of metal can be heated so that the solid melts to form a liquid. Further heating can change the liquid metal to vaporized metal, a gas. In general, then, matter can usually be changed from one state to another merely by changing its temperature. Such changes are called **physical changes.** A physical change is one in which *no* new substance is produced, although there may be a change in state. Simply, a physical change means a change in appearance as evidenced by the appearance of water in its three states. Other examples of physical change are melting glass, breaking glass, chopping wood, and raining.

Chemical Change

Pure substances are also known to undergo changes from which the original substances can no longer be recovered. For example, natural gas, which is largely the pure substance, methane, burns readily in air to give materials (gases) that do not burn. Interestingly, if a small flame of burning natural gas is directed against a cold surface, such as a ceramic plate or a sheet of metal, some liquid water will be observed to form on the plate. The burning of the natural gas has produced water, another substance quite different from the original methane. The burning of methane produces both water and carbon dioxide (Chapter 6).

Such transformations of substances are known as **chemical changes.** Other familiar chemical changes include burning of wood, burning of coal, rusting of iron, the flash of a flash cube, the action of antacids in the stomach, and the decay of foodstuffs and plant life. Before studying chemical changes (Chapter 6) though, we need to know something about the basic constituents of matter, atoms and molecules.

Atoms

Dalton's Atomic Theory

The observations of the behavior of substances, although very interesting and important in themselves, led many to ponder what it is that makes up pure substances. Perhaps one of the most notable contributors to chemistry was Antoine Lavoisier (1743–1794).[1] Lavoisier made many important observations during his quantitative studies of the behavior of substances in chemical changes.

Prior to Lavoisier, chemistry to most people was closely akin to magic. The alchemists of the Middle Ages sought to transform readily available,

---

[1] Lavoisier's experience illustrates an interesting relation between science and government. Lavoisier was a tax collector for the French King. This position gave him both time and money to pursue his interests in chemistry. Even though he made major contributions and was internationally recognized for his chemical work, when the French monarchy was overthrown, Lavoisier was first imprisoned and ultimately lost his life at the guillotine.

cheap metals to rare gold, or they sought to attain potions that would prolong life or convert old people to youths again. Lavoisier considered chemical changes with the instinctive feeling that there ought to be conservation of matter, that is, any material entering into a chemical change should be present at the end, though in altered form. He recognized the significance of the fact that air or gases have weight, and he sought ways to account for the materials involved in combustion. For his insight he became known as the father of modern chemistry (see Figure 5-3).

Lavoisier's careful experiments prepared the way for a concept, a theory, which is the cornerstone on which the science of chemistry is built, the **atomic theory** of John Dalton. Dalton (1766–1844), an English school teacher, became fascinated with the changes that matter undergoes and, after many years of study and thought, proposed a series of ideas in 1804 that provided a basis for understanding chemical changes. In a thoroughly modernized form, Dalton's atomic theory states:

1. All matter is ultimately composed of extremely small particles called atoms.
2. Atoms of a particular element are identical in size, shape, and form, and differ from other atoms in size, shape, and form.
3. Chemical change is the separation and/or union of atoms. Atoms unite to form molecules.
4. Molecules of substances are formed when atoms of one or more elements combine.
5. In a given pure compound, the relative numbers of atoms of the elements present will be definite and constant. In general, these relative numbers can be expressed as integers (law of definite proportions).

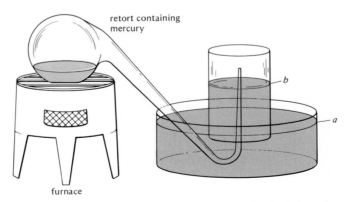

Figure 5-3. Lavoisier's experiment, which showed that air is one fifth oxygen. Heating the liquid metal mercury in the presence of air produces a red powder, mercuric oxide.

In these statements, we have a definition of the term, atom. Dalton's atom is the basic constituent of matter. However, the term element has also been introduced. A pure substance that is made up of only one kind of atom is known as an element or an elementary substance. A pure substance that is made up of two or more different kinds of atoms is known as a compound. About 90 different kinds of atoms or elements occur naturally, and another 15 are man-made. Some 3 million pure substances are known, all of which are made up of combinations, or multiple combinations, of these 105 elements.

The Rutherford-Bohr Atom

In Dalton's time, atoms were considered the ultimate (smallest) form of matter. Dalton considered atoms to be indivisible. In the late nineteenth century and early twentieth century, scientific studies of x–rays and of radioactive atoms, such as radium and uranium, suggested that atoms in fact are divisible under extraordinary circumstances. However, in changes such as combustion, as we will discuss in Chapter 6, atoms are not divisible.

The work (1911) of Rutherford and others resulted in the theory that atoms are made up of electrically charged units, protons and electrons, which are of opposite charge [1]. These units are called subatomic particles. The **proton** is a positively charged particle, whereas the **electron** is a negatively charged particle. The magnitude of the electrical charge associated with one electron is exactly equal and opposite to the charge of the proton.

In addition to protons and electrons as components of atoms, a third subatomic particle was discovered in 1932 by James Chadwick (1891–   ). Chadwick proposed that atoms also contain an electrically neutral particle whose mass is approximately the same as the proton. He called it the **neutron.** Subsequent studies have identified several additional subatomic particles, but, for practical purposes, most chemists focus attention on only these three: the proton, the electron, and the neutron. The characteristics of these three particles are summarized in Table 5-1.

Rutherford and Bohr proposed a nuclear model of an atom, a common but incomplete illustration of which is shown in Figure 5-4.

The essential features of the modern nuclear model for an atom are:

Table 5-1. Relation of Mass and Charge of Elementary Particles

*The mass of an electron is approximately $\frac{1}{2000}$ the mass of the proton.

Characteristic Particle	Proton	Electron	Neutron
symbol	p	e	n
relative electrical charge	+1	−1	0
relative mass	1	about 0*	1

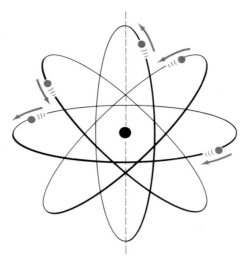

Figure 5-4. The Rutherford-Bohr model of an atom. Although this model is now used popularly and is useful, it does not provide a full modern concept of the atom.

1. An atom is composed of three fundamental particles: protons, electrons, and neutrons.
2. Atoms are electrically neutral; thus, they contain equal numbers of protons and electrons. The number of protons in an atom is called the atomic number (Z).
3. The neutrons and protons are located in the center of the atom, in a region known as the nucleus.
4. The electrons are located outside the nucleus. Each atom has a specific arrangement of electrons.
5. Because electrons have a very small mass and because the neutrons and protons provide nearly the entire mass of the atom, the sum of the masses of the neutrons and protons in an atom is commonly called the mass number of the atom (A).

Thus, the view of modern theory is that the atom is composed of a small, heavy nucleus (primarily protons and neutrons) with particles (electrons) in space about the nucleus, but at great distances from the nucleus. A simple model of the hydrogen atom gives the size of the nucleus as about 10^{-15} m as compared to an orbital radius of the electron of 5×10^{-11} m. That is, for the hydrogen atom, we would have to line up about 50,000 nuclei (protons) to reach out to the electron's position in the atom. Or, if the nucleus were the size of a basketball, then the electron would be a marble 6 mi (10 km) away. Hence, an atom is composed mostly of empty space. To further illustrate, if the nucleus of an atom were to be expanded to

Table 5-2. Composition of Selected Atoms

Element	Symbol	Number of Electrons	Number of Protons	Number of Neutrons	Atomic Number	Mass Number,
hydrogen	H	1	1	0	1	1
helium	He	2	2	2	2	4
lithium	Li	3	3	4	3	7
carbon	C	6	6	6	6	12
oxygen	O	8	8	8	8	16
aluminum	Al	13	13	14	13	27
sulfur	S	16	16	16	16	32
iron	Fe	26	26	30	26	56
uranium	U	92	92	146	92	238

be the size of the period at the end of this sentence, the size of the total atom would be equal to a house.

In Table 5-2, the composition of selected atoms is indicated.

Symbolism

In wide use today is a shorthand notation, or symbolism, which conveys information about atoms. One of the giants of early nineteenth century chemistry, J. J. Berzelius (1779–1848), a Scandinavian, introduced the symbolism of having letters stand for atoms of an element. Thus, H represents one atom of hydrogen, O one atom of oxygen, S one atom of sulfur (see Table 5-2). Sometimes the symbol for the atom is written with the atomic number, the number of protons (Z), as a subscript and the mass number, the number of protons plus neutrons (A), as a superscript. Thus, the general symbol is given as $_Z^A X$. The data for five of the atoms detailed in the above table is summarized as

$$_1^1 H, \quad _2^4 He, \quad _6^{12} C, \quad _{13}^{27} Al, \quad and \quad _{92}^{238} U.$$

The atomic number 13 for aluminum indicates that the aluminum atom has 13 protons and 13 electrons in the neutral atom. By subtraction of 13 mass units associated with the protons from 27 mass units for the entire atom, you are left with 14 mass units, which in our nuclear model for the atom must represent 14 neutrons. In general, the number of neutrons can be found by subtracting the atomic number of an element from its mass number. Thus, all the information in the table for aluminum is summarized by $_{13}^{27} Al$. Similarly, $_{92}^{238} U$ has 92 protons, 92 electrons, and 146 neutrons (Example 5-1). We will use this symbolism extensively in our later discussions of energy from the nucleus.

EXAMPLE 5-1

Calculate the number of protons, electrons, and neutrons in $^{238}_{92}U$.

$$\text{number of protons} = \text{atomic number} = 92$$
$$\text{number of electrons} = \text{number of protons} = 92$$
$$\text{number of neutrons} = (\text{mass number} - \text{atomic number}) = (238 - 92) = 146$$

Atomic Weights

We know that atoms are extremely small particles. Weighing individual atoms is an extremely difficult task. In fact, one atom cannot be weighed, even on the most sensitive weighing device. However, comparison of masses is possible by means of a device known as a mass spectrometer. As we have seen, different kinds of atoms have different masses (different mass numbers). In Table 5-2, comparison of mass numbers shows that a carbon atom (12) is 12 times as heavy as a hydrogen atom (1), or that a sulfur atom (32) is twice as heavy as an oxygen atom (16). Without elaborating the reasons why, scientists today have agreed to compare the masses of atoms to the mass of a specific carbon atom, $^{12}_{6}C$.

Interestingly, mass spectrometer analysis of a sample of carbon or of sulfur does not yield an integral number for the mass of the samples, as would be implied by our previous discussion. For example, the mass spectrometer analysis of a sample of sulfur shows that the sample contains four different kinds of sulfur atoms, each with different mass numbers. Most (95%) of the sulfur atoms have 16 neutrons and thus have a mass number of 32 ($^{32}_{16}S$). About 4% of the sulfur atoms have 18 neutrons or a mass number of 34 ($^{34}_{16}$)S. Somewhat under 1% have 17 neutrons or a mass number 33 ($^{33}_{16}S$), and a very small fraction have 20 neutrons or a mass number 36

EXAMPLE 5-2

Calculate the average mass number of a sulfur atom.

$$0.95 \times 32 = 30.40$$
$$0.04 \times 34 = 1.36$$
$$0.01 \times 33 = \underline{0.33}$$
$$32.09$$

(Ignore isotope 36 as of little significance.)

Thus: One atom (the average atom) has a mass number of 32.09 or 32.09 is the average mass number of a sulfur atom. (*Note:* The atomic weight of sulfur is 32.06 which is close, but not identical, to the 32.09 of this calculation.)

Figure 5-5. Table of atomic weights.

Element	Symbol	Atomic Number	Atomic Weight*	Element	Symbol	Atomic Number	Atomic Weight*
Actinium	Ac	89	(227)	Mercury	Hg	80	200.59
Aluminum	Al	13	26.9815	Molybdenum	Mo	42	95.94
Americium	Am	95	. . . .	Neodymium	Nd	60	144.24
Antimony	Sb	51	121.75	Neon	Ne	10	20.183
Argon	Ar	18	39.948	Neptunium	Np	93	237.05
Arsenic	As	33	74.9216	Nickel	Ni	28	58.71
Astatine	At	85	(210)	Niobium	Nb	41	92.906
Barium	Ba	56	137.34	Nitrogen	N	7	14.0067
Berkelium	Bk	97	. . . .	Nobelium	No	102	. . . .
Beryllium	Be	4	9.0122	Osmium	Os	76	190.2
Bismuth	Bi	83	208.980	Oxygen	O	8	15.9994
Boron	B	5	10.811	Palladium	Pd	46	106.4
Bromine	Br	35	79.909	Phosphorus	P	15	30.9738
Cadmium	Cd	48	112.40	Platinum	Pt	78	195.09
Calcium	Ca	20	40.08	Plutonium	Pu	94	(244)
Californium	Cf	98		Polonium	Po	84	(210)
Carbon	C	6	12.01115	Potassium	K	19	39.102
Cerium	Ce	58	140.12	Praseodymium	Pr	59	140.907
Cesium	Cs	55	132.905	Promethium	Pm	61	. . . .
Chlorine	Cl	17	35.453	Protactinium	Pa	91	. . . .
Chromium	Cr	24	51.996	Radium	Ra	88	226.0
Cobalt	Co	27	58.9332	Radon	Rn	86	(222)
Copper	Cu	29	63.54	Rhenium	Re	75	186.2
Curium	Cm	96	. . . .	Rhodium	Rh	45	102.905
Dysprosium	Dy	66	162.50	Rubidium	Rb	37	85.47
Einsteinium	Es	99	. . . .	Ruthenium	Ru	44	101.07
Erbium	Er	68	167.26	Samarium	Sm	62	150.35
Europium	Eu	63	151.96	Scandium	Sc	21	44.956
Fermium	Fm	100	. . . .	Selenium	Se	34	78.96
Fluorine	F	9	18.9984	Silicon	Si	14	28.086
Francium	Fr	87	(223)	Silver	Ag	47	107.870
Gadolinium	Gd	64	157.25	Sodium	Na	11	22.9898
Gallium	Ga	31	69.72	Strontium	Sr	38	87.62
Germanium	Ge	32	72.59	Sulfur	S	16	32.064
Gold	Au	79	196.967	Tantalum	Ta	73	180.948
Hafnium	Hf	72	178.49	Technetium	Te	43	. . . .
Helium	He	2	4.0026	Tellurium	Te	52	127.60
Holmium	Ho	67	164.930	Terbium	Tb	65	158.924
Hydrogen	H	1	1.00797	Thallium	Tl	81	204.37
Indium	In	49	114.82	Thorium	Th	90	232.038
Iodine	I	53	126.9044	Thulium	Tm	69	168.934
Iridium	Ir	77	192.2	Tin	Sn	50	118.69
Iron	Fe	26	55.847	Titanium	Ti	22	47.90
Krypton	Kr	36	83.80	Tungsten	W	74	183.85
Lanthanum	La	57	138.91	Uranium	U	92	238.03
Lawrencium	Lw	103	257	Vanadium	V	23	50.942
Lead	Pb	82	207.19	Xenon	Xe	54	131.30
Lithium	Li	3	6.939	Ytterbium	Yb	70	173.04
Lutetium	Lu	71	174.97	Yttrium	Y	39	88.905
Magnesium	Mg	12	24.312	Zinc	Zn	30	65.37
Manganese	Mn	25	54.9380	Zirconium	Ar	40	91.22
Mendelevium	Md	101	. . . .				

*The atomic weights are relative to $^{12}_{6}C$ as 12.0000. Numbers in parentheses are the mass numbers of the most stable or best-known isotopes.

Number above symbol = atomic weight
Number below symbol = atomic number

IA	IIA	IIIB	IVB	VB	VIB	VIIB	VIII	VIII	VIII	IB	IIB	IIIA	IVA	VA	VIA	VIIA	0
1.0079 H 1																	4.00260 He 2
6.94 Li 3	9.01218 Be 4											10.81 B 5	12.011 C 6	14.0067 N 7	15.9994 O 8	18.9984 F 9	20.179 Ne 10
22.9898 Na 11	24.305 Mg 12											26.9815 Al 13	28.086 Si 14	30.9738 P 15	32.06 S 16	35.453 Cl 17	39.948 Ar 18
39.098 K 19	40.08 Ca 20	44.9559 Sc 21	47.90 Ti 22	50.9414 V 23	51.996 Cr 24	54.9380 Mn 25	55.847 Fe 26	58.9332 Co 27	58.71 Ni 28	63.546 Cu 29	65.38 Zn 30	69.72 Ga 31	72.59 Ge 32	74.9216 As 33	78.96 Se 34	79.904 Br 35	83.80 Kr 36
85.4678 Rb 37	87.62 Sr 38	88.9059 Y 39	91.22 Zr 40	92.9064 Nb 41	95.94 Mo 42	98.9062 Te 43	101.07 Ru 44	102.9055 Rh 45	106.4 Pd 46	107.868 Ag 47	112.40 Cd 48	114.82 In 49	118.69 Sn 50	121.75 Sb 51	127.60 Te 52	126.9046 I 53	131.30 Xe 54
132.9054 Cs 55	137.34 Ba 56	57–71 *	178.49 Hf 72	180.9479 Ta 73	183.85 W 74	186.2 Re 75	190.2 Os 76	192.22 Ir 77	195.09 Pt 78	196.9665 Au 79	200.59 Hg 80	204.37 Tl 81	207.2 Pb 82	208.9804 Bi 83	(210) Po 84	(210) At 85	(222) Rn 86
(223) Fr 87	(226.0254) Ra 88	89–103 †	104	105													

232.0381 Th 90	231.0359 Pa 91	238.029 U 92	237.0482 Np 93	(242) Pu 94	(243) Am 95	(245) Cm 96	(245) Bk 97	(248) Cf 98	(253) Es 99	(254) Fm 100	(256) Md 101	(253) No 102	(257) Lr 103

Figure 5-6. Periodic table. The atomic weights (lower number) are based on $^{12}_{6}C$ as 12.0000. Numbers in parentheses are the mass numbers of the most stable isotopes of these radioactive elements. Most of the elements are metals, with the nonmetals to the right of the zig-zag black line.

($^{36}_{16}$S). Atoms of the same element (same Z) but differing in mass numbers (A) are known as **isotopes.** When we talk of a collection of sulfur atoms (for we almost always deal with large numbers of atoms), we have all these kinds of sulfur atoms present.

We think of the (weighted) average mass number of sulfur as 32.09. For reasons which will be discussed in Chapter 10, we do not use weighted average mass numbers for comparison of atoms, but rather data obtained by comparison of the masses of typical samples of an element to the specific isotope of carbon, $^{12}_{6}$C. These data are known as **atomic weights.**[2]

Similarly, when a sample of chlorine (Cl) atoms is compared with the same number of carbon atoms, we find that the assembly weighs 35.453/12 times as much or 2.954 times as much. Now clearly, 35.453 is not a simple integer. We find in the mass spectrometer that a sample of chlorine atoms has about 76% $^{37}_{17}$Cl atoms and about 24% $^{37}_{17}$Cl atoms. The average mass of a chlorine atom is 35.453 as compared to $^{12}_{6}$C. For convenience, the relative masses of all atoms (an average sample is usually taken) have been determined and this data is summarized in the **atomic weight table,** Figure 5-5 and in the **periodic table** Figure 5-6.

Molecules

In the Dalton view of matter, the union of atoms results in molecules (Dalton's laws). For example, two atoms of hydrogen (H) combine to form the hydrogen molecule, H_2. As we discuss molecules of substances, the subscript after the symbol for the atom indicates the number of that kind of atom to be found in the molecule. Also, two oxygen atoms comprise the oxygen molecule, O_2. A molecule containing only two atoms is known as a diatomic molecule. Many diatomic molecules containing two different atoms are also known, for example, HCl (hydrogen chloride), HF (hydrogen fluoride), NO (nitric oxide), and CO (carbon monoxide).

A pure substance that is made up of two or more different kinds of atoms is known as a **compound.** Water molecules contain two atoms of hydrogen and one atom of oxygen, that is, H_2O, a triatomic molecule (Figure 5-7).

---

[2] Again, atomic weights really are atomic masses. The older term, atomic weight, however, is still commonly used.

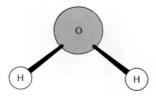

Figure 5-7.
Model (Ball and Stick) of
a water (H_2O) molecule.

Sulfuric acid is H_2SO_4. The molecule of sulfuric acid contains two atoms of hydrogen, one sulfur atom, and four atoms of oxygen. Protein and carbohydrate molecules are much larger and more complex, containing hundreds or even thousands of atoms.

We actually have been writing molecular formulas. The molecular formula represents the number of different kinds of atoms making up a molecule. A molecule, as we have described it, is simply a discrete and distinct cluster of atoms held tightly together. *The force or linkage holding one atom to another is called a* **chemical bond.** The more atoms in the molecule, the more chemical bonds exist in that molecule. Figure 5-8 presents the shapes of some important molecules that are discussed again later in this chapter.

The formation of molecules from atoms in a chemical bond is the single most crucial step in the architecture of matter and is directly related to the number of electrons and the arrangement of the electrons about the

Molecule	Formula	Structure
Oxygen	O_2	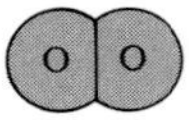
Carbon dioxide	CO_2	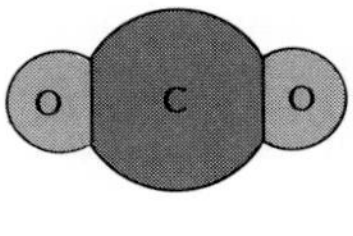
Sulfur dioxide	SO_2	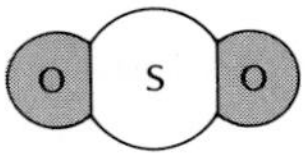
Water	H_2O	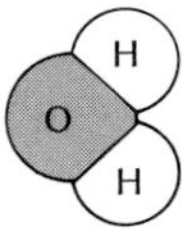
Ammonia	NH_3	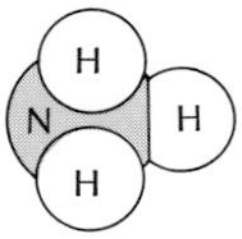
Methyl alcohol	CH_3OH	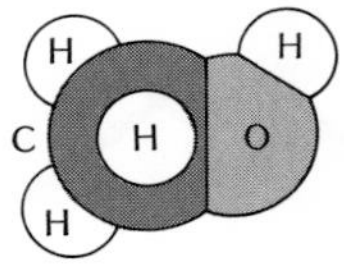

Figure 5-8. Space-filling models of some simple but important molecules. Space-filling models show the relative sizes of the atoms and the three-dimensional structure of the molecule.

nucleus. Chemists have studied extensively the reasons for the combination of atoms to form molecules, with special attention to the nature of particular combinations of atoms. Although this topic is very important to the scientist, we leave it to more extensive courses in chemistry.

Chemical bonds involve energy. Some of this energy can be obtained as heat energy when bonds are changed or broken during chemical changes (Dalton's law), as we shall see in the next chapter. It suffices to point out that each kind of matter—each pure substance—is made up of many, many molecules of exactly the same molecular formula (law of definite proportions) and has its own unique physical and chemical characteristics. For some materials, such as water, a relatively few physical observations may be sufficient to identify them, such as odor, taste, color, freezing temperature, boiling temperature, and density (mass or weight of $1 m^3$, $1 cm^3$, or $1 ft^3$). Some fairly common substances are listed in Table 5-3 along with selected physical properties of these substances. Casual inspection reveals easily measurable differences amongst the substances listed.

Table 5-3. Physical Properties of Selected Substances

Name	Common State at Room Temperature	Color	Density (g/cm^3)	Other
water	liquid	water-white	1.00	—
methane	gas	colorless	—	burns
air	gas	colorless	—	—
ethyl alcohol	liquid	water-white	0.79	burns, dissolves in water
octane	liquid	water-white	0.70	burns, does not dissolve in water
table salt (sodium chloride)	solid	white	—	dissolves in water
limestone (calcium carbonate)	solid	white	—	does not dissolve in water
mercury	liquid	silvery	13.6	—
iron	solid	silvery	7.86	—
lead	solid	silvery	11.3	—

Avogadro's Number and the Mole

We have mentioned that atoms, and of course molecules, are extremely small. For example, one carbon atom has a mass of about 2×10^{-23} g, that is, 0.00000000000000000000002 g. The smallest speck of charcoal or carbon black weighs much more than this, which means that a speck of carbon black contains many, many carbon atoms. One fly-speck contains some 10^{19} atoms.

Scientists today use an arbitrary but convenient system for working with molecules and collections of molecules. If we take 12.01 g of pure carbon (the isotopes of carbon are present in the average amounts found), we know that we have taken 6.02×10^{23} atoms of carbon. Because a carbon atom is 12 times as heavy as a hydrogen atom, only 1.008 g of hydrogen contains 6.02×10^{23} atoms of hydrogen. Similarly, when we take 32.06 g of sulfur, we have taken 6.02×10^{23} atoms of sulfur. This large number, 6.02×10^{23}, is known as Avogadro's number, in honor of Amadeo Avogadro (1776–1856), an Italian physicist, who found that a given volume of any gas (at the same temperature and pressure) contains the same number of molecules.

Today we call this number a **mole**. Thus, we have the following equivalent statements defining a quantity of material: 1 mole of atoms is 6.02×10^{23} atoms; 1 mole of atoms is an Avogadro number of atoms; 1 mole of atoms has a mass equal to the atomic weight of that element, with units given specifically in grams.[3] Thus, 1 mole of carbon atoms has a mass of 12.01 g; 1 mole of oxygen atoms has a mass of 16.00 g; and 1 mole of sulfur atoms has a mass of 32.06 g.

These same terms, the Avogadro number and the mole, are also used in counting and comparing masses of molecules. Molecules of oxygen consist of two atoms of oxygen. Then, one Avogadro number of O_2 molecules is 6.02×10^{23} molecules. Since each oxygen molecule is twice as heavy as an oxygen atom, 1 mole of O_2 molecules has a mass of 32.0 g. Similarly, one molecule of water is made up of two hydrogen atoms and one oxygen atom. Therefore 1 mole of water molecules has an approximate mass of 18 g (1 + 1 + 16). When more precise atomic masses are used, the mass is 18.016 g (1.008 + 1.008 + 16.00).

The mass of 1 mole of any substance is obtained by adding up the atomic weights of all the component atoms to obtain the molecular weight of the substance. For example, the approximate mass of 1 mole, or the molecular weight of H_2SO_4, is 98 g. Examples 5-3, 5-4, and 5-5 illustrate approximate molecular weight calculations (using rounded-off numbers).

---

[3]This definition of the mole as 6.02×10^{23} is most commonly used by chemists. The chemists' mole is often known as the gram-mole. Physicists usually use the kilogram as the mass unit. If the kilogram (kg) is the mass unit, then Avogadro's number becomes 6.02×10^{26}, which then is a kilogram-mole (kg-mole). Thus, 1 kg-mole of sulfur contains 6.02×10^{26} atoms and has a mass of 32.06 kg.

EXAMPLE 5-3

Calculate the approximate mass of 1 mole of H_2SO_4 molecules using rounded-off atomic masses (or atomic weights).

$$2 \text{ moles of hydrogen atoms} \times (\text{atomic weight}) = 2 \times 1 \ = \ 2$$
$$1 \text{ mole of sulfur atoms} \times (\text{atomic weight}) = 1 \times 32 = 32$$
$$4 \text{ moles of oxygen atoms} \times (\text{atomic weight}) = 4 \times 16 = \underline{64}$$
$$\text{Total} \qquad 98$$

That is, 1 mole of H_2SO_4 has a mass of 98 g and contains 6.02×10^{23} molecules.

EXAMPLE 5-4

Calculate the molecular weight of glucose, $C_6H_{12}O_6$.

$$\text{carbon } 6 \times 12 = \ 72$$
$$\text{hydrogen } 12 \times 1 = \ 12$$
$$\text{oxygen } 16 \times 6 = \ \underline{96}$$
$$\text{molecular weight of glucose} = 180$$

EXAMPLE 5-5

If the molecular weight of glucose is 180, what is the mass of 0.20 mole of glucose?

From Example 5-4, 180 g is the mass of 1 mole of glucose. Thus,

$$\frac{180 \text{ g}}{1.0 \text{ mole}} \times 0.20 \text{ mole} = 36.0 \text{ g glucose}$$

Sometimes the mass unit is left off the molecular weight, as shown in Examples 5-3 and 5-4, but the unit (gram) is understood. In summary, 1 mole of any substance contains 6×10^{23} molecules and has a mass in grams equivalent to the molecular weight. Finally, because 1 mole of a substance represents a very large number of molecules, we can think in terms of less than 1 mole—a fraction of a mole. To illustrate, if you have 0.20 mole of glucose, you have a mass of 36.0 g (see Example 5-5).

As we consider molecules, one of our concerns is how much of a certain element is present in the molecule. More specifically, in a molecule of O_2 or O_3, because only oxygen atoms are present, the molecule is 100% oxygen. But, in a molecule of H_2O, there is 11.1% by weight hydrogen and 88.9% by weight oxygen (see Example 5-6).

EXAMPLE 5-6

(a) Calculate the per cent by weight of hydrogen in water (H_2O). The molecular weight of H_2O is 18 g and contains $1 + 1$ or 2 g of hydrogen.

$$\frac{\text{mass hydrogen in } H_2O}{\text{mass water}} \times 100 = \text{wt. \% hydrogen}$$

or

$$\frac{2 \text{ g}}{18 \text{ g}} \times 100 = 11.1 \text{ wt. \% hydrogen}$$

(b) Calculate the per cent by weight of oxygen in water (H_2O).

$$\frac{\text{mass oxygen in } H_2O}{\text{mass water}} \times 100 = \text{wt. \% oxygen}$$

or

$$\frac{16 \text{ g}}{18 \text{ g}} \times 100 = 88.9 \text{ wt. \% oxygen}$$

The Periodic Law

Long before most atoms had been discovered, chemists recognized that some kinds of atoms behaved rather like other kinds of atoms. For example, magnesium and calcium are very similar; that is, they form similar compounds, magnesium carbonate ($MgCO_3$) and calcium carbonate ($CaCO_3$), which together comprise the rock, dolomite, named after the Dolomites, a mountain range in Italy. Both compounds have similar physical characteristics and behave similarly in chemical reactions. Chlorine and bromine are very similar in that they have salts, for example, sodium chloride (NaCl) and sodium bromide (NaBr) that are salty to the taste, are soluble in water, and are found together in salt wells or brine wells in Michigan and other places around the earth.

When one arranges all atoms in order of atomic numbers, we find that some atoms (and particularly compounds of these atoms) have similar features that appear at regular intervals. In Figures 5-9 and 5-10, we have presented in graphical form selected physical properties of the elementary substances of the atoms, such as the melting point of the solid, the normal boiling point, and the ionization energy (which is essentially the energy required to remove one electron from an atom of the element). The physical property (measurement) is plotted along the ordinate (vertical axis) and the atomic number is along the abscissa (horizontal axis).

Casual inspection of these figures shows that there is regular recurrence of properties (or periodicity to the properties) of the elements. The recurrence of properties is clearly present in the melting and the boiling points, with minimum numbers at elements of atomic numbers 2, 10, and 18 (cycles

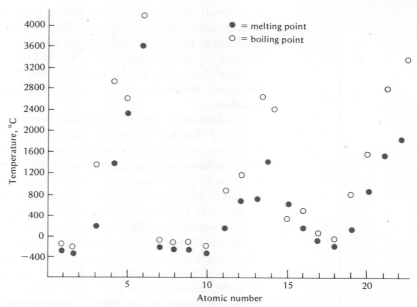

Figure 5-9. Variation of melting points and boiling points for the elements of atomic number 1 through 22.

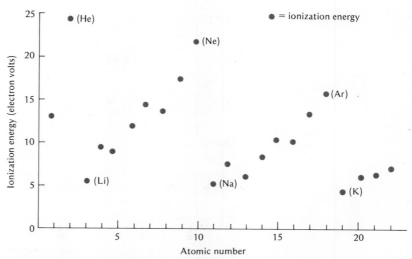

Figure 5-10. Variation of the ionization energy of the atoms with atomic numbers 1 through 22.

at intervals of eight). The most striking periodic character is shown by the ionization energy graph (Figure 5-10), which has minima (valleys) at atomic numbers 3, 11, and 19 and also has maxima (peaks) at atomic numbers 2, 10, and 18. Interestingly, the patterns between steps from number 3 to number 10 are repeated between numbers 11 and 18. This kind of periodic relation is observed with many other physical properties of the elementary substances.

In addition to regular recurrence of physical properties, regular recurrences of chemical properties are also observed. In Figure 5-11, data on the common, readily formed compounds of the first 22 elements with chlorine have been presented graphically. Several elements combine with chlorine to give compounds, usually known as chlorides (some chlorides are salts). As we have indicated, a molecule is made up of a specific number of atoms. In Figure 5-11, the atomic number is plotted along the abscissa and the number of chlorine atoms per molecule along the ordinate.

Several elements form more than one chloride. Those molecules (circled dots) do not detract from the remarkable regularity of the chemical composition of the first 22 elements. The recurrence of behavior shows a pattern of eight, as mentioned previously, with the physical properties of the elements. Similar graphs can be drawn for the carbonates of the elements. In all cases, the chemical characteristics are observed to recur regularly in this set of elements at intervals of eight elements.

An extension of this study to include all the elements reveals the existence of a similar, though somewhat more complex, periodicity of chemical and

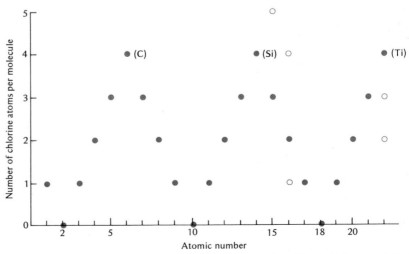

Figure 5-11. Number of chlorine atoms per molecule as observed in the known chlorides of elements (atomic numbers 1 through 22).

physical properties of all the elements. The experimental evidence that has been accumulated over many decades is indeed overwhelming. We summarize this data in the modern statement of the **periodic law:** *The elements, if arranged in order of their atomic numbers, exhibit a periodicity of properties.*

Some scientists suspected the existence of relations among the elements, which we have just been discussing, but the evidence even as late as 1860 was so sketchy that most scientists ignored the idea that such relations existed. Between 1869 and 1871 Mendeleev, a Russian chemist, wrote a series of papers on the periodic properties of the elements. He organized all the elements that had been discovered up to that time, about 70 in number, and used an approach similar to that presented above. Today the known elements number over 100.

Mendeleev grouped atoms of similar properties together, which grouping necessitated gaps in the arrangement, and was bold enough to assert that the gaps in his arrangement corresponded to elements undiscovered as of 1869. By noting the position of the gaps, Mendeleev was able to predict the properties of the undiscovered elements, thus enabling scientists to search and find them. These experimental confirmations of Mendeleev's predictions removed any doubt of the validity of the concept of the periodicity of the properties of the elements and their compounds. For this reason, Mendeleev is credited with the periodic law.

The Periodic Table

In order to emphasize the periodicity of the properties of the elements, Mendeleev arranged the elements in horizontal rows of such length that elements of similar chemical properties (see the figures above based on various properties) fell into vertical columns. This arrangement (somewhat altered and expanded) is used today and is called the periodic table (Figure 5-6). The elements in the vertical columns are known as families or groups of elements (Figure 5-12A). The horizontal rows of elements are known as periods of elements (Figure 5-12B).

The periods for the lighter elements are shown very simply in Table 5-4. There is a sequence of two elements followed by two very clear periods

Figure 5-12. Groups of elements in the periodic table. The elements in a vertical group (or column) **(A)** of the periodic table have similar properties, whereas those in a period (horizontal row) **(B)** have different properties.

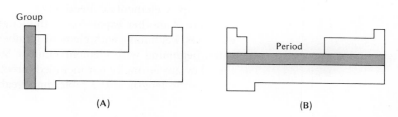

Group

Period

(A)

(B)

Table 5-4. The Lighter
Elements of the
Periodic Table

Group	IA	IIA	IIIA	IVA	VA	VIA	VIIA	VIIIA
1st period	H						(H)	He
2nd period	Li	Be	B	C	N	O	F	Ne
3rd period	Na	Mg	Al	Si	P	S	Cl	Ar
4th period	K	Ca						

of eight elements. Lithium, atomic number 3, begins one period of eight, and sodium, atomic number 11, begins the next period of eight. Hydrogen and helium comprise the first period. The next eight elements, including, for example, boron and oxygen, are in the second period, and the next following eight elements, including, for example, sodium, aluminum, and chlorine, are in the third period.

Examination of both Table 5-4 and the full periodic table (Figure 5-6) shows that helium, neon, and argon are in the same vertical family, group VIIIA, known as the rare or noble gases. Hydrogen has been placed in two places, group IA and group VIIA, because its behavior has similarity to both groups. In both tables, elements 19 and 20, potassium and calcium, have been placed in groups IA and IIA respectively, which is consistent with their chemical properties.

Although one might believe that the periodic table would be a single, rectangular array of elements arranged in periods of eight, this is not the case. Following calcium there are ten elements whose observed behavior does not correspond to any of the groups of eight designated in Table 5-4. For example, element number 31, gallium, behaves like aluminum and element 36, krypton, is a noble gas like neon. Thus, an expansion of the table nearly in the middle is necessary, as is shown in Figure 5-6. Also following krypton, rubidium, number 37, has properties like sodium and potassium; strontium, number 38, has properties like calcium; iodine, number 53, has properties like bromine and chlorine; and xenon, number 54, is one of the noble gases. Thus, period 4 and period 5 are 18 elements in length. The ten elements, by which the period of eight is thus expanded, are known as the transition elements or transition metals. Interestingly, the transition metals taken collectively, have more similar properties than do the group A elements collectively.

There is another expansion of the length of a period (to a period of 32 elements) beginning with element number 57, and the expansion is duplicated beginning with element number 89. These expansions are so large that they are normally not included directly in the periodic table, in order to keep the length of the table manageable. Thus, in our construction of

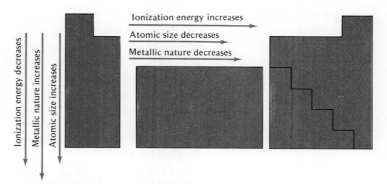

Figure 5-13. Trends in the periodic table.

the periodic table, the transition metals of the fourth and higher periods are inserted between groups IIA and IIIA. The rare earths, atomic numbers 58 and 71, would rightly appear between the first and second transition metals in the sixth period, but these are listed at the bottom of the table as a separate sequence, as is the related actinide sequence (atomic numbers 90–103) of the seventh period.

Without going into great detail, Figure 5-13 summarizes some of the trends observed when atoms are organized in the periodic table. Atoms to the left of the zig-zag line (see the full periodic table, Figure 5-6) are metals. Metals are generally shiny, malleable, hard solids, and good conductors of heat and electricity. Atoms to the right of the zig-zag line have very different properties, such as being gases or soft solids and poor conductors of heat and electricity. Due to the contrast in physical properties, atoms to the right of the zig-zag line are known as nonmetals (for lack of a better term). Atoms in the table along the zig-zag line have "in between" properties.

The study of chemistry includes in-depth analysis of the similarities and differences of elements belonging to the same vertical family, the comparison of different families, and the kinds of reactions that occur between elements of different families. In the discussion of radioactive wastes of Chapter 12, the compounds of strontium and calcium (found in bones) are shown to have chemical similarities. There is also great chemical similarity among the compounds of cesium, potassium, and sodium. The periodic law is a landmark in the development of chemistry. It is a key organizing principle for many, many pieces of data and is very useful for predictions of the chemical and physical behavior of atoms, molecules, and groups thereof. Such information can be found in any general chemistry textbook.

Some Interesting Molecules

These organizing principles of chemistry discussed so far—Dalton's atomic theory, the concept of molecules, the periodic law—provide the basis for a chemical view of matter. This view states that all matter is made up of molecules. The simplest molecules, conceptually, are gaseous molecules. We have already mentioned several gases: H_2 (hydrogen), O_2 (oxygen), CO_2 (carbon dioxide), SO_2 (sulfur dioxide), He (helium), H_2O (as steam), and CH_4 (methane). However, even with this list you recognize that merely calling all of these substances gases is only part of the story because you are familiar with some of the differences between liquid H_2O (water) and solid CO_2 (dry ice).

One of the fossil fuels of great importance in the United States is natural gas. Natural gas from Texarkana (Texas-Arkansas-Louisiana) is largely methane (CH_4). Natural gas from other sources contains significant quantities of ethane (C_2H_6) and propane (C_3H_8). Some of your gasolines are butane-primed in winter for better starting (butane is C_4H_{10}). The liquid gasoline dissolves the butane gas readily. Gasoline, oils, and greases are liquids made up of complex mixtures of simple molecules, such as octane (C_8H_{18}), benzene (C_6H_6), heptane (C_7H_{16}), nonane (C_9H_{20}), cetane ($C_{16}H_{34}$).

Petroleum is also a complex liquid mixture of molecules, most of which contain only atoms of carbon and hydrogen. Such molecules are generally known as hydrocarbons. In a petroleum refinery, the complex liquid mixture is separated into portions containing different sized molecules for different uses. The process of separating crude oil into its component hydrocarbons is known as **fractionation.** Table 5-5 summarizes some of the component fractions and their usefulness to us.

Table 5-5. Selected Petroleum Fractions

Approximate Boiling Range (°C)	Molecular Composition Range (number of carbon atoms per molecule)	Name of Portion	Use
−164–0	C_1–C_4	gas (natural gas)	gaseous fuel
30–175	C_5–C_{10}	gasoline	motor fuel
175–275	C_{12}–C_{16}	kerosene	fuel for stoves
250–400	C_{14}–C_{18}	gas oil; fuel oil; diesel fuel	Diesel engines; oil burners; power plants
above 400	greater than C_{16}	lubricating oils; petroleum grease; paraffin wax; pitch; and tar	

You may have noticed that all the molecules making up gaseous or liquid substances we have discussed so far consist of atoms from the right-hand section of the periodic table designated in Figure 5-13 as nonmetals.

A large number of relatively simple molecules are encountered in the world around us. Many of the some three million substances known to man are compounds containing carbon. The grain alcohol in liquor is ethyl alcohol (C_2H_6O). Rubbing alcohol is largely isopropyl alcohol (C_3H_8O). Ethylene glycol used as permanent antifreeze in your car's radiator is $C_2H_6O_2$. Carbon tetrachloride (CCl_4) is often a component of dry-cleaning fluids. Acetic acid ($C_2H_4O_2$) is the key component of vinegar. Decaying flesh gives off odors that consist of molecules made up of carbon, hydrogen, and nitrogen atoms; certain other disagreeable odors are caused by compounds containing carbon, hydrogen, and sulfur. On the whole, the molecular composition of gases and liquids based on carbon, while quite variegated, is not complex and is one major aspect of the field of science known as organic chemistry.

Pure solids are a bit more complex. The celluloses (carbohydrates) that are found in woody stems of trees and other plant life are made up of very, very large molecules containing only the atoms, carbon, hydrogen, and oxygen. Rather than discuss these fascinating complex molecules, let us look at one of the most deceptively complex substances, namely solid carbon.

Pure carbon is known to exist in three forms—the brilliantly clear, crystalline diamond; the black but at times shiny, crystalline graphite; and an amorphous, dull black powder sometimes called carbon black (soot). In all three of these forms there is no such thing as a molecule containing one carbon atom. So many carbon atoms make up "a molecule" with so many different numbers of carbons that chemists "cop out" and write all three just as C, although C_x or C_n might be more appropriate. Coal is even more complex than pure carbon, in that some quantities of other atoms, such as, hydrogen, oxygen, and sulfur are present. As we will see in the next chapter, the carbon in coal behaves essentially like the carbon in amorphous carbon. For convenience, then, we consider the carbon of coal to be just that.

The molecules in a substance such as coal (or tar or asphalt) are very large molecules—sometimes containing as many as a million atoms bonded together. The study of molecules containing many, many atoms lies at the heart of the work of the polymer chemist, the plastics chemist, the cellulose or wood chemist, the protein chemist or physicist, the macromolecular chemist or physicist—to name a few. Some of the many products of polymer research and development follow: paints, fibers (drip-dry clothing, wash and wear fabrics, carpets), vinyl records, vinyl sheeting, polyethylene products (bottles, sheets, and ropes), Lucite, Nylon (in its many forms),

Table 5-6. Some Inorganic Materials

Common Name	Chemical Name	Formula	Properties	Uses
water	water	H_2O	colorless liquid	covers 75% of the earth's surface; $\frac{2}{3}$ of body wt. is H_2O; dissolves many substances
common salt	sodium chloride	NaCl	white solid; m.p.[a] 801°C; soluble in H_2O	from NaCl, chlorine is made for bleach; found in seawater and brines; important in body fluids
dietetic salt	potassium chloride	KCl	white solid; m.p. 776°C; soluble in H_2O	used in low salt diets; important in body fluids
washing soda (soda ash)	sodium carbonate	Na_2CO_3	white solid; 851°C; soluble in water	base (reacts with acids)
limestone	calcium carbonate	$CaCO_3$	white solid; 1339°C (at 1025 atmospheres); insoluble in water	material of construction; heat limestone → lime used in mfg. of iron [CaO (lime) and CO_2]
epsom salts	magnesium sulfate	$MgSO_4 \cdot 7H_2O$	white solid; m.p. 1124°C; soluble in H_2O	medicinal purgative
gypsum	calcium sulfate	$CaSO_4 \cdot 2H_2O$	white solid; m.p. 1450°C; insoluble in water	building material; source of plaster of Paris
barite	barium sulfate	$BaSO_4$	white solid; m.p. 1580°C; insoluble in H_2O	x-ray diagnosis
oil of vitriol	sulfuric acid	H_2SO_4	oily liquid; b.p.[b] 338°C; reacts vigorously with water	as strong acid, reacts with bases; destroys fabrics and tissues; widely used by industry; cleans rust
dry ice	carbon dioxide	CO_2	gas; b.p. 78°C; very slightly soluble in water	product of combustion and respiration; used in photosynthesis
	sulfur dioxide	SO_2	gas; b.p. 10°C; slightly soluble in water	byproduct of combustion; used to make sulfuric acid; behaves like an acid

[a] m.p. means melting point.
[b] b.p. means boiling point.

Bakelite resins (black telephones), plastic dishes, and so on. The molecules of these solid substances are made up of carbon atoms, along with other atoms, and are also part of the science of organic chemistry. Many of these materials are made from molecules found in petroleum and are produced by the petrochemical industry.

Other solids, commonly discussed in first chemistry courses are the salts of inorganic chemistry—sodium chloride, calcium carbonate, calcium sulfate, potassium nitrate, and so on. Many of these salts are soluble in water; others are not soluble in water. We have written the molecular formulas of some of these salts earlier in this chapter. You might have noticed that salts are made up of atoms from both sides of the periodic table; that is, salts are made up of both metals and nonmetals.

For our purpose, salts are characterized by their high melting points. Furthermore, our principal concern with these interesting substances is that they are the dominant materials in the physical world. They make up ores from which we obtain materials for our civilization, and they are waste products, particularly from the combustion of coal and in radioactive wastes from nuclear power plants. A very brief collection of inorganic materials (not containing carbon) are listed in Table 5-6.

Although the chemical and physical behavior of these inorganic substances is of much interest and importance, salts are not fuels. Thus we leave this subject for you to explore further elsewhere.

Conclusion

Throughout this chapter, we have developed the scientists' molecular view of matter and, more particularly for our purposes, the concept that fossil fuels are molecules. The fossil fuels—coal, petroleum, and natural gas (also wood)—are made up of molecules containing carbon. Such molecules are made up of atoms that are held together by chemical bonds. These molecules have bonds whose energy can be released as heat during chemical changes. The chemical change known as combustion is the subject of the next chapter.

Questions

1. Define or describe:
   (a) physical property
   (b) state of matter
   (c) atom
   (d) molecule
   (e) mole
   (f) nucleus of a helium atom

2. Which of the following properties are physical and which are chemical?
   (a) odor
   (b) taste
   (c) boiling point
   (d) flammability

3. Illustrate in your own words at least five examples of chemical changes familiar to you.

4. Which of the following processes involves a physical change and which a chemical change?
   (a) breaking glass   (c) boiling water   (e) distilling gasoline
   (b) burning coal   (d) digesting food   (f) freezing water

5. What elements do the following symbols represent?
   (a) O   (d) K   (g) Pb
   (b) N   (e) Cl   (h) U
   (c) Na   (f) H   (i) Fe

6. What are the symbols for the following elements?
   (a) calcium   (d) phosphorus   (g) sodium
   (b) magnesium   (e) thorium   (h) helium
   (c) carbon   (f) sulfur   (i) potassium

7. What are considered to be the major constituent particles of atoms?

8. What is the name for the element for which
   (a) the nucleus contains 7 protons and 7 neutrons?
   (b) the nucleus contains 7 protons and 8 neutrons?
   (c) the nucleus contains 11 protons and 12 neutrons?
   (d) the nucleus contains 17 protons and 19 neutrons?
   (e) the nucleus contains 1 proton and 2 neutrons?
   (f) the nucleus contains 1 proton and 3 neutrons?

9. Practice assigning numbers of protons and numbers of neutrons to the nuclear structure of the following elements by completing the following table. Identify the atomic masses from a periodic table.

Atomic Number	Symbol	Number of Protons	Number of Neutrons	Mass Number
2	He			4
6	C		6	
8	O		8	
10	Ne		10	
10	Ne (isotope)			
11	Na			23
	Mg		12	
	Al		14	
16	S			32
18			22	
	K	19	20	
20	Ca			40
26	Fe		30	
		82	125	
88	Ra			226
	U		143	
	U		146	

10. How do isotopes differ from one another? How are they alike?

11. Where is the mass of the atom concentrated?

12. What is atomic number? What is mass number?

13. In terms of protons, neutrons, and electrons, what does each of the following represent?
   (a) ^1_1H (b) ^3_1H (d) $^{235}_{92}U$
   (b) ^2_1H (c) $^{14}_6C$ (e) $^{238}_{92}U$

14. Explain why atoms are considered to be the fundamental building blocks of matter even though smaller particles (subatomic particles) exist.

15. What information is embodied in the molecular formula of a compound?

16. How many hydrogen atoms are found in each of the following molecules?
   (a) H_2O (c) C_2H_5OH (e) $(NH_4)_2SO_4$
   (b) H_2SO_4 (d) $C_6H_{12}O_6$ (f) $Ca(H_2PO_4)_2$

17. How do molecules of elements and compounds differ from each other?

18. In the periodic table, where are the metals located? The nonmetals? The noble gases?

19. What are atoms in the vertical columns in the periodic table called?

20. What are atoms in the horizontal rows in the periodic table called?

21. Gases are compounds made up of atoms from what section of the periodic table?

Numerical Exercises

1. Calculate the molecular weight of:
   (a) CH_4 (e) NaCl   (i) C_6H_{12}
   (b) $CaCl_2$ (f) C_2H_6 (j) H_3PO_4
   (c) H_2O (g) H_2SO_4 (k) SO_2
   (d) NaOH   (h) C_2H_6O (l) CO_2

2. What is the mass in grams of 1 mole of methane (CH_4)? 0.1 mole of methane? 10^{-6} mole of methane?

3. If you have 180 g of glucose ($C_6H_{12}O_6$) how many moles of glucose do you have? 1800 g of glucose? 30 g of glucose?

4. How many molecules are in 2 moles of CO_2?

5. How many nitrogen atoms are there in each of the following?
   (a) $\frac{1}{2}$ mole of N_2 (c) 28 g of N_2
   (b) 1 mole of NO   (d) 60 g of NO

6. Ammonium nitrate (NH_4NO_3) is probably the most important nitrogen containing fertilizer. Calculate the per cent by weight of nitrogen in this compound.

7. Calculate the per cent by weight of sulfur in sulfur dioxide (SO_2).

8. If you have 100 g of sulfur (S), how many grams of SO_2 can be formed?

9. If you have 100 g of carbon (C), how many grams of CO_2 can be formed?

10. **(a)** How many metric tons of carbon dioxide are formed from 1 metric ton[4] of carbon?
    (b) How many kilograms of carbon dioxide are formed from 1 metric ton of carbon?
    (c) How many metric tons of carbon dioxide are formed from 12 metric tons of coal, assuming the coal is 100% carbon? This amount of coal is the amount used by a typical home per winter in the northeastern part of the United States.
    (d) How many metric tons of carbon dioxide are formed from 12 metric tons of coal, assuming the coal is only 92% carbon?

Reference

[1] For a discussion of electrostatics, the reader is referred to V. H. Blooth and M. C. Bloom: *Physical Science.* The Macmillan Company, New York, 1972.

[4] A metric ton is 1000 kg.

6

Reactions
Provide Heat

In Chapter 5, we examined one of the two major aspects of chemistry, specifically the structure of matter. Chemistry is the science that deals with the structure of matter *and* the changes that matter undergoes. We surveyed some of the important substances or molecules of these substances with particular reference to materials either familiar to you or important to our theme of energy in society.

Only certain combinations of atoms go to make up molecules, and some 3 million different kinds of molecules are specifically known to exist either in nature or have been synthesized by man. Some of these molecules comprise the substances known as the fossil fuels. Mention has been made that molecules are made up of atoms held together by forces known as chemical bonds. You recall that our version of Dalton's atomic theory stated that chemical change is the separation or union of atoms. Because chemical changes are involved in the production of heat energy, we now turn to the chemist's view as to how this heat energy is obtained from chemical reactions.

In this chapter, we elaborate specifically the concept that chemical change is the separation or union of atoms as we discuss combustion reactions, which are generally illustrative of chemical reactions. We work with chemical equations as shorthand for describing the changes that substances undergo. Chemical equations are viewed from a molecular point of view and also from a mass point of view in that the law of conservation of mass holds true throughout chemical processes. In discussing masses, we will use the chemist's convention of measuring mass in units of grams rather than kilograms.

Chemical change, particularly combustion reactions, is related quantitatively to the heat produced in a chemical reaction. The scientist's measurement of the heat from a chemical reaction is then correlated with the heat value of various fossil fuels. We digress to some nutritional heats for our calorie-conscious society, and then return to the fossil fuels by summarizing briefly the resources of fossil fuels available to us on earth.

Combustion Reactions (Chemical Reactions)

Burning is probably the most familiar chemical transformation in your experience. Fire is spectacular, sometimes terrifying, and very useful. As the material burns, the flame is a source of heat, light, and often sound. Furthermore, when the flame is extinguished, the burned material is completely changed in appearance. We find that the properties of the combustion products are different, particularly in the fact that they do not burn.

Burning was probably the most extensively studied process by chemists up until 1800. The burning of some solid materials results in the disappearance of the solid, leaving a little ash, that is most of the material seems to be "lost." The burning of other materials gives solids that have more mass than the materials you started with. With fuels, such as wood, oil, and coal, one observes a loss of solid or liquid matter, that is, what is left behind as solid (ash) or liquid has a mass much less than the starting masses of these fuels. However, if a strip of magnesium or aluminum is burned (bright white light as in a flash bulb) (Figure 6-1), the mass of solid after burning is greater than the original mass of metal. The rusting of iron (Figure 6-2) is related to burning. Actually, rusting of iron and burning of iron (say steel wool) are the same process chemically, but rusting proceeds very much more slowly than burning. In both cases, the material left after burning has a greater mass than the original iron metal.

You probably already know that burning, rusting, respiration, and oxidation are processes involving combination of a material with the oxygen of air to give a class of materials known as oxides. The role of oxygen in

Figure 6-1. Magnesium burning in air. Magnesium and aluminum are used as the element in flash-bulbs. Magnesium oxide is a white solid.

Figure 6-2. The iron in the steel from which the automobile is made has combined with oxygen to give the powder known as rust. The dollar loss from rusting of iron objects in the United States was about $30.00 per person per year in 1972.

Table 6-1. Gaseous Composition of the Atmosphere (Clean Dry Air Near Sea Level)

Substance	Percentage
nitrogen	78.09
oxygen	20.94
argon	0.93
carbon dioxide	0.0318
noble gases (other than argon)	0.0024
other gases	0.0058

these processes was not understood until oxygen had been prepared in a relatively pure state or "discovered." Although oxygen was discovered in 1775 by Joseph Priestley, it was Lavoisier in 1780 who postulated that combustion is the combination of the substance that burns with oxygen. He also found that air is only about 20% oxygen (Figure 5-4) and he named the remainder azote, which is today the French name for nitrogen gas, the major component of air. Table 6-1 records the principle constituents of clean dry air near sea level.

Lavoisier's interpretation has been confirmed in all combustions with air. The chemical changes known as combustion are usually accompanied by the evolution of heat and light (Figure 6-3). Burning of both metals and nonmetals produces oxides. The oxides of metals, such as aluminum oxide, magnesium oxide, and iron oxide, are usually solids; the oxides of nonmetals, such as sulfur dioxide, carbon monoxide, and carbon dioxide, are usually gases at room temperature.

Figure 6-3. A match burning. This chemical reaction gives off carbon dioxide, water vapor, heat, and light. Note the formation of the moisture ring as water vapor condenses on the cool surface.

Chemical Equations

In Table 6-2, some combustion processes are presented schematically, utilizing some of the symbol notation developed earlier.

Let us look more closely at the meaning of these chemical equations. A chemical equation represents a reaction occurring between substances or, more specifically, between molecules of substances. Therefore, correct molecular formulas must be used. The law of conservation of mass states that in chemical reactions, matter is neither created nor destroyed. In terms of atoms involved in the combustion of methane, the oxygen atoms, which are originally present in oxygen molecules (O_2), are found after burning either as oxygen atoms in carbon dioxide molecules or as oxygen atoms in water molecules. Thus, in word form, the last reaction in Table 6-2 would read: methane plus oxygen yields carbon dioxide plus water, or

$$CH_4 + O_2 \longrightarrow CO_2 + H_2O \qquad (6\text{-}1)$$

The convention of the chemist is to put the original substances (**reactants**) **on the left** and to put the final substances of the chemical change (**products**) **on the right.** You will see that, as written, there are four H atoms in the CH_4 molecule on the left of the arrow but there are only two H atoms in the water molecule to the right of the arrow. So, equation (6-1) as written is not an equality, that is, the left hand side does not equal the right hand side. The law of conservation of mass and energy requires that the left side equals the right. Because the arrow (by convention) signifies an equality, we must place appropriate coefficients in front of each term as was done in Table 6-2.

This procedure of determining the appropriate coefficients is called **balancing by inspection** and is as follows. By writing equation (6-1) with a coefficient of two in front of the H_2O, as in equation (6-2)

$$CH_4 + O_2 \longrightarrow CO_2 + 2\,H_2O \qquad (6\text{-}2)$$

Table 6-2. Selected Combustion Processes (Burning Reactions)

Process	Equation	
burning of coal or charcoal	$C + O_2 \longrightarrow CO_2$	carbon dioxide
incomplete combustion	$2\,C + O_2 \longrightarrow 2\,CO$	carbon monoxide
burning of sulfur	$S + O_2 \longrightarrow SO_2$	sulfur dioxide
metal oxidation	$2\,Mg + O_2 \longrightarrow 2\,MgO$	magnesium oxide
	$2\,Ca + O_2 \longrightarrow 2\,CaO$	lime or calcium oxide
	$3\,Fe + 2\,O_2 \longrightarrow Fe_3O_4$	iron oxide, rust
	$4\,Al + 3\,O_2 \longrightarrow 2\,Al_2O_3$	aluminum oxide
burning of hydrogen	$2\,H_2 + O_2 \longrightarrow 2\,H_2O$	hydrogen oxide, water
burning of methane (natural gas)	$CH_4 + 2\,O_2 \longrightarrow CO_2 + 2\,H_2O$	carbon dioxide and water

we suggest that the four H atoms of the methane molecule appear as two molecules of H_2O, each containing two H atoms ($2 \times 2 = 4$). Thus, the same number of H atoms appear in the molecules on the *left* as there are H atoms in the molecules on the *right*.

From the outset, we had the same number of C atoms on each side of the equation, but it is not yet balanced as to O atoms. Equation (6-2), which is balanced as to H and C atoms, has two O atoms on the *left* in an O_2 molecule whereas on the *right* there are a total of four O atoms (two in a CO_2 molecule and one in each of the two molecules of H_2O). Placing a two in front of the O_2 on the left provides four O atoms (two in each of two molecules of O_2).

Thus, equation (6-3) is totally balanced. That is, on the left and right are the same number of C atoms, H atoms, and O atoms. Note that in all cases, the atoms are found in molecules.

$$CH_4 + 2\,O_2 \longrightarrow CO_2 + 2\,H_2O \qquad (6\text{-}3)$$

This balancing of equations by inspection introduces the use of coefficients in front of the molecular formulas to indicate the number of a given kind of molecule in the reaction. No coefficient at all obviously implies one molecule of that substance in the balanced equation. Please note again that *correct molecular formulas* must be used and the final chemical equation *must balance*, in order to satisfy the law of conservation of matter. Example 6-1 develops, similarly, a chemical reaction which is not a combustion reaction.

The significance of a balanced equation arises from the fact that each correct molecular formula communicates specific information about the chemical composition of each substance involved and that the equation relates the reactants and products in a uniquely characteristic ratio (the coefficients). So, let us look further at these chemical sentences for additional meaning. Consider the combustion of coal, assuming that coal is pure carbon,

$$C + O_2 \longrightarrow CO_2 \qquad (6\text{-}4)$$

EXAMPLE 6-1

Write a balanced equation for the reaction of steam with hot iron to give iron oxide and hydrogen.

Preliminary equation: $H_2O + Fe \longrightarrow Fe_3O_4 + H_2$

Balance Fe atoms: $H_2O + 3\,Fe \longrightarrow Fe_3O_4 + H_2$

Balance O atoms: $4\,H_2O + 3\,Fe \longrightarrow Fe_3O_4 + H_2$

Balance H atoms: $4\,H_2O + 3\,Fe \longrightarrow Fe_3O_4 + 4\,H_2$

In equation (6-4) showing the combustion of carbon to give CO_2, we say one atom of C has reacted with one molecule of O_2 to give one molecule of CO_2. What happens when two atoms of C are used? A reasonable interpretation is that two molecules of O_2 are needed and that two molecules of CO_2 will be produced.

Suppose 6.02×10^{23} atoms of C are taken, then 6.02×10^{23} molecules of O_2 will be needed and 6.02×10^{23} molecules of CO_2 will be produced. You recall that 6.02×10^{23} is the Avogadro number. One Avogadro number of anything is by definition, 1 mole. Thus, another way of reading our equation is to say, 1 mole of C atoms reacts with 1 mole of O_2 molecules to give 1 mole of CO_2 molecules.

Furthermore, we can now think in terms of masses of material. You recall that 1 mole of carbon atoms has a mass of 12 g, one mole of oxygen molecules has a mass of 32 g, and 1 mole of carbon dioxide molecules has a mass of 44 g. Table 6-3 summarizes these important relationships in the total combustion of carbon. These different ways of expressing a chemical equation (data in Table 6-3) in terms of quantities of materials are shown using equation (6-4) again for our example.

$$C \quad + \quad O_2 \quad \longrightarrow \quad CO_2 \qquad (6\text{-}4)$$
$$1 \text{ atom} + 1 \text{ molecule} \longrightarrow 1 \text{ molecule} \qquad (6\text{-}4a)$$
$$\text{or} \quad 1 \text{ mole} + \quad 1 \text{ mole} \quad \longrightarrow \quad 1 \text{ mole} \qquad (6\text{-}4b)$$
$$\text{or} \quad 12 \text{ g} + \quad 32 \text{ g} \quad \longrightarrow \quad 44 \text{ g} \qquad (6\text{-}4c)$$

We can do a similar analysis for all balanced chemical equations to obtain the definite proportion of materials that go into chemical reactions. Equation (6-4c) also clearly shows the principle of conservation of mass in a chemical reaction. The total mass of reactants has the same total mass as the products, namely 44 g. The fact that 12 g of C requires 32 g of O_2 for reaction is not quite as important as the ratio of masses appearing in the equation, that is, 12:32:44 which is the same as 1:2.67:3.67. Thus, we find that 12 lb of C will react exactly with 32 lb of O_2 to give 44 lb of CO_2. Similarly, 1 lb of C plus 2.67 lb of O_2 gives 3.67 lb of CO_2. Thus, equation

Table 6-3.

Substance	Number of Moles	Number of Atoms or Molecules	Atomic Weight or Molecular Weight	Mass (number of grams)
C	1	6.02×10^{23}	12	12
O_2	1	6.02×10^{23}	32	32
CO_2	1	6.02×10^{23}	44	44

(6-4) may be used with any mass as long as mass unit consistency and ratio consistency is maintained.

$$C \; + \; O_2 \; \longrightarrow \; CO_2$$
$$1 \text{ g} \; + \; 2.67 \text{ g} \; \longrightarrow \; 3.67 \text{ g}$$
$$1 \text{ lb} \; + \; 2.67 \text{ lb} \; \longrightarrow \; 3.67 \text{ lb}$$
$$1 \text{ ton} \; + \; 2.67 \text{ ton} \; \longrightarrow \; 3.67 \text{ ton}$$

The same procedure for methane combustion yields the following:

$$CH_4 \; + \; 2 \; O_2 \; \longrightarrow \; CO_2 \; + \; 2 \; H_2O$$
$$16 \text{ lb} + \; 64 \text{ lb} \; \longrightarrow \; 44 \text{ lb} \; + \; 36 \text{ lb}$$

or $\quad 1 \text{ lb} + \frac{64}{16} \text{ lb} \; \longrightarrow \; \frac{44}{16} \text{ lb} + \frac{36}{16} \text{ lb}$

or $\quad 1 \text{ lb} + 4.0 \text{ lb} \; \longrightarrow \; 2.75 \text{ lb} + 2.25 \text{ lb}$

or $\quad 1 \text{ ton} + 4.0 \text{ tons} \longrightarrow 2.75 \text{ tons} + 2.25 \text{ tons}$

At this point, we have three ways of reading a balanced chemical equation.

1. In terms of molecules.
2. In terms of moles.
3. In terms of masses.

Thus, it is always true of any chemical reaction, and the balanced equation describing it, that the mass and the number and identity of the atoms involved must remain unaltered. A chemical change, then, is merely a rearrangement of atoms. Reactant molecules are made up of certain kinds of atoms that become product molecules, which are made up of the same atoms but with molecules that are obviously put together differently. Chemical bonds holding atoms together in the reactant molecules are broken, whereas new bonds are formed comprising the product molecules.[1]

Heats of Reaction

The most frequently observed reaction of oxygen is combustion. Why are combustion reactions useful? Very simply, combustion reactions are useful because the energy released primarily as heat energy, when harnessed for practical purposes, is used to provide goods and services for mankind. Combustion of the fossil fuels, coal, oil, and natural gas, are used in the heating of man's shelters, the preparation of man's food, and in making all kinds of man's clothing. Combustion of fossil fuels provides the energy used to power our land and air transportation system and most of our sea transportation. Combustion of fossil fuels is the principal means by which electricity is generated in power stations.

[1] In Chapter 10, we discuss nuclear reactions in which an atom is transformed into another kind of atom. Such atomic transformations are not observed in ordinary chemical reactions, such as combustion.

Processes that release heat energy are known as **exothermic processes.**

$$\underset{\text{coal}}{C} + O_2 \longrightarrow CO_2 + \text{heat} \qquad (6\text{-}5)$$

$$2\,H_2 + O_2 \longrightarrow 2\,H_2O + \text{heat} \qquad (6\text{-}6)$$

$$\underset{\text{methane}}{CH_4} + 2\,O_2 \longrightarrow CO_2 + 2\,H_2O + \text{heat} \qquad (6\text{-}7)$$

All of these reactions produce heat and are exothermic reactions. We are very interested in knowing quantitatively how much heat is produced from some specific quantity of material being burned. Why? At a very practical level, how much coal, oil, or natural gas will you need to heat your home this winter?

Scientists view **heats of reactions** (source of heat energy) in terms of balanced chemical equations. The study of heat of reaction is one branch of the area of scientific study known as thermochemistry (thermodynamics). The heat of reaction is known as ΔH (read delta H), and for exothermic reactions ΔH is arbitrarily defined by scientists as having a negative sign. Some reactions are heat absorbing (**endothermic**). For endothermic reactions ΔH has a positive sign.

This convention is not as mysterious as you might first think. During an exothermic reaction, heat is lost or given up by the reactants, in which case the $(-)$ minus sign seems appropriate. The heat produced in a chemical reaction is directly proportional to the amount of substance that reacts or that is produced. For example, in equations (6-8), (6-9), and (6-10), the heat of reaction recorded is that observed when the equation is read in terms of moles. For equation (6-8), the 94.4 kcal is produced when 1 mole of C reacts with 1 mole of O_2 to produce 1 mole of CO_2. Furthermore, when 2 moles of C react with 2 moles of O_2 to give 2 moles of CO_2, then twice as much heat or 188.8 kcal is produced. One can make similar inferences about equations (6-9) and (6-10). The heats of reaction are usually determined in a very carefully controlled experiment.

$$C + O_2 \longrightarrow CO_2 \qquad \Delta H = -94.4\,\text{kcal} \qquad (6\text{-}8)$$

$$2\,H_2 + O_2 \longrightarrow 2\,H_2O \qquad \Delta H = -116\,\text{kcal} \qquad (6\text{-}9)$$

$$CH_4 + 2\,O_2 \longrightarrow 2\,H_2O + CO_2 \qquad \Delta H = -210.8\,\text{kcal} \quad (6\text{-}10)$$

A very important extension of this chemical arithmetic permits you to determine the relation of masses of material combusted and the amount of heat produced. Let us consider equation (6-10): 16 g of CH_4 or 1 mole of CH_4 on complete combustion produces 210.8 kcal of heat energy; 160 g of CH_4 (10 moles) produces 2108 kcal of heat energy. As calculated in Example 6-2, 1 lb (454 g) of CH_4 produces 5980 kcal of heat energy.

EXAMPLE 6-2

How many kilocalories of heat energy are produced when 1 lb (454 g) of methane (CH_4) are combusted?

1 mole of methane (16 g) produces 210.8 kcal of heat.

$$\frac{210.8 \text{ kcal}}{16 \text{ g}} = \frac{13.2 \text{ kcal}}{1 \text{ g}}$$

$$\frac{13.3 \text{ kcal}}{1 \text{ g}} \times \frac{454 \text{ g}}{1 \text{ lb}} = \frac{5980 \text{ kcal}}{1 \text{ lb}}$$

Thus, for any amount of methane, then, the heat produced can be calculated from the thermochemical data for the *heat of reaction*.

Calorific Values of Fuels

Heats of Combustion

A large number of chemical compounds have been totally combusted. Heats of combustion are simply another term for heats of reaction but specifically for combustion reactions. Heats of combustion of some selected compounds consumed in various oxidation processes are recorded in Table 6-4. The heat of any reaction is the chemical energy left over as heat energy after certain bonds between atoms in the reactants have been broken and after certain bonds in the products have been formed. The chemical bond is the energy reservoir sometimes referred to as chemical potential energy.

The whole subject of chemical bond energy, although interesting, is not needed in any more detail for our development of the principle of heat energy derived from fossil fuels. Thus, we leave this topic and turn to combustion data obtained by engineers and technologists who have studied

Table 6-4. Heats of Combustion

Substance	State	Heat of Combustion (kcal/mole)	(kcal/kg)
methyl alcohol (CH_4O)	liquid	170.9	5,350
ethyl alcohol (C_2H_6O)	liquid	327.6	7,120
propane (C_3H_8)	gas	526.3	11,900
benzene (C_6H_6)	liquid	782.3	10,000
glucose ($C_6H_{12}O_6$)	solid	673.0	3,700
sucrose* ($C_{12}H_{22}O_{11}$)	solid	1,349.6	3,950
stearic acid† ($C_{18}H_{36}O_2$)	solid	2,711.8	9,520

* Table sugar.

† A major constituent in saturated fats.

Table 6-5. Calorific Value of Fuels

SOURCE: Data adapted from J. H. Perry: *Chemical Engineer's Handbook* 4th ed. McGraw Hill Book Company, New York, 1963, pp. 9-2 to 9-11.

*Heat data for oil is usually given per gallon, that is, 35,000 kcal/gal (each gallon weighs 7.2 lb).

†Gas heat data is usually reported per cubic foot. Texarkana gas has a heating value of 0.244 kcal/ft³.

‡Green wood provides 5–20% less heat per pound than does air dried wood.

Substance	Heat of Combustion	
	(kcal/lb)	(kcal/kg)
anthracite coal (86.7% C), Pennsylvania	3,450	7,500
bituminous coal, West Virginia	3,660	8,050
bituminous coal, Indiana	2,880	6,350
lignite (42% C, 43% O), North Dakota	1,820	4,000
oil,* domestic heating oil	4,920	10,800
natural gas† (96% CH₄), "Texarkana"	5,550	12,200
air dried wood‡		
red maple	2,400	5,300
white pine	2,140	4,700
white oak	3,030	6,700

the combustion of the fossil fuels (mixtures of molecules) from various sources (Table 6-5). Selected comparisons of Tables 6-4 and 6-5, in the kcal/kg columns, provide some subtle insight into the relative usefulness of various substances (fuels) in providing heat energy.

The data presented records the maximum amount of heat obtainable from the source indicated. How much useful heat—useful for what you are doing, that is, heating a house, driving a car or locomotive—will depend on the system converting the heat energy and how it is designed such that losses up a chimney or by radiation are minimized. However, one can

---

EXAMPLE 6-3

(a) Calculate the heat produced on combustion of 1 metric ton (1000 kg) of anthracite and of lignite.

Anthracite:

$$7500 \frac{kcal}{1 \ kg} \times 1000 \ kg = 7.5 \times 10^6 \ kcal$$

Lignite:

$$4000 \frac{kcal}{1 \ kg} \times 1000 \ kg = 4.0 \times 10^6 \ kcal$$

(b) Compare these two substances as to their relative heating value per ton.

$$\text{anthracite:lignite} = \frac{7.5 \times 10^6}{4.0 \times 10^6} = \frac{1.875}{1}$$

calculate that 1 metric ton (1000 kg)[2] of anthracite will produce 7,500,000 (7.5×10^6) kcal of heat energy, whereas 1 metric ton of lignite is a much poorer source of heat energy as it provides only 4,000,000 (4.0×10^6) kcal of heat energy (as in Example 6-3). More specifically, anthracite coal provides 1.875 times as much heat energy per metric ton (or per any quantity) as does lignite.

Photosynthesis—An Endothermic Reaction

Our attention has focused on the fossil fuels (coal, oil, and natural gas) which account for 96% of the energy generated and consumed each year in the United States. The fossil fuels are chemical compounds in which energy has been stored and which can be released on combustion. The energy so stored has its origin in the sun. A large amount (large when compared to terrestrial sources of energy) of solar energy is temporarily stored in living plants as a result of the photosynthetic process occurring in green plants.

$$6\,CO_2 + 6\,H_2O + \text{solar radiation} \longrightarrow \underset{\text{glucose}}{C_6H_{12}O_6} + 6\,O_2 \quad (6\text{-}11)$$

The glucose molecules become incorporated into the carbohydrates (starch, cellulose) of plants.

$$n\,\underset{\text{glucose}}{C_6H_{12}O_6} \longrightarrow \underset{\substack{\text{cellulose, starch, or} \\ \text{carbohydrates}}}{H-(C_6H_{10}O_5)_n-OH} + (n-1)\,H_2O \quad (6\text{-}12)$$

These reactions are *endothermic processes* for which the sun's energy provides the driving force. Most living matter (carbohydrates) is gradually oxidized (fast or slow combustion) back to CO_2 and H_2O with release of energy as heat. Decaying matter gives off heat; for example, hay in barns sometimes ignites spontaneously. Remember that the decaying matter is continually exposed to air which contains 21% oxygen.

However, at some point in time some plant matter has, in a few favored places such as swamps and bogs, been exposed to a *reducing environment* (the opposite of an oxygen containing, oxidizing environment). In such environments, our fossil fuels are believed to have been formed. Thus, the fossil fuels are reservoirs of combustible compounds or stores of energy—the sun's energy. Actually, a very small portion of the sun's energy has been stored in these materials—approximately 7 days worth. As we use the fossil fuels, we must be aware that additional fossil fuel reserves are either not being created or developed in the earth, or the rate of formation of fossil fuels from plant matter is so slow as to be negligible. Thus, we must consider fossil fuels to be a nonrenewable source of energy.

[2] The metric ton is 1000 kg or 2204 lb.

Fuels in the Life Process

Before discussing fossil fuels further, let us examine briefly some of the materials that man metabolizes (burns). We have discussed the photosynthesis reaction by which plants convert carbon dioxide into stored energy as carbohydrates. We have also noted that burning of wood (essentially cellulosic carbohydrate) provides heat as the wood combines with oxygen to give back carbon dioxide and water and heat energy. Man metabolizes a number of materials to provide energy for living and for his activities. Heat and muscle energy is furnished in the body by the oxidation (combustion) of food materials such as sugars by oxygen of the air.

The amount of food energy needed by a person varies according to his weight and the amount of his muscular activity. In the United States, the daily per capita consumption of food provides about 3000 kcal.[3] Because individual needs for daily food energy vary according to both body size and activity, the United Nations Food and Agricultural Organization (FAO) has established standard "reference" body weights and standard daily per capita caloric (heat energy from food) requirements. On the basis of these FAO standards, the President's Science Advisory Committee has estimated the current world average needs at 2354 kcal per capita per day.

This figure, as with food production figures, takes no account of whether a proper balance of protein, fat, and carbohydrate is available, nor does the figure refer to any other special nutrient. Starvation level is considered to be much lower than 2000 kcal/day, perhaps about 1000 kcal. A knowledge of the caloric food values is useful in nutrition. Tabulated below (Table 6-6) is the portion of food having a food (fuel) value of approximately 100 kcal (100 Calories). Reference to Table 6-4 shows that for the same mass (or weight), fats, for example, stearic acid, provide about twice as many calories as do carbohydrates, for example, sugar.

Actually, the metabolism of foodstuffs and the combustion of fuels have two common features. First, the exothermic heat of reaction or heat of combustion provides the energy either for the human (or animal) system or for the system needing heat energy, such as a building, a boiler, a power

[3]The kilocalorie of the scientist is the Calorie (capital "C") of the weight-intaker or the nutritionist. When one counts Calories of foods, one is actually counting "scientific" kilocalories, and not the calorie (small "c") as defined earlier.

Table 6-6. Portion of Food Having Fuel Value of Approximately 100 kcal

1 large egg	hamburger $2\frac{1}{2}$ in. in diameter,
$\frac{2}{3}$ cup of milk	$\frac{7}{8}$ in. thick
1 slice of buttered toast	$\frac{2}{3}$ oz candy bar
$\frac{1}{3}$ cup of cream of tomato soup	6 oz of beer
piece of apple pie, $1\frac{1}{2}$ in. at edge	

plant, an engine, and so on. In general, man's use of energy far exceeds his metabolic requirements. Energy from stored sources (fossil fuels) has been consumed in the early 1970s at a rate of some 100 times that of his metabolic consumption, and some 20 times that of the combined metabolic consumption of man and his livestock.

The other feature in common between foodstuff metabolism and fuel consumption is the production of carbon dioxide, which must go somewhere. We concern ourselves in later chapters (Chapters 8 and 13) with the effect of both heat, and carbon dioxide on the environment. Suffice it to say at this point, that both the amount of heat and the amount of carbon dioxide can be determined quantitatively if you know what is being burned and what its heat of combustion is. Both are summarized in the balanced chemical equation and its heat of reaction.

Fossil Fuel Resources and Reserves— A Summary

Natural gas, petroleum, and coal serve principally as energy producing fuels in that the chemical energy of the molecules is released on combustion. These same materials are the sources of nearly all the organic chemicals, natural and synthetic, that are used by society in such forms as drugs, dyes, synthetic fibers, and plastics. In this section, we examine the locations and quantities of these fossil fuels, which were formed in the earth many millions of years ago.

Man discovered these substances and found them useful. He first used the most easily available supplies. The best supplies are the most concentrated and the closest to the surface. When these readily available supplies were exhausted, man sought supplies accessible with more difficulty, which means that the supplies are of lower concentration, or are deeper in the earth, or are more remote from man's usual habitats. You can imagine a situation in which the energy expended to obtain a pound of coal or a gallon of petroleum or a cubic foot of natural gas, might approach or exceed the amount of energy that could be recovered by burning the material as fuel.

A Question for the Reader: When the energy cost of obtaining the fuel is as much as its energy value, will it then be worth going after the material?

We use three terms to describe the supplies of fossil fuels and other materials in the earth. **Total stock** is the sum of everything known to exist even though the substance desired may be present in trace quantities and, therefore, would be highly unlikely to be worth trying to extract. **Resources** comprise that portion of the total stock that man can make available under some technological and economic conditions *different* from those that currently prevail. For example, with today's technology it is generally agreed that from a given oil well only half of the oil is recoverable, which means that half of the oil is not currently obtainable using present technology. With improved methods, perhaps the recovery rate of oil from an oil well might be significantly raised, although almost certainly the recovery rate will not come close to 100%. Thus, the term **reserve** refers to that proportion of a resource that is known with reasonable certainty to be available under current technological, economic, and other societal conditions.

Coal

Paul Averitt of the U.S. Geological Survey has estimated that an initial supply (resource) of 7.6×10^{12} metric tons of coal was available to man, say at the beginning of recorded history. He then defines as mineable (reserve), coal in beds as thin as 14 in. (36 cm) and extending to depths of 4000 ft (1.2 km) [1]. Table 6-7 shows the distribution of the coal resources by region. The world reserve of coal, then, is about 3.8×10^{12} metric tons of coal [1]. For your further understanding of the geographical distribution of coal, Figure 6-4A shows the geographical distribution of known world coal reserves, and Figure 6-4B shows the major locations of coal in the United States.

The principal constituents of coal are carbon, hydrogen, and sulfur, although coal also contains certain minerals. The term coal refers to anthracite (hard coal) and bituminous coal (soft coal). These coals derive from lignite, which in turn is considered to come from peat. Peat is partially decayed plant matter found in swamps and bogs. Although little carbonization has taken place in its formation, peat is dried and used as a fuel in some countries. Lignite is the first stage of carbonization of peat and still contains about as much oxygen as carbon. The coals, on the other hand, have rather low amounts of oxygen in them. A typical coal might contain

Table 6-7. World
Resources of Coal*

Area	10^9 Metric Tons	Per Cent of Total
United States	1486	19.5
North America (outside U.S.A.)	601	7.9
Western Europe	377	4.9
U.S.S.R.	4310	56.5
Asia (outside U.S.S.R.)	681	8.9
Africa	109	1.4
Oceania including Australia	59	0.7
Latin America	14	0.2
Total	7637	100.0

* SOURCE: Adapted from M. K. Hubbert: *Sci. Amer.*, **224**(3): 61 (1971).

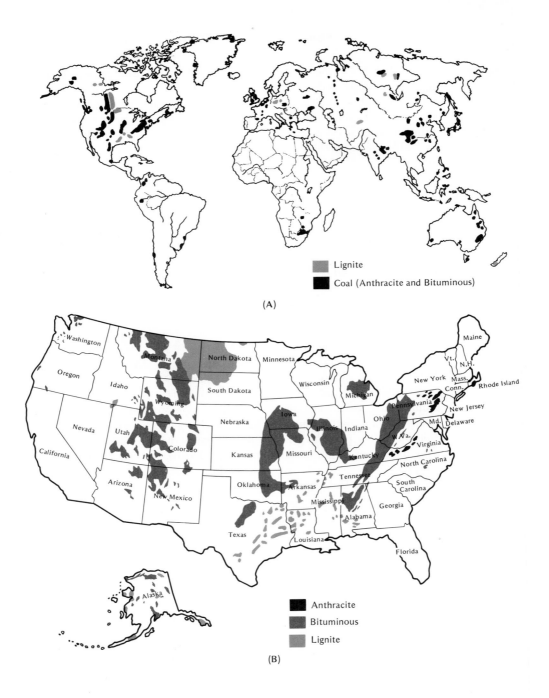

Figure 6-4. **(A)** Geographic distribution of known world coal reserves.
(B) Geographic distribution of coal reserves of the United States.

Table 6-8. Combustion Products from 1 Metric Ton of Coal

Combustion Reaction	Mass of Product (kg)
$C + O_2 \longrightarrow CO_2$	3370 CO_2
$4H + O_2 \longrightarrow 2H_2O$	270 H_2O
$S + O_2 \longrightarrow SO_2$	60 SO_2
minerals $\longrightarrow$ smoke	10 smoke
Thus, 1000 kg of typical coal gives	3710 reaction products

92% carbon, 3% hydrogen, 3% sulfur, and 2% minerals (all percentages are by weight). Thus, such a metric ton of coal would contain:

> 920 kg of carbon
> 30 kg of hydrogen
> 30 kg of sulfur
> 20 kg of minerals

Combustion of 1 metric ton of this coal produces the quantities of materials shown in Table 6-8.

Petroleum and Natural Gas

Table 6-9 records the distribution of "original" petroleum reserves by region. Figures 6-5 and 6-6 show the major locations of petroleum worldwide and in the United States, respectively. The estimates for petroleum and natural gas are somewhat more difficult to make than for coal. These fluids occur in restricted volumes of space within the earth at depths

Table 6-9. Original World Reserves of Petroleum Liquids *

* Adapted from Hubbert [1] who quotes W. P. Ryman, Deputy Exploration Manager, Standard Oil Company of New Jersey. Ryman's estimates include: Produced + Proved + Probable + Future Discoveries.

†A barrel of petroleum is 42 gal.

‡One barrel of petroleum on the average delivers the same amount of heat energy on combustion as does 0.203 metric tons of hard coal (203 kg).

	Barrels × 10^9 †	Metric Tons of Hard Coal Equivalent × 10^9 ‡	Per Cent of Total
United States	220	40.6	9.6
Canada	95	19.3	4.5
Latin America	225	45.7	10.8
Europe	20	4.1	1.0
Africa	250	50.7	12.0
Middle East	600	121.8	28.7
Far East	200	40.6	9.6
U.S.S.R. and China	500	101.2	23.8
Total	2090	424.0	100.0%

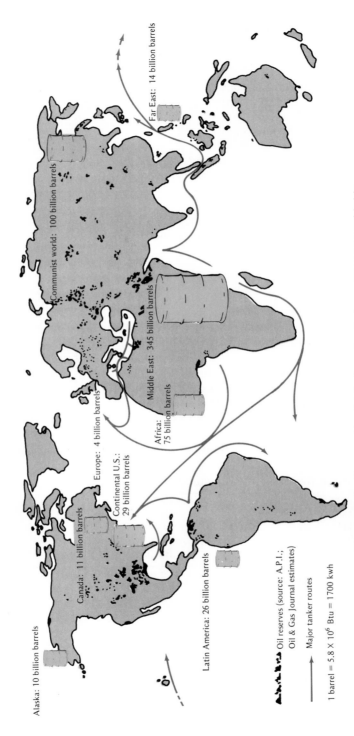

Figure 6-5. Geographic distribution of petroleum reserves worldwide.

Far East: 14 billion barrels

Communist world: 100 billion barrels

Middle East: 345 billion barrels

Africa: 75 billion barrels

Europe: 4 billion barrels

Continental U.S.: 29 billion barrels

Canada: 11 billion barrels

Alaska: 10 billion barrels

Latin America: 26 billion barrels

Oil reserves (source: A.P.I.; Oil & Gas Journal estimates)

Major tanker routes

1 barrel = 5.8 × 10⁶ Btu = 1700 kwh

180

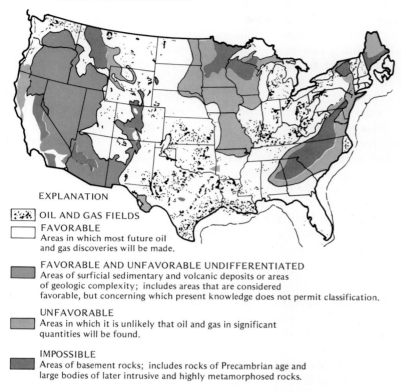

DEPARTMENT OF THE INTERIOR
UNITED STATES GEOLOGICAL SURVEY

EXPLANATION

OIL AND GAS FIELDS

FAVORABLE
Areas in which most future oil
and gas discoveries will be made.

FAVORABLE AND UNFAVORABLE UNDIFFERENTIATED
Areas of surficial sedimentary and volcanic deposits or areas
of geologic complexity; includes areas that are considered
favorable, but concerning which present knowledge does not permit classification.

UNFAVORABLE
Areas in which it is unlikely that oil and gas in significant
quantities will be found.

IMPOSSIBLE
Areas of basement rocks; includes rocks of Precambrian age and
large bodies of later intrusive and highly metamorphosed rocks.

OIL AND GAS FIELDS IN THE UNITED STATES AND UNPRODUCTIVE AREAS
CLASSIFIED AS TO THEIR RELATIVE LIKELIHOOD OF YIELDING COMMERCIAL
QUANTITIES OF OIL OR GAS

Figure 6-6. Geographic distribution of petroleum reserves in the
United States.

ranging from a few hundred meters to more than 8 km. The mass of
exploration data developed in the oil producing regions of the United States
is used to give estimates for other areas of the world.

In addition, on the average 6500 ft^3 of natural gas is discovered for each
barrel of oil [1]. Thus, the ultimate natural gas production calculates to
about $13 \times 10^{15} (6500 \times 2090 \times 10^9)$. Since 10^6 ft^3 of natural gas is equivalent in heating value to 30 metric tons of coal, this ultimate figure for
natural gas is equivalent to 93×10^9 metric tons of coal as compared with
424×10^9 metric tons of coal equivalent available in total petroleum
resources.

Hubbert [2] discusses these figures for petroleum and natural gas at considerable length, especially by comparing them with other estimates. In general, the estimates used here seem reasonable, but must be used with the caution that these figures include all material produced, proved to exist, probably existing, and include future discoveries. You should also be aware that with current technology for every barrel of petroleum produced, another barrel is left in the ground.

In Chapter 7, we will look at the consumption of energy, 96% of which has its origin in fossil fuels. World oil and gas reserves represent 424×10^9 metric tons equivalent of coal as compared to 3800×10^9 metric tons of coal reserve. Oil and gas represent only a little more than 10% of the world's fossil fuel reserves. Yet, today these more convenient liquid and gaseous fuels are being consumed preferentially. For example, in the United States in 1970 [3], 75% of the energy consumed was supplied by oil and gas with only 22% supplied by coal. An additional concern in the United States is the fact that, since 1945, the gas reserves have been consumed at an ever increasing rate. It would appear that our consumption habits must change (the subject of Chapters 8 and 15).

Conclusion

We began this chapter with a discussion of combustion reactions, especially as represented by chemical equations. The heats of these straight forward chemical processes have been related to metabolism of foods, photosynthesis, and, most importantly, to the generation of heat energy from fossil fuels. We turn now in the next chapter and the two chapters following to the utilization of heat energy from fossil fuels to power our society.

Questions

1. What is an exothermic reaction? An endothermic reaction? What is meant by a chemical equation?

2. Balance each of the following equations:
   (a) $CH_4 + O_2 \longrightarrow CO_2 + H_2O$
   (b) $H_2 + N_2 \longrightarrow NH_3$
   (c) $C_2H_6 + O_2 \longrightarrow CO_2 + H_2O$
   (d) $H_2S + O_2 \longrightarrow H_2O + SO_2$
   (e) $H_2SO_4 + KOH \longrightarrow H_2O + K_2SO_4$
   (f) $H_2 + Cl_2 \longrightarrow HCl$
   (g) $C_6H_6 + O_2 \longrightarrow CO_2 + H_2O$
   (h) $NO + O_2 \longrightarrow NO_2$

3. Slaked lime, $Ca(OH)_2$, used in plaster, absorbs carbon dioxide from the air to form a molecule that has the same formula as marble, $CaCO_3$; water also is formed. Write a balanced equation for this reaction which occurs in the hardening of plaster.

4. Starches and cellulose are sometimes written as the formula $(C_6H_{10}O_5)_n$, where n is a large number. Write the burning reaction for such molecules.

5. When Humphrey Davy visited Florence in 1813, he was given a large lens to use as a burning glass. With this he focused the sun's rays intensely on a diamond enclosed in an oxygen atmosphere. The diamond gradually disappeared. Write the chemical reaction that occurred. (*Hint:* What is diamond?)

6. Balance each of the following:

   (a) $C_3H_8 + O_2 \longrightarrow CO_2 + H_2O$
   propane

   (b) $C_{12}H_{22}O_{11} + O_2 \longrightarrow CO_2 + H_2O$
   sucrose

   (c) $N_2H_4 + O_2 \longrightarrow N_2 + H_2O$
   hydrazine
   (rocket fuel)

7. Two major air pollutants produced by automobiles are unburned gasoline (hydrocarbons) and carbon monoxide (CO). One device for the abatement of this pollution is an air-injection system by which air is introduced into the very hot exhaust gases emerging from the cylinder. Suggest the essential chemistry of the operation of the device. [*Hint:* Write equation of combustion for CO and for gasoline (C_8H_8).]

8. A typical formula for a fat is $C_{54}H_{110}O_3$. A comparable carbohydrate would be $C_{54}H_{92}O_{46}$. When fats burn (including being used in your body), about twice as much energy is released as when an equal amount of carbohydrates burns. Can you anticipate a reason for this?

9. State several reasons why the air in rural regions is likely to contain more oxygen than the air in urban areas.

10. What is the source of the black smoke emerging from the exhaust of a diesel powered truck that is accelerating? Diesel fuel is approximately $C_{16}H_{34}$.

11. Which would give the higher temperature, a gas-air flame or a gas-oxygen flame? Why?

12. Compare the fossil fuels as to heating value per kilogram and as to availability (quantities).

13. In the process of photosynthesis in plants, CO_2 and H_2O are combined to make a sugar. Is this an exothermic or an endothermic process? Make your reasoning clear.

14. As you examine the fossil fuel "geography" of the United States, what states must import fossil fuels for home heating or for fueling electric power plants?

15. Weight for weight, how does the calorie value of fats compare with sugars (carbohydrates)?

Numerical Exercises

1. For the balanced chemical equation:

$$C_2H_4 + 3O_2 \longrightarrow 2CO_2 + 2H_2O$$
ethylene

   (a) one molecule of C_2H_4 requires exactly how many molecules of O_2 and produces how many molecules of CO_2 and H_2O?

   (b) 1 mole of C_2H_4 requires exactly how many moles of O_2?

(c) 2 moles of C_2H_4 produces exactly how many moles of CO_2?
(d) 0.25 mole of C_2H_4 produces exactly how many moles of H_2O?
(e) 28 g of C_2H_4 produces exactly how many grams of CO_2?
(f) 1 metric ton of C_2H_4 produces exactly how many metric tons of CO_2?

2. If the heat of combustion of ethane (C_2H_6) is 372.7 kcal/mole (g-mole), how much heat will be liberated
   (a) when 10 moles of ethane is combusted?
   (b) when 30 g of ethane is combusted?
   (c) when 1.0 g of ethane is combusted?
   (d) when 1.0 lb of ethane is combusted?
   (e) when 1 metric ton of ethane is combusted?

3. If kerosene were pure $C_{10}H_{22}$, whose heat of combustion is 1625 kcal/mole, and if kerosene costs 15 cents per gallon (each gallon weighs 3000 g), calculate the cost of obtaining 10,000 kcal of heat by burning kerosene.

4. Compare the percentage of carbon in sucrose with that in stearic acid. Compare the percentage of oxygen in those two substances. Why does one substance give off more heat per pound than the other?

5. In walking a mile, a man expends about 0.55 kcal/lb of body weight. How much extra food does a 180-lb man require for each mile he walks? How many eggs are needed to provide the energy?

6. How many pounds of carbon dioxide must a plant absorb from the air to produce 10 lb of glucose?

7. If a ton of coal has only 87% carbon, how many tons of carbon dioxide will be produced on total combustion of 1 ton of coal? If, as in the United States during 1970, 7.5×10^8 tons of coal are burned in a year, how many tons of carbon dioxide are emitted to the atmosphere?

8. If a metric ton of coal has 3% sulfur, how many kilograms of SO_2 will be produced on total combustion of this ton of coal? If 7×10^8 tons of coal are burned in 1 year, how many kilograms of SO_2 are emitted to the atmosphere? How many metric tons of SO_2?

References

[1] Estimate made in 1969, U.S. Geological Survey, quoted by M. K. Hubbert: *Sci. Amer.*, **224**(3):61 (1971).

[2] Committee on Resources and Man, National Academy of Sciences-National Research Council: *Resources and Man*. W. H. Freeman and Co., 1969, Chap. 8.

[3] U.S. Department of the Interior.

Heat Is Transformed and Transferred

Combustion of molecules of fossil fuels, or more specifically the rearrangement of the chemical bonds in these molecules, releases the stored chemical energy to give us heat energy. We use this heat energy to do useful work and ultimately transform this work into the goods and services for our consumption. The fact is that only some of the heat energy (the amount depends on the machine process as we shall see) goes into these goods and services while much of it is wasted. In this context, C. M. Summers [1] has compared our modern industrial society with a "complex machine, for degrading high quality energy into waste heat while extracting the energy needed for creating an enormous catalogue of goods and services."

The term **waste heat** refers to the fact that no heat engine or other thermodynamic system (physical, chemical, or biological) can be 100% efficient; that is, no heat engine can convert a quantity of heat entirely (100%) into work. Thus, the inevitable result of energy use

185

or consumption is the degradation or loss of some part of that ener
converted. Usually, the waste heat or energy is immediately discarded ir
the environment in the vicinity where the heat is being produced. T
amount of heat being discarded as waste and its effects on the environme
are subjects for the following chapters. Is there any reason why 100
conversion of heat into work cannot be attained? Is there some compelli
force in nature that tells us that our society and its machines must *wa*
in order to function? These questions can only be answered by a stu
of the laws governing heat flow, the laws of thermodynamics.

In this chapter, we first introduce the concept of efficiency and ine
ciency of machines in our society. A discussion of the energy tra
formations and the inefficiencies inherent in the operation of a fossil-fuel
electrical power generating plant follows as our "model" machine. We th
introduce the laws of thermodynamics as the basis of inefficiency, follow
by a brief discussion of Carnot's "ideal" heat engine. Finally, we disc
the various means by which heat is transferred.

Inefficiency

In Chapter 6, we found that the total heat available from burning 1
of domestic heating oil is about 4900 kcal (10800 kcal/kg). However, t
"useful" heat from the fuel depends on how we use it. Different types
machines convert different fractions of the maxium heat from combusti
of fuels to useful heat or work. The fraction not converted to useful he
of course, is wasted and is known as waste heat. This leads us to a scientis
definition of efficiency, and, conversely, inefficiency. **Efficiency** is the use
heat or work (output) divided by the maximum heat (input).

$$\text{efficiency } (\eta) = \frac{\text{work (output)}}{\text{heat (input)}} = \frac{W}{Q_{\text{in}}} \qquad (7$$

where the useful heat is to be identified with work (output) and the he
input with maximum heat.

Equation (7-1) is a fraction. By multiplying the fraction by 100%
obtain the efficiency in per cent. Similarly, inefficiency is the waste he
divided by the maximum heat. We have not provided an equation for p
cent inefficiency since it is just $100\%-\eta(\%)$. Table 7-1 records the efficienc
of representative energy converters that are important in the maintenan
of our society.

For example, a home oil furnace is about 65% efficient in its conversi
of the chemical energy in the oil to heat for the home. This means th
35% of the energy of the fuel burned is discarded out the chimney as h
gases and unburned hydrocarbons. A steam power plant converts t
chemical energy of the fossil fuel to electrical energy with an over

Table 7-1. Efficiency
of Energy Converters

Device	Efficiency(%)
electric generator	99
dry cell battery	90
steam boiler	88
home gas furnace	85
storage battery	72
home oil furnace	65
steam turbine	47
steam power plant	41
diesel engine	38
auto engine	25
fluorescent lamp	20
Wankel engine	18
steam locomotive	8
incandescent lamp	4

efficiency of 41%. Much of the waste heate, 59%, is discarded into plant cooling waters, such as rivers (Figure 7-1A, p. 188).

An interesting comparison to make is that of comparing the efficiency of the incandescent light bulb for converting electrical energy into light with the fluorescent lamp. About 96% of the electrical energy consumed by the incandescent lamp goes into heat with only 4% into light. In the fluorescent lamp, 80% of the electrical energy goes into heat and 20% into light. Many new buildings utilize this waste heat from the lighting systems in designing the capacity of heating plants. Ultimately, one could project an end to the use of the incandescent lamp in favor of the more efficient fluorescent lamp for lighting.

Overall, our society operates at an average efficiency of about 50%.[1] That is, half of the fuel consumed produces useful work whereas the other half is wasted immediately or discarded as waste heat into the environment. Refer again to Table 7-1. If we average these representative energy converters for society, we obtain about 50% for the efficiency for these devices, which is consistent with the overall efficiency for society mentioned above.

In order to understand the full implications of fuel consumption and the growth thereof, let us look at what we do with heat energy in our society. Economically speaking, the amount of fuel consumed is directly related to the quantity of goods and services produced. The value of these

---

[1] The efficiency of an incandescent light bulb is obtained by comparing the visible emission with the total energy emission. The efficiency of society in producing goods and services, and other similar examples lie outside our basic definition in equation (7-1), but it is common use to refer to these other efficiencies in a like manner.

Figure 7-1(A). Twin 700-ft stacks of Potomac Electric Power Co., Morgantown, Maryland. Stations loom high above the Potomac River. Generating at 1251 Mw, the facility burns coal (and residual oil) at the rate of 2.5×10^6 tons of coal/year. [Courtesy of the National Coal Association, photograph by Len Henig, 1973.]

Figure 7-1(B). A steam turbine being installed in an electrical generating facility. Note the various stages of blades needed for transforming a maximum amount of the steam's kinetic energy.

goods and services in our economy is essentially our gross national product (GNP) (see Chapter 1). In 1970, the United States had a GNP of about $1000 billion, achieved with the help of 20×10^{12} kwh or 65×10^{15} Btu of energy. Fossil fuels provided 95.9% of this energy while only 3.8% was provided by hydroelectric power. Nuclear power was responsible for the remaining 0.3%. Other sources, such as combustion of wood, were insignificant by comparison.

In the twentieth century, until 1967, the GNP grew at an average rate of 3.2% per year as compared to an annual growth of fuel consumption of 2.7%. An economic growth—growth in the GNP—which is greater than the growth in fuel consumption, generally means that we have been getting more goods and services each year per unit of fuel. This favorable ratio of growth in GNP to growth in fuel consumption has been made possible by advancing technology or improved technology, and, in particular, by increasing the efficiency of converting energy to goods and services. One of the most dramatic increases in the efficiency of machines has been made by the electric power generating industry. The increase has been from about 5% in 1900 to an average efficiency of about 33% in 1970.

Unfortunately, this favorable situation has changed since 1967. Fuel consumption since then has been growing faster than the GNP. Thus, to get a unit of goods, more energy must be used. Ultimately, this means that we are paying more for the same units of goods (Figure 7-2).

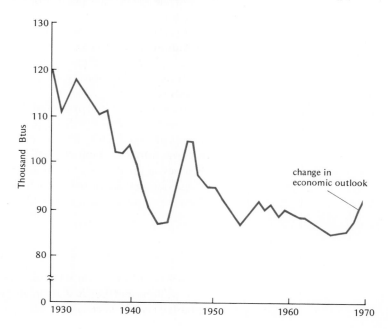

Figure 7-2. Energy consumed in the United States per dollar of GNP (in 1958 dollars). [Source: U. S. Department of the Interior.]

What is the reason for this unfavorable change or "turn around"? Some people blame inflation. Although inflation is prevalent in the "workings" of our complex machine, the problem has more fundamental bases. The rate of advance of technology has slowed. Until the late 1960s, we were able to cover up our fantastic increase both in use and in waste of energy with correspondingly larger increases in efficiency of use.

This is no longer the case. For example, we find ourselves, in the United States today, with a rapidly increasing demand for oil and natural gas but with inadequate supplies at home and with the political front somewhat "clouded" abroad (where the largest untapped reserves are available). Whereas, at one time a few drillings were only necessary for an oil well discovery, now drillings numbering about 70 in the U.S.A. are necessary to bring in a single significant, producing oil well. More effort must be expended to maintain the supply of oil. The situation is even more acute with natural gas. As a result, in many areas of the country, no new natural gas customers are being accepted because of tightened supplies.

Furthermore, many of us consumers have been converting our home heating systems from the more highly efficient oil and gas systems to less efficient electrical heating systems for reasons of convenience and cleanliness. Figure 7-3 compares the efficiency of a home heated directly with oil and one heated by electricity. As is shown the oil furnace is about twice as efficient for home heating as electrical heating due to the large inefficiency of the generating plant. For this reason alone, the cost of electrical heating is considerably greater than directly heating with oil (or gas), not to mention the effect on energy consumption. For the home heating plants, the pollution (heat and combustion products) is widely distributed because a "little" is produced from each home, as compared with the pollution concentrated in the vicinity of a large power generating station serving thousands of homes.

By the year 2000, it is projected that a far greater proportion, near 50%, of our electrical power will be provided by nuclear generating stations, operating at about 30% efficiency as compared to fossil fuel generating stations operating at about 40% efficiency. Since 30% efficiency means 70% inefficiency, a greater proportion of our energy is to be degraded immediately and dumped as waste heat, which may represent a severe heat burden on the environment, particularly in the vicinity of the power plant.

A Question for the Reader: The electric automobile has been suggested as the answer to the air pollution problem in an area such as Los Angeles. What effect might this have on our already overburdened energy supplies?

What has all this to do with us? Very simply, this will put more stress on our wallets. Putting it a little more strongly, we are running out of inexpensive, easily accessible, high quality energy. We have been living

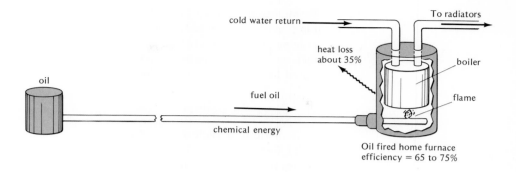

cold water return

To radiators

heat loss
about 35%

boiler

oil

fuel oil

flame

chemical energy

Oil fired home furnace
efficiency = 65 to 75%

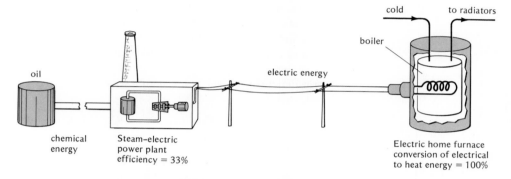

cold

to radiators

boiler

oil

electric energy

chemical
energy

Steam–electric
power plant
efficiency = 33%

Electric home furnace
conversion of electrical
to heat energy = 100%

Figure 7-3.   Efficiencies of direct use of fuel versus indirect use. A large loss of energy occurs at the electrical power generating plant (67%) as compared to the loss when directly using fuel for home heating (about 30%). Thus, heating the home with electricity consumes at least twice as much fuel energy as does direct use of fuel for heating.

"cheaply" off our environment, but in the future we will have to pay for it, almost certainly through an increased percentage of our incomes. One ecologist has expressed the seemingly careless disregard of our society toward growth in energy consumption and its effects on the environment when he said, "there is no such thing as a free lunch." Let us, for the benefit of present and future generations, try to understand why, when, and how the price must be paid.

**A Fossil-Fueled
Steam Power Plant**

In keeping with our central theme, we have chosen to discuss some of the intricacies of operation of the fossil-fueled steam power plant as a typical complex machine. Besides its central importance to the operation of our industrial society, its workings illustrate very clearly the specific meaning of the terms energy consumption, efficiency, waste, thermo-dynamic engine, and so on, which are so often used in drawing a parallel between society and a complex machine. In the beginning of our discussion, we would like to pose two questions for the reader to keep in mind:

1. Why must our complex machine be inefficient?
2. Where does the result of inefficiency, namely waste, go?

A schematic of a typical fossil-fuel steam electric generating plant is shown in Figure 7-4. Such a plant is a **thermodynamic engine** and is sometimes referred to as a **heat engine.** The word **thermodynamic** means heat flow. A heat engine takes heat energy and converts some of this energy into work. Some of the heat energy is also discarded as waste, specifically wasted heat energy. Interestingly, spaceship earth can be considered as a heat engine which takes in energy from the sun, using the energy to do work, and then discards (reradiates) heat back into space.

As shown in Figure 7-4, the first step in the operation of the steam electric generating plant is the combustion (burning) of a fossil fuel, such as coal, that is,

$$C + O_2 \longrightarrow CO_2 + heat$$

Figure 7-4. Schematic of a fossil-fuel steam electric generating plant. The efficiency of the various stages were obtained from Table 7-1.

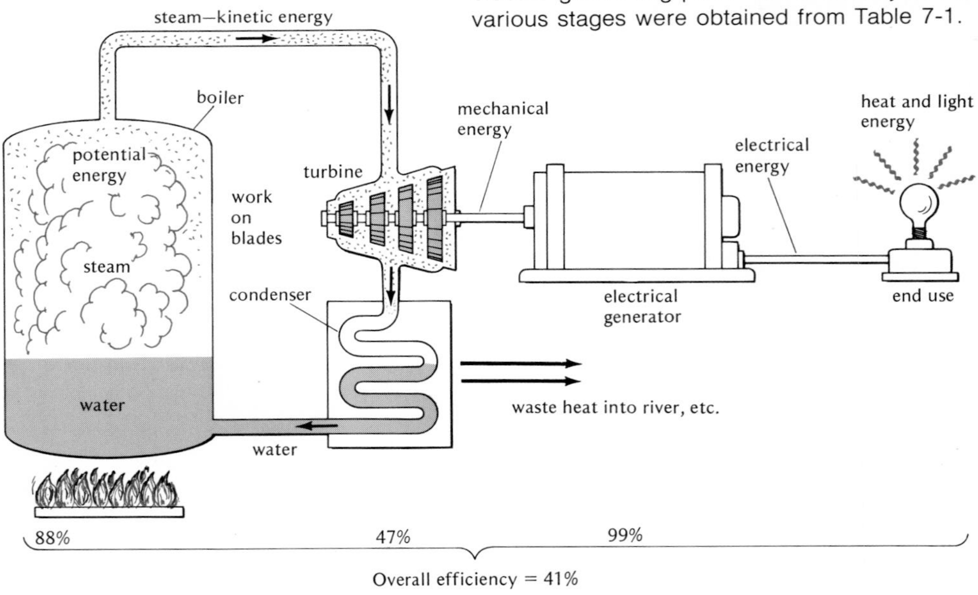

because carbon is the major constituent of coal. The heat of combustion raises the temperature of the water in the boiler to the boiling point and converts the water to steam. Fossil-fuel fired steam boilers today are able to use steam under pressure up to a temperature of about 1000°F or about 538°C. This "hot" steam—steam with high heat content—has energy available for doing work. We roughly may even consider the hot steam as having potential energy relative to the cold water being circulated back to the boiler from the condenser.

The steam is allowed to escape from the boiler through a nozzle at high speeds. You will recall that high speed means high kinetic energy and, hence, high capacity to do work. The steam or water molecules impinge on the turbine blades exerting a force on them and moving the blades through a distance (analogous to blowing on a pinwheel). The work done on the blades moves them and is thus transformed into mechanical energy of the spinning blades. The turbine blades are mechanically linked through the shaft holding the blades to an electrical generator which converts the mechanical energy (spinning motion) into electrical energy for our use (often called **end use**). We should mention that, although we show a simple turbine here, often a series of turbines or turbine stages is used for obtaining as much work as possible out of the steam (Figure 7-1B, p. 188).[2]

The final stage of operation is the condensing stage where the cool steam, having lost most of its kinetic energy in doing work on the turbine blades, is condensed to water. Water, usually river water, then serves as the coolant by being circulated across the condenser coils to remove the remaining heat energy of the steam. The heat of condensation, which is the same as the latent heat or heat of vaporization of steam, is 540 kcal/kg. The coolant nominally remains at about 100°F or 32.2°C. Because a given volume of steam (gas) when condensed to water (liquid) occupies considerably less volume, a partial vacuum is created which pulls on the turbine blades, giving a little extra "oomph" (energy) to the blades and hence to the electrical generator. The cooled water is now pumped back (recirculated) to the boiler for reheating.

Our description of the boiler–steam turbine system represents one completed **thermodynamic cycle** of operation, that is, we are back to the initial pressure, temperature, and volume for the working substance, the water in the liquid state. One of the most important points to take note of is that removing some heat in order to condense the steam into water is absolutely necessary for efficient completion of the thermodynamic cycle.

This waste heat is carried away by the coolant, which is then discharged

---

[2] A large number of turbine stages is desired so as to decrease the temperature (or pressure) drop between input and output, which means the system is operating with less thermodynamic irreversibility and, hence, higher overall efficiency. See the following discussion on irreversibility.

EXAMPLE 7-1

Calculate the overall efficiency of steam electric power plants from the data given in Table 7-1 and Figure 7-2.

Stage	Inefficiency(%)	Efficiency(%)	Fraction
boiler	12	88	0.88
turbine	53	47	0.47
electrical generator	1	99	0.99

Overall efficiency (fraction): $0.88 \times 0.47 \times 0.99 = 0.41$
Overall efficiency (percent): $0.41 \times 100\% = 41\%$

---

back to the source from whence it was drawn, namely into rivers, lakes, and estuaries. On discharge, the warm coolant spreads, and cools as it spreads, throughout the receiving body of water. Thus, the waste heat is now no longer available for producing work. Ultimately, some part of this heat will be reradiated into space. Whether some remains to raise the average surface temperature of the earth is open to question.

Also shown in Figure 7-4 is the efficiency of each stage of energy conversion. From Table 7-1, we obtain the following efficiencies for the various stages: boiler, 88%; turbine, 47%; electrical generator, 99%. The net or overall efficiency of 41% is obtained by taking the product of the efficiencies of each stage. See Example 7-1.

About 12% of the heat of combustion goes up the chimney as hot gases and ash. The electrical generator loses about 1% due to frictional heat losses. From our analysis, we find that the turbine assembly, the heat engine, is the major source of inefficiency. There are several reasons for this inefficiency. The most important consideration is the inherent loss caused by the necessity to waste in order to complete the operating cycle. As we will see in the section on the Carnot engine, there is no practical or theoretical way to eliminate totally this loss! Here we have a physical limit placed by nature, that is, one of nature's laws!

Other sources of heat losses are those due to friction, radiative heat losses (see the section, Modes of heat Transfer, in this chapter) and a loss often called **irreversibility** (we will later include friction, radiative losses, and so on, as irreversible losses). A complete discussion of irreversibility is beyond our scope, but a brief explanation is in order (also see the section, Carnot Engine, in this chapter).

The maximum available work from a heat engine is obtained only when the thermodynamic process takes place very slowly and can be reversed (without doing work on the turbine system) at any point in the operating cycle. This is simply not the practical case wherein a large quantity of steam impinges on the turbine blades at very high speeds. To operate the turbine reversibly, we would have to allow the steam to hit the blades molecule by molecule. Thus, in reality, reversible engines are not possible. The Carnot engine, to be discussed later, is a thermodynamically reversible engine. In that discussion, we will differentiate between the ideal or theoretical efficiency and the practical efficiencies of heat engines.

In summary, this complex machine (steam electric power plant) converts the high quality energy (concentrated) of the coal or oil partly into work for producing goods and services and partly into waste heat. Unfortunately, the greater part is waste while the lesser part is useful. During operation, the energy transformations are as follows: conversion of chemical energy (coal) into heat energy, heat energy into "potential energy" of the steam, potential energy into kinetic energy of the steam, kinetic energy into work on the turbine blades, work into mechanical energy, mechanical energy into electrical energy, and electrical energy into end use energy (light, heat, and work). In the sections following immediately, we will discuss the laws that govern the operation of these complex machines on which our society is very dependent.

Nature's Guidelines: The Laws of Thermodynamics

"Mother (or father) nature" has guidelines that we are compelled to follow (ultimately!) in our course of living and working together on space-ship earth. These guidelines have been formulated into laws by scientists (men and women). We do not know why these laws exist. We do not know for sure whether they are valid on spaceships or stars other than our own, although we would assume so until we have reason to believe otherwise. Trying to answer these questions is the same as trying to answer the questions of how and why the matter in the universe came together in that first big fireball and what is the matter's origin. However, these laws are based on observations that have been made over centuries, particularly in the last century, and we believe from our observations that these laws will never be overthrown. Furthermore, we believe so completely in them that they have become more than just facts, they seem to have become truths.

Zeroth Law of Thermodynamics

The Zeroth law may or may not be considered as one of the fundamental laws of thermodynamics, depending on your point of view. However, since the concept of temperature is basic in discussions of heat flow, we include this law as a law of thermodynamics. The Zeroth law does not actually establish a temperature scale but it defines what we mean by **temperature equilibrium.** The Zeroth law can be stated as follows:

> If we choose three bodies, A, B, and C, which are in thermal (heat) contact with each other, and if B and C are at the same temperature as A, then B and C are at the same temperature.

This law is always assumed in every temperature measurement, and we believe that, so long as two bodies are at the same temperature and are measured by the same method, there will be no change in the measurable properties (pressure, volume, or temperature) when the two bodies are brought into contact. That is, no heat will flow to cause a change in properties when these bodies at the same temperature are brought into contact.

Now that we have defined what we mean by bodies at the same temperature, consider the case where A is at a higher temperature than B (or C). This immediately poses two rather basic questions: (1) how much higher in temperature is A than B, and (2) how do we measure it? The most obvious answer to us, as members of a scientific-technological world, is to use a thermometer. For eighteenth century man,[3] the answer was not so obvious.

The type of thermometer familiar to most is the small glass capillary (a tube with a very small lengthwise hole) with a small amount of mercury or alcohol trapped inside. We know from experience that when fluids are heated they expand, and, as they are cooled, they contract. Thus, as the small alcohol-filled bulb or reservoir is heated on the bottom, the alcohol will move up the capillary (the lengthwise hole), and, conversely, will move down on cooling. The heating or cooling and hence the rising or falling of the column of liquid, is related to the temperature of the thermometer.

In order to tell how much higher A is than B, we must mark (calibrate) the thermometer and establish a temperature scale. The three most commonly used scales are shown in Figure 7-5.

Two fixed points are chosen for the calibration, the ice point (mixture of pure ice and pure liquid water), and the boiling point of pure water at sea level. The level of alcohol, in the thermometer, is marked for each of these two points. The ice point is 32 degrees on the Fahrenheit (F) scale, 0 degrees on the Celsius (C) scale, and 273 degrees on the Kelvin (K) scale. The boiling points on the three scales are 212°F, 100°C, and 373°K, respectively. It is easy to see that the temperature interval between the boiling and ice points on the Fahrenheit scale is 180 degrees as compared

---

[3] The greatest problem of thermometer construction in the 1600s and 1700s was obtaining glass tubes with uniform capillaries so that thermometer readings could be compared. It was Daniel Fahrenheit (1686–1736) (German) who accomplished the construction. However, it was Anders Celsius (1701–1744) who used the fixed points of matter for calibration. The scale that we call Celsius is also referred to as the centigrade scale.

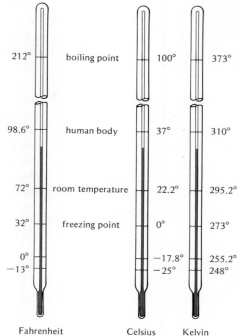

	Fahrenheit		Celsius	Kelvin
	212°	boiling point	100°	373°
	98.6°	human body	37°	310°
	72°	room temperature	22.2°	295.2°
	32°	freezing point	0°	273°
	0°		−17.8°	255.2°
	−13°		−25°	248°

Figure 7-5. The three most commonly used temperature scales.   (A)   (B)   (C)

to 100 degrees for each of the other two scales. Thus, a change of 1°C or 1°K is a change of 1.8°F. The comparison of degree intervals can be used in converting between scales, keeping in mind that the Fahrenheit scale begins at 32°.

The conversions between the scales are summarized by the equations (7-2), (7-3), and (7-4).

$$T(°C) = \tfrac{5}{9}[T(°F) - 32°] \qquad (7\text{-}2)$$
$$T(°F) = \tfrac{9}{5}T(°C) + 32 \qquad (7\text{-}3)$$
$$T(°K) = T(°C) + 273° \qquad (7\text{-}4)$$

(*Note:* $\tfrac{9}{5} = 1.8$.) For example, a room temperature of 77°F is 25°C or 298°K (see Example 7-2).

EXAMPLE 7-2

Convert 77°F to the appropriate temperature on the Celsius and Kelvin scales.

$$T(°C) = \frac{5}{9}(77° - 32°) = 25°C; \qquad T(°K) = 25° + 273° = 298°K$$

Perhaps one of the most significant points to remember is that the Fahrenheit and Celsius scales were derived by man, for the convenience of man and have little fundamental relation to processes in nature. However, the Kelvin scale, which establishes the concept of an "absolute" zero of temperature, does appear to have a fundamental meaning in our models of how nature behaves. For example, the "zero" on the Celsius scale was arbitrarily set at the ice point. As temperatures drop below the ice point, as they frequently do during winter in the temperate zones of earth, we begin to start counting in negative degrees.

The question of how low, in the negative direction, temperatures can go has intrigued man for a century. Both theoretically and experimentally, the absolute minimum in temperature appears to be about $-273°C$. Thus, we call this limit **absolute zero.** The significance of absolute zero is that at that temperature no further heat can be extracted from a substance. Some low temperatures of importance to the reader and for energy considerations are as follows: nitrogen liquefies at $77°K$, and helium liquefies at $4.2°K$. Physicists have produced temperatures as low as $0.000001°K$ $(10^{-6}°K)$.

Certain metals, such as lead, are called superconductors because they dissipate no heat energy when an electrical current flows through them if they are cooled below about $7°K$. Interestingly, even at the temperature of liquid nitrogen, $77°K$, the heat lost in transmission of electricity through power lines is considerably less than at room temperature. The loss of electrical energy as heat (due to the electrical resistance of the wire) while being transmitted along power lines is very serious and costly. Thus, the study of materials at low temperatures may have practical use if economic feasibility for keeping materials at these low temperatures can be achieved.

Another aspect of temperature that has significance for us is the model (view) of a gas[4] called the **kinetic model.** The kinetic model of gas considers the molecules of gas like little balls moving around like angry bees (Figure 7-6). These imaginary bees (or molecules) only interact through billiard-ball-like collisions. The model of the gas is described by the average speed and the average kinetic energy of the molecules. Obviously, just after collision, some molecules will be moving more slowly while others will be moving more rapidly. Because collisions are continually happening, we would observe a whole distribution of speeds and kinetic energies (Figure 7-7). However, the properties of the system of molecules as a whole are described by an average kinetic energy which depends (not too unexpectedly) on the absolute or Kelvin temperature.

$$\text{average KE} = \tfrac{1}{2}\overline{mv^2} = \tfrac{3}{2}k_B T. \tag{7-5}$$

---

[4]In contrast to the traditional approach, the authors have chosen to discuss an "ideal gas" in Appendix B because it is not necessary for our purposes.

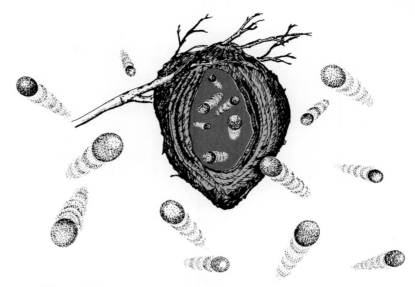

Figure 7-6. Molecules of a gas are like busy bees in a hive.

The factor of $\frac{3}{2}$ is appropriate for a monatomic gas, such as helium, and k_B is a constant called Boltzmann's constant. The bar over mv^2 means average.

Thus, *the Kelvin temperature is a measure of average molecular kinetic energy.* The higher the temperature, the faster the atoms or molecules move. For example, at room temperature, about 295°K, the average speed of a

Figure 7-7. Molecular distribution of speeds. The number of molecules having a "particular" speed are shown plotted. At room temperature, the speed at the peak of the curve is about 500 m/sec.

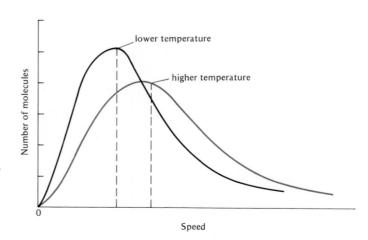

gaseous molecule is that of a high-powered rifle bullet, about 500 m/sec. When the temperature is lowered, the molecules buzz around more slowly, and, ideally, at absolute zero (0°K), the ideal molecules collapse on the floor.

However, molecules of a gas at very low temperatures do not act ideally because the gas liquefies before absolute zero is reached; for example, steam liquefies at 373°K and helium liquefies at 4.2°K. Furthermore, most liquids eventually solidify, for example, water at 273°K. Helium has not been observed to solidify (at atmospheric pressure). Nonetheless, the Kelvin temperature scale is also basic for discussions of the properties of liquids and solids. For atoms and molecules of a solid that are not free to move around as they are in gas, the Kelvin temperature is related to the vibrational kinetic energy of the atoms about their fixed positions in the solid. As we approach absolute zero, the vibrational energy approaches a small fixed value, called the **zero point energy.**

From observations, via experiments, this kinetic model of matter appears to be well verified, at least with respect to its general characteristics. Hence, even though the absolute temperature scale, too, was devised by man, it does seem to have some fundamental basis in our attempt to characterize nature's behavior.

First Law of Thermodynamics

Let us return now to our example of systems A, B, and C in thermal contact. With systems B and C at a higher temperature than A, we conclude from our personal experience that heat flows from B and C to A. This simple example of heat flow includes many processes in nature: spontaneous heat flow from a hot pan to a container of cold water touching the pan, spontaneous heat flow from a steam condenser to some coolant, or, in general, processes that involve work done by systems on surroundings or vice versa.

The first law of thermodynamics expresses one point in common about all of these processes, that the heat or other forms of energy involved may be simply transferred from one body to another or it may be transformed in type, but it is never destroyed. For our steam turbine example, even though the efficiencies of the conversion at each step were less than 100%, all of the energy could be accounted for in some way. Thus, the first law is hardly more than a statement of the conservation of energy[5]: *Energy can neither be created nor destroyed but can be transformed from one form to another.*

[5]To the scientist or engineer, the form of the first law is a little more mathematical in that it expresses a relationship between the heat flow into or out of a system, work done on or by the system, and the change in internal energy of a system. The internal energy of a monatomic (ideal) gas is given by equation (7-5). However, our statement is adequate for our purpose.

In using the first law of thermodynamics, we must carefully define the system that we are considering lest we end up with contradictions. In considering the boiler of the power plant, Figure 7-4, we have found that the practical efficiency in the transfer of the fuel's chemical energy to heat energy in the boiler is about 88%. If we restrict our attention to the boiler alone, we find that 12% of the energy is lost, apparently contradicting the first law of thermodynamics. However, if we consider the surroundings or the environment, we find the missing 12% in the form of hot gases discharged out the chimney stack.

This example may seem to be too obvious, but for the last 100 years or so people have been finding contradictions that, in fact, stem from a poor or inadequate description or definition of the system and its energy. In retrospect, when their apparent contradictions have been resolved, these too have seemed obvious. Hence, an equivalent statement of the first law is as follows: *the energy of a system, including its surroundings, remains constant* (see Figure 7-8).

Ultimately, when we consider the earth as "one big heat engine," we must include the universe as our complete system to avoid contradictions. It is believed that the law of conservation of energy holds for the universe, although we cannot rule out the possibility of our universe exchanging energy with another universe much in the same way that our earth system exchanges energy with the sun and the space around the earth.

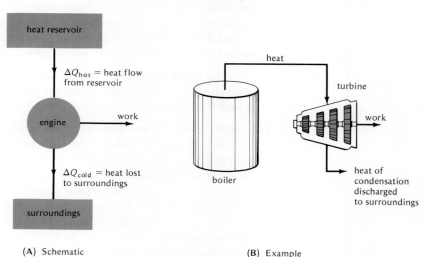

(A) Schematic   (B) Example

Figure 7-8. A thermodynamic system plus surroundings. In testing real processes, the choice of system is important. Most apparent violations to the laws of thermodynamics are due to inappropriate choice of systems. Probably the only true isolated system is the universe.

Also, in avoiding contradictions, we generalize our concept even further and call it the law of conservation of mass and energy: *mass-energy can neither be created nor destroyed.* Apparently, at some points in our universe, mass is continually being converted into energy and probably vice versa. The sun and other stars (in the earlier stages of their evolution) obtain their energy from a fusion of hydrogen nuclei with a resulting change of some nuclear mass into energy. The equivalence between mass and energy has been expressed by Albert Einstein in the relation, $E = mc^2$, where E is energy, m is mass, and c is the speed of light. With the recognition of Einstein's equivalence, a new era of study and understanding of the physical world has evolved which is of direct concern to all of us. We will return to this point in Chapter 10.

The Second Law of Thermodynamics

In our discussion of the Zeroth law, we observed that one expects heat to flow from a system of higher temperature to one of lower temperature.

A Question for the Reader: Can you think of any reason why the heat should flow from the colder system to the hotter system, making the cooler system colder and the warmer system hotter?

Further, can you imagine the heat from the power plant cooling water flowing into the plant's condenser coils causing them to become warmer rather than cooler? There is nothing to prevent it from happening. The fact that we have not seen it happen and do not expect it to happen is expressed in the second law of thermodynamics. Rudolf Clausius in 1852 expressed this law in one of its many guises.

> No process is possible whose sole result is the removal of heat from a reservoir (system) at one temperature and the absorption of an equal quantity of heat by a reservoir at a higher temperature.

Put simply, heat always flows *spontaneously* from a system at a higher temperature to one at a lower temperature (Figure 7-9).

Your suspicious face immediately lights up and you ask, "But what about a refrigerator?" Obviously, as the refrigerator operates, heat flows from the cool interior to a warm exterior, the room. The process of heat flow from the cooler interior to the warmer exterior of a refrigerator does not represent a *spontaneous* flow of heat. The electric motor, as it drives the refrigerator cooling system, provides work in order to reverse the natural direction for heat flow, that is, from a hot region to a cold region.

In principle, we could imagine using the now cool interior of the refrigerator as a low temperature reservoir and the warmer exterior (room) as a high temperature reservoir (a reservoir of heat) to operate a heat engine. After having used the flow of heat to obtain work, we could store

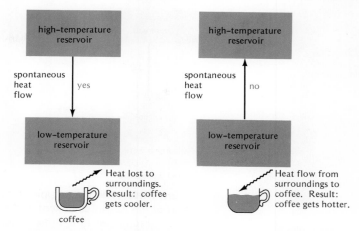

Figure 7-9. Heat flows spontaneously from hot to cold—the Clausius statement of the second law of thermodynamics.

this work, perhaps as potential energy of a weight, and then reuse it to run the refrigerator. Perhaps we might further imagine that for an initial energy investment to get the process started, the process would be self-sustaining, forever. This is a rather nice idea, is it not?

A Question for the Reader: Can you find the fallacy in our imagined self-sustaining refrigerator-heat engine?

This has been appropriately called a **perpetual motion machine,** and people have been trying to design such machines for centuries.

Our thoughts of a perpetual motion machine brings us to another statement of the second law, one which includes thermodynamic processes involving work done on or by a system, for example, the steam-turbine system of a power plant, an automobile engine, or perhaps the cooling system of a refrigerator. This formulation, appropriate for work processes, was provided by Lord Kelvin about the same time as the Clausius statement.

> No process is possible whose sole result is the abstraction of heat from a single reservoir and the performance of an equivalent amount of work.

More simply, a quantity of heat from a reservoir at some temperature cannot be converted solely into work (Figure 7-10). In the conversion process some heat must be discarded to a reservoir at a lower temperature. This is illustrated by the steam engine shown in Figure 7-11. To violate this form of the second law means to operate an engine at 100% efficiency. This has never been observed to happen. This statement is actually one of the guises assumed by the second law and is worth emphasizing again and again.

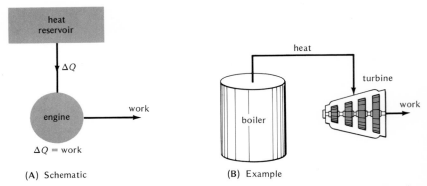

(A) Schematic   (B) Example

Figure 7-10. The conversion of heat entirely into work. This is a clear violation of the Kelvin statement of the second law. There must be a quantity of heat discharged to the cold reservoir—usually the surroundings as shown in Figure 7-8.

No engine may operate at an efficiency of 100%. It is fundamentally for this reason that our perpetual motion refrigerator will not work. The electric motor, which drives the cooling system, is less than 100% efficient. Any heat engine that would operate between the room and cool refrigerator interior would be much less than 100% efficient at converting heat into work, which could in turn operate the refrigerator. Thus, in all cases, we must continually add energy to sustain the machine's operation!

The application of the second law of thermodynamics to real systems and processes is quite straightforward in many cases. In our earlier description of the steam turbine, we first observed that heat flows from the high temperature boiler to the low temperature condensers—the Clausius statement. Secondly, we observed that part of the heat is converted into work

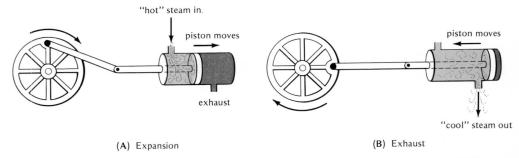

(A) Expansion   (B) Exhaust

Figure 7-11. The operation of a steam engine (piston type). The cool steam is discharged into the surroundings to allow the piston to return to its starting point for a new cycle. The discarded heat is waste.

and this ultimately into electricity. The rest of the heat from the steam is then discarded because of the necessity of a complete thermodynamic cycle of operation—the Kelvin statement.

In summary, water is turned into steam, circulated through the turbine, and then condensed back into water in the condenser coils. The heat of condensation is carried off by the cooling water. The heated cooling water, containing this waste heat, is then discharged into a river, lake, stream, and so on, to remove the waste heat from the plant. Thus, the thermodynamic cycle of the steam turbine is completed by removing the remaining heat from the used steam and then discarding it. Therefore, only part of the heat input to the boiler is used in producing work (turning the turbine blades). This thermal cycle alone necessarily has reduced the efficiency from 100% to about 55% or 60%, without even considering other heat losses (irreversible losses).

The laws of thermodynamics express what we observe and what we intuitively feel about processes involving heat flow in nature. Earlier, it was mentioned that we believe that the first law of thermodynamics holds not only for our small spaceship and our solar system, but for the universe as well. We believe the same about the second law in its many forms. However, generalizing what we observe to be true here on earth to the universe is like a child staring out his tiny window at his immediate surroundings and saying to himself, "the rest of the world must look just as I see it here."

In Chapter 2, we concluded that man has been on earth for such a minute period of time that our thoughts and ideas would surely change as our civilization matures (advances technologically). Thus, the application of our ideas to the universe seem highly speculative at this time and must surely be based on faith. This implicit faith was well expressed by Albert Einstein,

> . . . therefore, the deep impression which classical thermodynamics made on me. It is the only physical theory of universal content which I am convinced, that within the framework of applicability of its basic concepts, will never be overthrown.

The Carnot Engine— The Best We Can Do

When discussing cyclic thermodynamic processes and the general topic of heat engines, the most important example or reference point is the Carnot engine proposed in 1820 by Sadi Carnot, a French military engineer, when he was still a young man. Carnot showed that the maximum efficiency of a steam engine is determined by the temperature of its boiler and the temperature of its condenser (refer to our discussion of the steam power plant).

The full merit of Carnot's work was not recognized, however, until about 1848 when the Englishman, William Thomson, showed that, in addition to Carnot's observations about the efficiency of steam engines, Carnot's conclusions were sufficient to establish a temperature scale that is independent of the physical properties of substances, for example, the freezing and boiling points of water. Thomson (later Lord Kelvin) called it the absolute temperature scale—the Kelvin scale. This is the same scale that we introduced earlier on an **ad hoc** basis. It also is the scale necessary for the description of matter, such as the average kinetic energy of molecular motion (equation 7-5). Thus, the absolute scale is fundamental to our discussion of the second law of thermodynamics which, in itself, is a description of "how" nature behaves.

The importance of Carnot's accomplishments can hardly be overstressed. The development of the steam engine (Chapter 1) was effectively the "mother" of the Industrial Revolution, which marked the transition from labor provided by the backs of men, horses, and oxen to more powerful energy converters. Without the steam engine, England would not have been able to develop and maintain her coal production, and hence her superiority (economic and political) over other countries. However, advances in steam technology were very slow. There were very serious questions about the maximum work obtainable from a quantity of available heat. The observations of Carnot turned the art of making steam engines into a science. Figure 7-12 shows a chart illustrating the improvements in heat engine efficiencies.

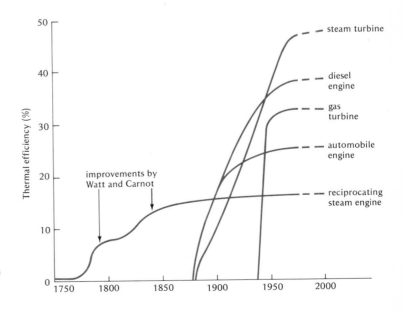

Figure 7-12. Efficiencies of engines. [Sources: M. W. Thring: *J. Inst. Fuel,* **27:**401 (1954); and C. M. Summers: The conversion of energy. *Sci. Amer.,* **224**(3):149 (1971).]

The rapid improvement of the reciprocating steam engine in the middle 1800s is evidence of Carnot's contribution. The development of the steam turbine about 1900 with its improvements is central to our interest in electrical power generation.

Carnot's studies of the steam engine led to the concept of an **ideal, reversible heat engine.** By ideal, we mean that there are no heat losses either due to radiation or to work against friction. A reversible process or engine (as mentioned earlier) is one that can be made to go in the reverse direction by an infinitesimal (very, very small) change in the conditions. For example, a very slow compression of a spring is practically a reversible process. If the compressing force is slightly decreased, the spring expands and does an amount of work exactly equal to the work done in compressing it. In contrast, hitting the spring with a mighty blow of a hammer will waste energy in the form of heat (work against air friction and in deformation of the spring). The work obtainable from the spring compressed in this manner is measurably less than that work used in compressing it.

A similar observation can be made for compressing a gas in a cylinder or for steam impinging on the blades of a steam turbine. This raises the question of why do we not design engines that operate slower and more reversibly. Besides the practicality of designing a truly reversible engine, we are actually more interested in the rate of obtaining work, or the power, rather than in obtaining the most from the energy available for producing work. Thus, real or practical engines are neither ideal nor reversible. Real processes are then **irreversible.** In summary, all changes that occur suddenly are inherently irreversible.

It further follows from the second law (although not immediately obvious) that a Carnot heat engine has a greater efficiency than any other heat engine operating between a given pair of heat reservoirs or between two specific temperature conditions. Put another way, a Carnot heat engine operating between a pair of reservoirs is the most efficient of all heat engines known or possible. These statements may be taken as other statements of the second law of thermodynamics.

The efficiency of the Carnot engine represents the maximum efficiency any heat engine can have while operating within given temperature limits.

Therefore, the Carnot engine represents the "best we can do."

General schematic diagrams for a heat engine and a refrigerator are shown in Figure 7-13. Let us concentrate for the moment on the heat engine. There are three basic components: the high temperature reservoir, the engine, and the low temperature reservoir. The engine may be a valve, a piston, or perhaps a turbine. Also, let us assume that the engine is operating reversibly between the two reservoirs. From previous discussions

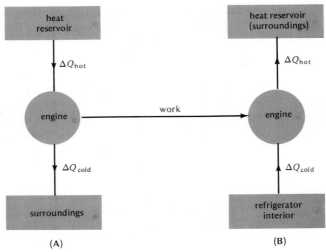

Figure 7-13. Schematic diagrams for a heat engine **(A)** and a refrigerator **(B).** The work from the heat engine can be used to run the refrigerator.

the efficiency is

$$\eta = \frac{\text{work}_{\text{out}}}{\text{heat}_{\text{in}}} = \frac{W_{\text{out}}}{Q_{\text{in}}} \qquad (7\text{-}6)$$

where W is the useful work output for an amount of heat, Q_{in} flowing into the engine. From the diagram, it also follows that the work out (energy put to use) is the difference between what we put in, Q_{in}, and what we discard into the low temperature reservoir, Q_{out}.

$$W_{\text{out}} = Q_{\text{in}} - Q_{\text{out}} \qquad (7\text{-}7)$$

Substituting this relation for W_{out}, we obtain for the efficiency

$$\eta = \frac{Q_{\text{in}} - Q_{\text{out}}}{Q_{\text{in}}} = 1 - \frac{Q_{\text{out}}}{Q_{\text{in}}} \qquad (7\text{-}8)$$

For a Carnot engine, the following relationship exists [2] between the heats transferred and the reservoir temperatures

$$\frac{Q_{\text{in}}}{T_{\text{hot}}} = \frac{Q_{\text{out}}}{T_{\text{cold}}} \qquad (7\text{-}9)$$

where T_{hot} and T_{cold} are the temperatures of the hot and cold reservoirs, respectively. Thus, the efficiency of the Carnot engine in terms of the temperatures of the reservoirs is

$$\eta = \frac{T_{hot} - T_{cold}}{T_{hot}} = 1 - \frac{T_{cold}}{T_{hot}} \qquad (7\text{-}10)$$

where the temperatures are expressed in Kelvin degrees. Equation (7-10) may be multiplied by 100% to give η in per cent.

The Carnot efficiency is usually referred to as the **maximum theoretical efficiency.** To distinguish between ideal efficiency and real efficiency, we will call the real or practical efficiency the **conversion factor** and **conversion efficiency.** From equation (7-10), we make the following important observations:

1. So long as T_{cold} is other than absolute zero, the efficiency is less than 1. or the per cent efficiency is less than 100%.
2. So long as the T_{hot} is finite (not infinitely large), the efficiency is less than 100%.
3. The most practical way of increasing the efficiency is to increase T_{hot}, which means increasing the steam temperature. The nature of materials limits this possibility.
4. If $T_{hot} = T_{cold}$, the efficiency is necessarily zero and the engine cannot operate. This particular condition precludes, for example, utilization of the large quantities of heat stored in ocean water. Near its surface, the ocean is a "sea of constant temperature" and no heat can be made to flow (the second law).

Therefore, in reality, the efficiency for heat engine operation is always less than 100%. After having determined the maximum theoretical efficiency for a given condition, a rule of thumb (of the authors') is that the conversion factor (practical efficiency) may be estimated by taking about two thirds (times $\frac{2}{3}$) of the maximum theoretical efficiency.

In Example 7-3, we determine the maximum theoretical efficiency for a fossil-fuel steam power plant. Present metallurgical technology allows operation of the hot reservoir (the boiler) at about 811°K (1000°F or 538°C), whereas the cold reservoir (the condenser) operates nominally at 311°K (100°F or 38°C). The maximum theoretical efficiency is then 0.62 or 62%.

Thus, even for an ideal, reversible (Carnot) engine operating between these specific temperature limits, only 62% of the heat used from the hot reservoir is actually convertible to work. The remaining 38% must be discarded as waste. You recall from our explanation of steam turbine operation earlier that the waste of heat energy is inevitable in order to complete the thermodynamic operating cycle. This is also apparent from the steam engine in Figure 7-11.

EXAMPLE 7-3

(a) Determine the maximum theoretical efficiency in per cent of a fossil-fuel steam plant with a boiler temperature of 1000°F (811°K) and a condenser temperature of 100°F (311°K).

$$\eta\ (\%) = \frac{T_{hot} - T_{cold}}{T_{hot}} \times 100\% = \frac{811 - 311}{811} \times 100\%$$

$$= \frac{500}{811} \times 100\% = 62\%.$$

(b) Calculate the minimum percentage of the heat from part a that is waste. Equation 7-3 relates the useful heat or work out, the maximum available heat, Q_{in}, and the heat discarded (waste heat), Q_{out}. Thus

$$Q_{out} = Q_{in} - W_{out}$$
$$= 100\% - 62\%$$
$$= 38\% \text{ minimum waste}$$

From our calculation, we see theoretically that the useful heat is considerably greater than the waste heat. But, in reality, irreversible heat losses reverse the results; the useful heat is more like 40% as compared to 60% waste.

In reality, the steam turbine does not operate exactly according to a Carnot cycle (even an imagined reversible turbine), so we expect maximum efficiency of this process to be somewhat less than the 62% calculated for the maximum theoretical efficiency; the maximum efficiency just for the thermodynamic operating cycle of the steam turbine is actually only about 54%. When the heat losses due to irreversibility of the processes are included, the maximum efficiency is reduced to a real efficiency or conversion factor (or conversion efficiency) of about 41%, as we recorded earlier. A similar calculation for a nuclear-fueled power plant gives a maximum theoretical efficiency of about 50% with an actual conversion efficiency of about 30% (of heat into electricity). In the nuclear case, the materials technology limits boiler operation to a temperature of about 623°K as compared to 811°K for the boiler of a fossil-fuel fired system.

In Carnot's engine, we see the essence of all of these three laws of thermodynamics.[6] First, his analysis of the ideal engine provided a basis for the absolute temperature scale, the Zeroth law. Second, his engine takes in a quantity of heat and transforms a portion to work with the remainder necessarily discarded as waste. All of the energy is accounted for, which

---

[6]There is another law of thermodynamics, known as the third law. This third law plus the first and second laws described above, are commonly known to scientists as the three laws of thermodynamics. The Zeroth law is usually not included.

is consistent with the requirement that energy be conserved, the first law. Third, only part of the heat taken in is converted to work, the Kelvin statement of the second law. Thus, even Carnot's ideal engine is less than 100% efficient (again the second law). Further, if $T_{hot} = T_{cold}$ (observation 4 above), no heat will flow and the efficiency is zero, essentially the Clausius statement of the second law. Thus, Carnot's work, published in 1820, provided the basis for Kelvin's and Clausius' later contributions during the middle 1800s.

The Carnot Refrigerator (Useless Devices)

If the arrows which indicate direction of heat flow and work in Figure 7-13A are reversed, as shown in Figure 7-13B, the Carnot heat engine becomes a Carnot refrigerator. Whereas the engine takes heat from a reservoir at high temperature and discards heat to a reservoir at low temperature consistent with the natural direction for heat flow, the refrigerator removes heat from the cold reservoir and transfers it to the hot reservoir. The second law precludes any possibility of a spontaneous flow of heat from cold to hot, so work must be done to effect the heat transfer in the unnatural direction, that is from cold to hot.

Let us assume that the Carnot engine provides the work for operating the refrigerator. A mathematical analysis [3] shows that the quantity of heat transferred **to** the cold reservoir in the engine in Figure 7-13A is the same as the heat transferred **from** the cold reservoir in the refrigerator in Figure 7-13B. Similarly, the heat discharged into the hot reservoir in 7-13B is the same as that removed in 7-13A. Thus, once started, the heat discharge from the refrigerator could supply heat for running the engine which, in turn, drives the refrigerator, and so on. The result is a machine that seems to run perpetually at no extra cost other than the cost for starting the cycle.

The fallacy, of course, is that our imagined machines and their apparent success is due to their reversible operation. Unfortunately, in reality, *all* machines are irreversible. Such an uncomfortably large number of machines operating in this manner have been presented for patents, that the U.S. Patent Office will no longer consider them. This action expresses the faith of the U.S. Patent Office in scientists and in the laws of thermodynamics.

Irreversibility means that a significant portion of the energy our machines consume in turning out goods and services is discarded as waste heat into the environment. This waste heat has the peculiar property of having little or no capacity for producing work. So, our definition of energy in Chapter 4 as the capacity to do work is not entirely appropriate for waste heat energy. Thus, in running our complex machine, we have "degraded" high quality energy supplies. We will return to this point for discussion from a more philosophical point of view in Chapter 9.

The Meaning of Heat—A Summary

A very basic question that puzzled eighteenth century man was the relationship between heat and temperature and how they manifest themselves in matter: solids, liquids, and gases. From equation (7-5), we see that the absolute temperature of a gas is a measure of the average kinetic energy of its molecules. For a simple gas, a monatomic gas such as helium, the kinetic energy is that of translational motion. However, for a diatomic gas, such as oxygen, or for a more complex molecule such as carbon dioxide, the molecules can rotate about some axis and have rotational kinetic energy, too.

The other extreme is a solid, where the molecules are not free to move around, and, in this case, the kinetic energy is related to the vibrational motion of the atoms about some average position in the solid's lattice structure. Molecules in a liquid may undergo motions that are combinations of the motions exhibited by solids and gases. When heat is added to matter, this heat simply goes into molecular motion. This view is part of the kinetic molecular view of matter. That is,

Heat manifests itself in the average kinetic energy of the molecules in matter.

As seen for the case of a gas, temperature is a measure of the average kinetic energy of molecules, which means that temperature must be a measure of the amount of heat in a body. But, it most certainly is not the only measure. For example, there is more heat in a bucket of water (or gas molecules) at room temperature ($22°C$) than in a cup of coffee (or gas molecules) at $100°C$. Here is a chance for you to be a scientist. Find out how many ice cubes the cup of coffee will melt as compared to the bucket of water.

The result of your observations will indicate that the amount of mass (in your case, measured simply by the number of ice cubes and liquid) is also important in considering the amount of heat in a body. The ocean then is a tremendous heat reservoir. Thus, why not use the limitless heat energy in the ocean to run society's machines? Unfortunately, nature does not allow this. The second law of thermodynamics says that heat flows spontaneously only from a region of higher temperature to a region of lower temperature. Thus, the **motive power** (the availability of work from heat) of heat depends on temperature differences, not on the absolute temperature. For practical purposes, the ocean (near the surface) is a sea of constant temperature. We note that proposals have been made for use of the temperature gradients (differences) between the surface and the ocean depths for generating power. We will return to this in Chapter 15.

Specific Heat Capacity

The property of a body that determines the amount of heat lost or gained during a change in temperature for a given mass is the specific heat capacity. As we mentioned in our discussion of the Joule experiment, 1 kcal of heat when added to 1 kg of water will increase the temperature of the water 1°C. Also, 1 Btu added to 1 lb of water will raise the water's temperature by 1°F.[7]

Contained within the term, specific heat capacity, is the definition of a kilocalorie and a British thermal unit of heat. The amount of heat required to raise the temperature of 1 kg of water by 1°C is defined as 1 kcal of heat, and similarly, using pounds and degrees Fahrenheit, for 1 Btu. The specific heat capacities of other substances are then measured relative to that of water. Table 7-2 gives specific heat capacities for selected substances.

Among common fluids water has one of the largest specific heat capacities, which makes it particularly good for cooling purposes or for heat storage. An amount of heat added to water, produces a smaller rise in temperature as compared to the same amount of heat being added to a substance such as copper. The concept of heat capacities is important when considering the advantages of various coolants used in nuclear reactors (Chapter 11).

For an example of the use of specific heat capacities, consider the simple case of adding 1 kcal of heat to 1 kg of water or to 1 kg of copper. If each substance is initially at 22°C, then the increase in temperature for the water and the copper will be 1°C and 10.9°C, respectively. The final tempera-

---

[7]We should also add that the specific heat capacity of 1 kcal/kg °C = 1 cal/gm °C is defined for the temperature interval between 15.5°C and 16.5°C. The reason is that, in general, the heat capacity of liquids varies slightly with temperature, but for the values of temperature we are working with, the heat capacity is essentially constant.

Table 7-2. Specific Heat Capacities at 25°C

Substance	Specific Heat (kcal/kg °C, or Btu/lb °F)
alcohol (C_2H_6O)	0.60
aluminum	0.215
coal (anthracite)	1.5
copper	0.092
helium	1.24
hydrogen	3.41
ice	0.51
iron	0.106
lead	0.038
mercury	0.033
oil	0.40
sodium	0.29
water (by definition)	1.00

EXAMPLE 7-4

(a) If 1 kcal of heat is added to 1 kg of water at an initial temperature of 22°C, what is the final temperature of the water? (Change in temperature is ΔT.)

$$\Delta T = \frac{1 \text{ kcal}}{1 \text{ kg}} \times \frac{1}{1 \text{ kcal/kg °C}} = 1°C$$

$$T_{\text{final}} = T_{\text{initial}} + \Delta T$$
$$= 22°C + 1°C = 23°C$$

(b) Repeat part a for copper.

$$\Delta T = \frac{1 \text{ kcal}}{1 \text{ kg}} \times \frac{1}{0.092 \text{ kcal/kg °C}}$$

$$= 10.9°C$$

$$T_{\text{final}} = 22°C + 10.9°C = 32.9°C$$

Note: (1) $\Delta T = T_{\text{final}} - T_{\text{initial}}$. (2) The use of the specific heat capacity and the cancellation of units are described by the equation, $\Delta T = \Delta Q/(m \times C)$.[8]

---

tures are then 23°C for water and 32.9°C for copper. See Example 7-4 for details of the calculation.

Thus, the change in temperature of a given quantity of water for a given amount of heat is small as compared to those substances that have smaller specific heat capacities. A second example of using specific heat is as follows. We would like to raise the temperature of a cup of coffee that has a mass of 0.10 kg (100 g) from 22°C (room temperature) to 62°C (just right for drinking). How much heat should be added? From Example 7-5, 4 kcal of heat must be added.

[8]The specific heat capacity is often illustrated with the formula, $C = \Delta Q/m \, \Delta T$, where ΔQ is the heat added into the mass, m, and ΔT the change in temperature. However, our approach in using the conversion factor, C (specific heat capacity), and a dimensional analysis does not require the student to use such formulas.

EXAMPLE 7-5

How much heat must be added to a cup of coffee having a mass of 0.10 kg and at 22°C raise its temperature to 62°C? Note that the change in temperature is 40°C.

$$\text{heat added} = \frac{1 \text{ kcal}}{\text{kg °C}} \times 0.10 \text{ kg} \times 40°C = 4.0 \text{ kcal}$$

Again, note the use of the specific heat capacity and the cancellation of units ($\Delta Q = C \times m \times \Delta T$).

Latent Heats

At the melting point of a substance, for example, water, the substance can absorb large quantities of heat with no increase in temperature. For example, to melt ice at 0°C, requires 80 kcal/kg. This quantity of energy is also called the **heat of melting** or, conversely, when liquid water freezes, this same quantity of energy is known as the **latent heat of fusion.** Similarly, to boil water (at atmospheric pressure) requires 540 kcal of heat per kilogram of liquid evaporated to gas or water vapor. This quantity of energy is called the **latent heat of vaporization** or, conversely, when the gas changes back to a liquid (condensing), the **heat of condensation.** Note that the difference between evaporation and boiling is that evaporation takes place at the surface, whereas, boiling occurs within the liquid as the necessary heat is added. Interestingly, melting ice at 0°C is a reversible process because a small change in conditions, a very slight lowering in temperature, could reverse the melting process and cause refreezing.

The Modes of Heat Transfer

From the first law of thermodynamics we found that heat energy flows from warmer bodies to cooler bodies. That is, when bodies are in thermal contact, heat will flow so long as there are temperature differences. As a result of the heat flow, the warmer bodies get cooler and the cooler bodies get warmer. The flow of heat can be compared analogously with the flow of water. As water always flows spontaneously downhill, so does heat flow spontaneously down the "temperature hill" (Figure 7-14). The steeper the

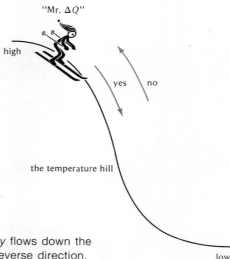

Figure 7-14. Heat *spontaneously* flows down the "temperature hill," never in the reverse direction.

hill, that is, the greater the temperature difference, the faster the heat flows. When the temperature difference between the bodies becomes zero, we have temperature equilibrium between the bodies (the Zeroth law of thermodynamics), and heat ceases to flow. We cannot push our analogy of heat flow with water flow too far, but the analogy is particularly useful when considering heat flow by **conduction.** The other methods of heat transfer are **convection** and **radiation.**

Conduction

All of you are familiar with heat flow by conduction. Perhaps your first startling experience happened when you grasped the handle of your mother's solid iron skillet while she was frying bacon. If the fire was under the skillet bottom, why was the metal handle so hot? Or, perhaps you have placed marshmallows on the end of a metal rod (since wooden ones burn) for toasting and, within a minute or so, the metal handle became too hot to hold (Figure 7-15).

You might also have noticed that some metals conduct heat more readily than others, for example, copper and aluminum (skillets) as compared to iron or steel (skillets). Interestingly, metals that are good heat (thermal) conductors are also good conductors of electricity. Other materials like wood, cork, wool, glass, and so on are poor thermal conductors and also poor electrical conductors. Such materials are called insulators. If you want to insulate against heat loss from your body on a cold day, you wear thick, wool clothing. You should note that materials like wool, which have many trapped air spaces, are particularly good insulators.

The phenomena of heat conduction is believed to take place by molecu-

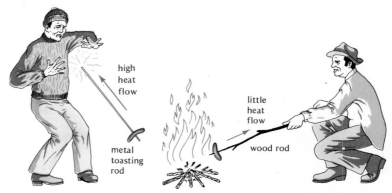

high
heat
flow

little
heat
flow

metal
toasting
rod

wood rod

Figure 7-15.   Conduction of heat by a metal. Metals are good thermal conductors and electrical conductors. Nonmetals, wood, glass, and so on, are poor thermal conductors and electrical conductors.

lar collisions. For example, in the metal toasting rod, the molecules at the heated end have a high average kinetic energy. They collide with adjacent cool molecules and give up some of their kinetic energy to these cool molecules, making them warmer. The collision process continues along the rod with the result that the heat travels to the other end of the rod. The heating process we have just described depends on the type of material, the cross-sectional area of the rod, the length of the rod, and the temperature differences between the ends. Such a relationship is usually expressed as the rate of heat (energy) transfer, Q/t, which is power.

$$\text{power} = \frac{(\text{constant}) \times \text{area} \times \text{temperature difference}}{\text{length of rod}}$$

or
$$P = \frac{Q}{t} = \kappa \frac{A \, \Delta T}{L}. \tag{7-11}$$

The constant, κ (kappa), is called the thermal conductivity and has units of kilocalories per meter-second-Celsius degree (kcal/m sec °C) or watts per meter-Celsius degree (w/m °C). Kappa, κ, is a number that depends on the type of material and the temperature; for copper, κ is large (0.092 kcal/m sec °C) whereas for wood, κ is small (about 10^{-3} kcal/m sec °C). That is, wood conducts heat 100 times poorer than copper. Glass wool, used for building insulation, conducts heat 1000 times poorer than wood ($\kappa \sim 10^{-5}$ kcal/m sec °C). These thermal conductivity values are given for room temperature. Equation (7-11) can be read as follows: the rate of heat transfer, for a given material, is larger for a greater area perpendicular to the direction of heat flow for a greater temperature difference and smaller length of the flow path.

There are many practical uses of equation (7-11), but of particular interest to us here is the reduction of heat loss through the walls of your house (Figure 7-16). Very simply, thick walls, insulation of the walls with a material with a small κ, and a minimal exterior wall area, will reduce the heat loss from your house. With the exception of using good thick insulating material, our society's building practices and habits limit the other considerations. However, a very important point to remember in the interest of conserving fuel energy, is to lower your thermometer setting to perhaps 66 or 68°F from 72°F or higher, and put on a sweater. The obvious effect is to reduce ΔT (the difference between the temperature inside and the temperature outside) and hence the rate of heat loss (see Figures 1-6B and C).

Convection

The transfer of heat by convection means the circulation of warm air or liquid. The movement of masses of air, air currents, is most commonly brought about by changes in air density. We mentioned earlier (Chapter 3) that warm air is less dense than cool air. Thus, the warm air heated at the surface of the earth rises while the cool air descends. This sets up a circulation pattern which we called "cells" in discussing atmospheric circulation.

Convection is also important in home heating. The air above baseboard heaters rises after being heated. The cool air descends, is then heated, and then rises. This sets up air currents as shown in Figure 7-17 and is a very effective means of heat transfer. Air currents also facilitate a significant

Figure 7-16. Heat loss by conduction. Insulation, thicker walls, and a cooler interior, will reduce heat loss from your home.

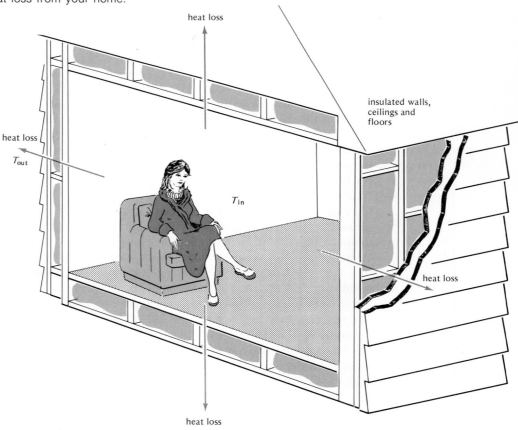

heat loss

heat loss

T_{out}

insulated walls, ceilings and floors

T_{in}

heat loss

heat loss

$$\text{rate of heat loss} \propto \frac{T_{out} - T_{in}}{\text{wall thickness}}$$

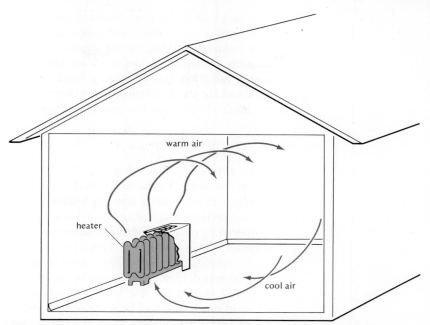

Figure 7-17. Heat flow by convection. Convective heat flow involves a flow of matter.

heat loss in the home. If there is no insulation in the walls, air currents in the walls and between floors transfer large quantities of heat to the outside. When insulation is added in the walls, there is a reduction of heat loss that occurs both by convection and conduction.

Radiation

Radiation, or radiant energy, is a form of electromagnetic energy. Thus, in contrast to the previous two methods of heat transfer, radiation does not require any medium for the transfer. For example, light, heat, and so on, are transferred from the sun to the earth by radiation because space is mostly empty. All bodies emit radiation with wavelengths that are characteristic of the body's temperature. We have noted that the sun, at $6000°K$, emits energy mainly in the visible spectrum, $0.6\,\mu$ (microns) in wavelength. Whereas, the earth, at an ideal radiation temperature of $255°K$, radiates energy mainly in the heat region of the electromagnetic spectrum, $12\,\mu$ in wavelength. The human body also radiates energy in the heat portion of the spectrum, as most of you are aware. Upon absorption of radiation or radiant energy by a body, this energy becomes energy of molecular motion (heat).

Bodies that are good absorbers of radiation also are good emitters. Black bodies are good absorbers and good emitters. This is the reason for wearing light colored clothing in the summer. Lighter clothing reflects greater amounts of heat than does black or dark clothing. In general, the rate of energy radiation (power) from a black body is related to the area of the body and the absolute temperature (Kelvin scale). The relationship is called the Stefan-Boltzmann law and is written as

$$\text{power} = \text{constant} \times \text{area} \times (\text{temperature})^4$$
$$P = \sigma\, A\, (T)^4 \tag{7-12}$$

The constant σ, has a value of $5.7 \times 10^{-8}\, \text{w/m}^2(^\circ\text{K})^4$. Again, the power unit is the watt. It is to be noted that the rate of radiation increases very rapidly with increases in absolute temperature. Black-body radiation curves for various temperatures are shown in Figure 7-18. As can be seen, a higher temperature means a shorter peak emission wavelength. This observation is called Wien's law. From Wien's law we see why the sun's hotter surface emits wavelengths shorter than those from the earth's cooler surface.

Heat transfer, in general, can be by any one of the three methods just discussed or can be due to any combination of these methods. These different mechanisms of heat transfer, along with cooling by evaporation, are of particular interest to us in our discussions of power plant cooling and thermal pollution.

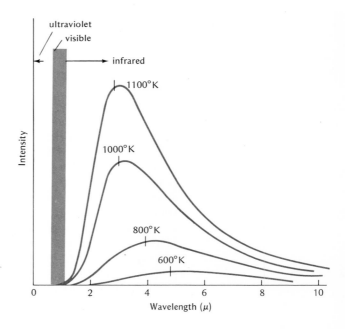

Figure 7-18. Black-body radiation curves. Intensity is shown plotted against wavelength for different temperatures. For higher temperatures, the peak intensity shifts to shorter wavelength or high frequency (Wien's law). For example, the sun is hotter than the earth; thus it has its peak in the visible spectrum, whereas the earth's peak is in the infrared or heat spectrum. [After C. J. Adkins: *Equilibrium Thermodynamics*, McGraw-Hill Book Co., London, 1968.]

Cooling by Evaporation

Evaporation is the process by which molecules leave the surface of a liquid and escape into the air above the surface. Liquids left uncovered gradually disappear. The temperature of a liquid is a measure of the average kinetic energy of the molecules in the liquid. And, some molecules have kinetic energies higher than the average. These molecules have adequate energy to overcome the forces which hold the molecules to the surface of the liquid and escape into the air. This leaves the slower moving molecules, which then will have a lower average kinetic energy, and hence a lower temperature (cooler). Thus, evaporation is a cooling process. Clearly, heat energy is involved in evaporation. A good estimate of this heat is the heat of vaporization. For water 540 kcal of heat are needed to evaporate 1 kg.

More rapid evaporation can be effected by spraying the liquid. This procedure provides a larger surface area from which the "hot" molecules can escape. Also, ventilation is important. Circulating air above the liquid to remove the escaped molecules (as wind does for water in the hydrologic cycle) will keep the air immediately above the liquid from becoming "saturated" with vapor which would stop further evaporation (or further loss of liquid). The cooling effect is very readily noticed by coming out of a pool on a windy day or by rubbing the skin with alcohol which evaporates rapidly.

Conclusion

Around 1820, Sadi Carnot published his brochure entitled, *Reflections on the Motive-Power of Heat, and on Machines Fitted to Develop That Power.* He wrote,

> Everyone knows that heat can produce motion. That it possesses vast motive-power no one can doubt, in these days when the steam-engine is everywhere so well known. The study of these engines is of the greatest interest, their importance is enormous, their use is continually increasing, and they seem destined to produce a great revolution in the civilized world.
>
> The [greatest][9] service that the steam-engine has rendered to England is undoubtedly the revival of the working of the coal-mines, which had declined, and threatened to cease entirely . . . To take away today from England her steam-engines . . . would be to dry up all her sources of wealth, to ruin all on which her prosperity depends, in short, to annihilate that colossal power.
>
> . . . the possible improvements in steam-engines have an assignable limit,—a limit which the nature of things will not allow to be passed by any means whatever; or whether, on the contrary, these improvements may be carried on indefinitely . . .

[9] Words in square brackets are the author's words substituted for clarity.

. . . The production of motive power is then due in steam-engines not to an actual consumption of [heat], but to its transportation from a warm body to a cold body. According to this principle, the production of heat alone is not sufficient to give birth to the impelling power: it is necessary that there should also be cold; without it, the heat would be useless . . .

Carnot addressed himself to the question of the best way to build a steam engine in order to get the maximum possible work out of a quantity of heat. Carnot's engine was the "ideal" and set the limits to the machines (driven by heat) that man could contrive with a given set of conditions. Carnot's observations and the resulting laws of thermodynamics are so fundamental in governing heat energy transfers and transformations that we have called them "nature's guidelines," for they seem to express universal truths rather than merely to wear the cloak of scientific facts.

Underlying Carnot's words is the philosophy that a society must have a continuing source of energy along with energy converters to produce the goods and services necessary for the maintenance of that society. The absence of either means the crumbling of the very foundations on which societies are built—energy. In short, energy means power! Thus, even though it may seem cruel and inhuman to consider our society as a complex machine that degrades high quality energy, it is a way to see how conservative or profligate we are as we consume energy. Any striving towards the "ideal" in energy conversions ultimately must mean a greater longevity for our energy supplies and, certainly just as important, less waste. A knowledge and general awareness of the ideal can only be to the benefit of our present and future generations.

Until now, our society has had at its disposal large supplies of inexpensive, high quality energy. Our growth and industrialization since the beginning of the twentieth century has been based on the consumption of this inexpensive energy. Fortunately, advances in energy conversion technology have offset our rapidly growing energy appetites. However, as the twenty-first century—Millenia Three—approaches, advances in energy conversion technology have slowed and we are faced with energy shortages. An energy shortage can only lead to a reduction in the growth that we have assumed in the past and that most people largely still do assume to mean both progress and improvement in the general quality of life. Present and future generations almost certainly will have to pay more for less.

In the following chapter, we present an overview of our energy consuming habits along with projections to the year 2000 and a discussion of the consequences—resource depletion and pollution. In Chapter 9, we then return to the second law of thermodynamics and a more philosophical discussion of energy degradation and its relation to the evolution of spaceship earth and to the universe.

Questions

1. Discuss reasons for the analogy between society and a complex machine.

2. Define efficiency and inefficiency in terms of the "maximum" heat available from a combustion reaction, the "useful" heat, and the "waste" heat.

3. What do we mean when we say that society as a whole is 50% efficient?

4. Discuss three reasons for the unfavorable situation of the ratio of growth in GNP to growth in fuel consumption becoming less than one.

5. (a) In what way is the fuel consumption related to the GNP? (b) What is the significance of the GNP leveling off but fuel consumption continuing to increase?

6. Describe the energy transformations taking place during operation of a steam turbine power plant.

7. What are the major reasons for inefficiency evident in power plant operation?

8. How does the steam turbine complete a cycle of operation?

9. (a) What is waste heat? (b) Is waste heat necessarily waste? (c) How might waste heat be utilized?

10. List the three laws of thermodynamics that are pertinent to power plant operation.

11. Room temperature is normally about 22°C. What is it on the Kelvin scale?

12. Discuss the pertinence of the word "spontaneous" in the Clausius statement of the second law of thermodynamics.

13. "The earth may be considered as a large heat engine." Discuss.

14. Give five examples of how the second law of thermodynamics manifests itself in your everyday experience.

15. Since it takes 1 kcal of heat to raise the temperature of 1 kg of water 1°C, the ocean must contain an enormous amount of heat. It has been suggested that a ship be designed which takes in the warm water (consider the surface layer to be about 20°C) in the front of the ship, removes the heat for engine operation, and then discharges the cooler water out the back. Can heat energy of such a process be utilized for propelling the ship?

16. What are the basic differences between the Kelvin, Celsius, and Fahrenheit temperature scales? (*Hint:* Sketch the scales.)

17. What is the meaning of temperature using the kinetic model of a gas?

18. Discuss what is meant by a thermodynamic operating cycle?

19. What do we mean by a perpetual motion machine and how does it violate the laws of thermodynamics?

20. It has been suggested that we rate the efficiency of air conditioners by taking the British thermal unit rating and dividing it by the power requirement in kilowatts. Explain.

21. Compare the maximum theoretical efficiency with the real or practical efficiency.

22. Why does the Carnot engine represent the best we can do with regard to machine operation?

23. Is the electric refrigerator a violation of the second law of thermodynamics? Why or why not?

24. The first and second laws of thermodynamics have been restated in the following way:
    First Law: You can't win the game.
    Second law: You can't break even in the game.
    Third law: You can't get out of the game.
    The usual third law has been added here for general interest. In your own words, relate these statements, in particular, to the first and second laws of thermodynamics as stated in this chapter.

25. Clearly illustrate the three methods of heat energy transfer: conduction, convection, and radiation.

26. Describe three different ways to reduce heat loss by conduction. In particular, reference these ways to the home.

27. Discuss how a liquid cools by evaporation.

28. What is meant by a black-body radiator?

29. Why is heat loss by radiation so critically dependent on the temperature of a body?

30. What must you do to reduce heat loss by convection?

31. How is the Gulf Stream an excellent example of heat flow by convection?

32. Why is the global circulation of air so important in so far as climate is concerned?

Numerical Exercises

1. A power plant is producing for "users," 500 Mw (megawatts) (500×10^6 w) of electrical power. The plant's rate of energy consumption from fossil fuels is 1500 Mw. At what efficiency is the plant operating?

2. An engine uses 100 kcal of heat to produce 50 kcal of work. What is the efficiency of this engine?

3. (a) Assume that an electric car utilizes a dry cell battery for power. Neglecting frictional losses in the car itself, find the overall efficiency of using fuel. Refer to Table 7-1 for the efficiencies, and see Example 7-1.
   (b) Keeping in mind that the power transfer to the wheel in the automobile is itself inefficient, do you think that the electric automobile is the answer to the energy shortage problem?

4. (a) Convert the following temperatures in degrees Fahrenheit to degrees Celsius.
   (1) 32°F (freezing point of water)
   (2) 212°F (boiling point of water)
   (3) 98.6°F (average body temperature)
   (4) 105°F (dangerous fever)
   (5) 140°F (pasteurization temperature for milk)
   (6) 68°F (room temperature)
   (7) 1000°F (temperature of steam boiler in power plant)
   (8) −40°F (Arctic cold)
   (b) Convert each of your answers in part a to degrees Kelvin.

5. By using the fact that absolute zero on the Celsius scale is −273°, find absolute zero on the Fahrenheit scale.

6. Liquid nitrogen at its boiling point has a temperature of about 77°K.
   (a) Find the equivalent temperature on the Celsius scale.
   (b) Find the equivalent temperature on the Fahrenheit scale.

7. From the equation relating average kinetic energy of molecules to absolute temperature ($\frac{1}{2}mv^2 = \frac{3}{2}k_BT$), find the speed of oxygen molecules (O_2) at room temperature. The oxygen molecule has a mass of about 5×10^{-27} kg and the Boltzmann constant, k_B, has a value of 1.38×10^{-23} joule/°K. *Note:* A joule is a kg-m^2/sec^2.

8. A high temperature reservoir supplies heat at 227°C to an engine. The low temperature reservoir is at a temperature of 27°C. What is the Carnot efficiency (maximum theoretical) of this engine?

9. (a) The cooling water from a fossil-fueled power plant has a temperature of 100°F or 311°K. It is suggested that we use this heat to operate an eingine with the low temperature reservoir (the air) at 77°F or 298°K. What is the Carnot efficiency of such an engine?
   (b) What would you guess for the real or practical efficiency of this engine? What do you think of its feasibility?

10. Calculate the amount of heat in calories needed to raise the temperature of 1 liter (1000 g) of water from the cold water tap (15°C) to almost boiling (95°C). How much heat is this in kilocalories? In British thermal units? Remember: 1 cal will raise the temperature of 1 g of water 1°C.

11. A cup of coffee contains 100 cm^3 (that is, approximately 100 g) of coffee. To a cup of coffee at room temperature (22°C), 500 cal of heat are added by setting it on a hot plate. Assuming that in the heating all the heat is transferred to the coffee, what is the final temperature of the coffee in degrees centigrade?

12. How much heat in kilocalories is necessary to raise 1 kg of copper from room temperature (20°C) to the boiling point of water (100°C)? The specific heat of copper is 0.092 kcal/kg °C.

13. 2 kg of water at 100°C are added to 1 kg of water at 20°C. What is the final temperature of the mixture? The specific heat capacity of water is 1 kcal/kg °C.

14. How much heat is required to vaporize 10 kg of water at 100°C? (Heat of vaporization of water is 540 kcal/kg.)

15. How much heat must be removed from 10 kg of water at 0°C to freeze it? (Heat of fusion of water is 80 kcal/kg.)

16. If 10 kg of water is at room temperature, 20°C, how much heat must be removed to freeze it at 0°C?

17. If 10 kg of water is at room temperature, 20°C, how much heat must be added to boil it away (at sea level)?

18. (a) A wall in the home (not insulated) probably has the thickness equivalence of about 2 in. (5 cm or 0.05 m) of wood. If the wall is basically 2.5 m high and 7 m long (area = 17.5 m^2), what is the rate of heat loss in kilocalories per second. Assume the difference in temperature between exterior and interior to be 10°C.
   (b) From part a, how much heat is lost during 24 hrs.

19. When considering the heat radiated by earth, we characterized earth as having a black body temperature

of 255°K. Find the rate of radiation in kilowatts per square meter of surface. How does this compare with the solar constant (Chapter 3)? How does this compare with the insolation?

References

[1] C. M. Summers: *Sci. Amer.*, **224**(3): 148 (1971).
[2] F. W. Sears: *Thermodynamics*. Addison-Wesley Publishing Co., Inc., Boston, Mass., 1964, p. 85.
[3] H. A. Bent: *The Second Law*. Oxford University Press, New York, N.Y., 1965, p. 53.

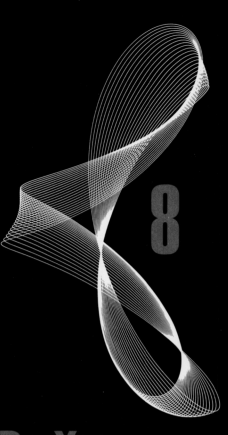

Do You Consume Power?

Throughout our consideration of energy, we have been concerned for quantities of energy, such as the calorie, the kilocalorie, and the British thermal unit (Btu). As we look at our electric bills, we find that we pay for kilowatt-hours of electricity. One kilowatt-hour (Figure 8-1) is a quantity of energy which is equivalent to 860 kcal or to 3412 Btu. As we consume energy, we usually are interested in the energy being used within a certain time period, for example, electricity used for a month. The rate at which we consume this quantity of energy is what we call **power.** In the metric system, a unit of power is the watt, or more commonly, the kilowatt, that is, 1000 j/sec (joules per second) or 0.239 kcal/sec. The kilowatt, as a unit of power, expresses

$$\frac{\text{quantity of energy}}{\text{time}}$$

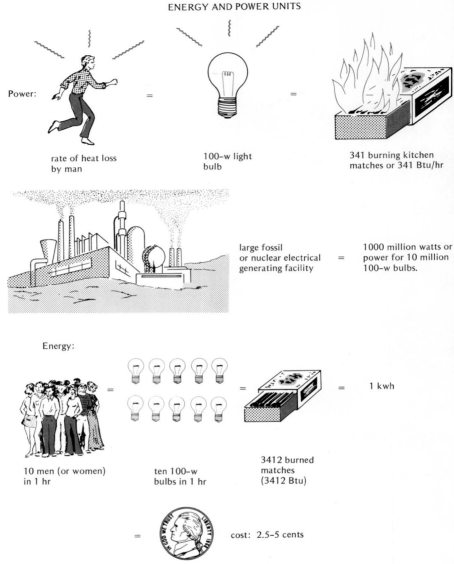

Power:

rate of heat loss by man = 100-w light bulb = 341 burning kitchen matches or 341 Btu/hr

large fossil or nuclear electrical generating facility = 1000 million watts or power for 10 million 100-w bulbs.

Energy:

10 men (or women) in 1 hr = ten 100-w bulbs in 1 hr = 3412 burned matches (3412 Btu) = 1 kwh

= cost: 2.5–5 cents

Figure 8-1. Some graphic representations of what 1 kwh is worth!

Then, the kilowatt-hour (the measure of electrical energy) is simply a quantity of energy:

$$\text{kilowatt} \times \text{hours} = \text{quantity of energy}$$

or

$$\frac{\text{quantity of energy}}{\text{time}} \times \text{time} = \text{quantity of energy}$$

You should note that when we write kilowatt-hour (kw-hr or kwh), in fact, we mean kilowatts *times* hours.

Well, what about our energy consumption or more basically, the pattern of increase of our energy needs, or more so, our energy desires? The trend in American society, particularly since the early 1800s, has been and continues to be the consumption of more energy per person each year. That such a trend exists suggests that there may be no limit to this growth in energy consumption. Our society has been one which has taken growth to be synonomous with a quality of life, and, in many cases, such growth clearly has been at the expense of the environment.

Over much of recent history of the United States, growth in energy consumption has been paralleled by an increase in efficiency of energy use, which means that today we obtain more goods and services per unit of energy consumed than people did 20, 50, or even 100 years ago. Increases in efficiency do translate into a higher standard of living. However, now we seem to be approaching a "new era" in our society when growth in per capita energy consumption is no longer being paralleled by a similar growth in gross national product (GNP). This simply means that we are paying more for units of goods and services. This is illustrated in Figure 8-2 by the steeper slope (= higher growth rate) of the GNP in the early to middle 1960s and then the somewhat leveling-off trend in the late 1960s

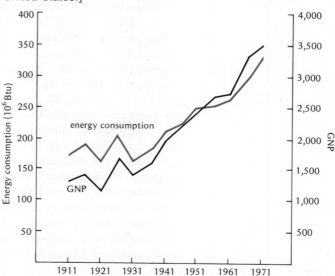

Figure 8-2. Per capita annual energy consumption and per capita GNP (in 1958 dollars). [Source: Statistical Abstract of United States.]

and early 1970s. Correspondingly, the line showing the energy consumption continues to slope upwards, which designates an increasing rate.

This very recent trend also means that growth can no longer be directly related to a higher standard of living. Thus, with prices rising partially as a result of the higher cost of energy and along with evidence of some sort of energy supply crisis, particularly imminent by the year 2000, there has developed a widespread concern for the future of society as we know it today—the closing years of Millenia Two. Many people are also showing increasingly deep concern about the prolonged impact of industrialization and the quantity of energy necessary to keep various industries growing. Interestingly, the daily average per capita food energy requirement is about 2400 kcal and will remain unchanged. Incidentally, this is about 1% of our total daily per capita consumption. Thus, continued growth in total energy consumption means simply that we are doing things in a more costly way (more costly in energy).

The particular sector of the economy of the United States which, in recent years, has had the fastest growth rate for energy production and consumption is the electrical power industry. This industry's growth evidences the changing pattern of consumption of energy. Electrical energy in the early 1970s comprised at least 22% of the total energy consumed in the United States. What will be the consumption pattern in the future? As we look towards the year 2000 A.D., various projections of electrical energy consumption have been made that predict electrical energy may require 50% of the total energy required in the United States. Such a change, coupled with increased demand for energy, will have a profound effect on us and our environment. Here, then, is the reason for selecting the generation of electrical energy as our central theme. We are all involved, and we will all be affected by these trends.

The purpose of this chapter is to survey patterns of energy consumption for the United States and the world in the past and the present and then to project these patterns for the future. We begin by studying growth rates from past to present and characterize them as being exponential (fast growing) in behavior. We then project into the future and derive two limits, the worst case—exponential growth based on present rates—and the best case—no growth or linear consumption. A "real" picture is then attempted, which draws heavily on the ideas of M. K. Hubbert, National Academy of Sciences. The demand for electrical energy is treated similarly due to its particular significance for us. We discuss the controversy between Professors Commoner and Ehrlich as to whether the problem facing society is due to overgrowth in energy consumption or is due to overgrowth in population? Finally, we present an introduction to the problem of waste heat disposal from the generation of electrical energy (elaborated in Chapter 13) and the potential carbon dioxide problem known as the "greenhouse effect."

Energy
Consumption:
Past and Present

Any analysis of society's use of energy—**energy consumption**—must of necessity start with data that summarizes in a meaningful way the past and present use patterns. This section surveys the past, whereas the following section looks into the crystal ball and attempts to see into the future. Our interest here is mainly in the energy use of fuels because such use constitutes the greatest proportion of total fuel use. The "energy use" of fuels includes fuel combustion to provide heat, to move our transportation system, to drive machinery, and to generate electricity. Nonenergy uses of fuels (about 6% of the total in 1970) include the use of fossil fuels as raw materials for producing substances such as resins, plastics, and synthetic fibers. Some caution must be exercised when studying data or fuel consumption published by certain sources, particularly industrial complexes, since such data often includes both energy and nonenergy uses. We will try to specify clearly what is included in our discussions. Although the information presented comes principally only from the time period up through 1970, such data is adequate for our purposes of identifying the overall energy situation we face for the forseeable future.

United States

An interesting summary of energy consumed by the various sectors of the economy of the United States is shown in Table 8-1. Similar data is shown pictorally in Figure 8-3. The total fuel energy consumed is 16.82×10^{12} kwh. Only energy uses of fuels are included in this number. Each type of fuel (coal, oil, and so on) is shown as a percentage of the total along with the consumption of each sector (electrical utilities, transportation, and so on). Figure 8-3 also illustrates pictorally the major energy sources. Two important numbers to note are that in 1968, oil (and natural gas) produced the greatest part of our total fuel energy needs, about 75%,

Table 8-1. U.S. Energy Consumption, 1968* (units of 10^{12} kwh)

Fuel	Utilities	Industry	Transport.	Residential and Commerical	Total	Per cent
coal total	2.09	1.60	0.00	0.17	3.86	22.95
oil	0.35	0.84	4.21	1.70	7.10	42.21
natural gas	0.95	2.58	0.18	1.89	5.60	33.29
hydropower	0.22	—	—	—	0.22	1.30
nuclear power	0.04	—	—	—	0.04	0.25
Total	3.65	5.02	4.39	3.76	16.82	100.00
Per cent	21.70	29.85	26.10	22.29	100.00	

*After the MIT Study Group: *Man's Impact on the Global Environment.* Massachusetts Institute of Technology Press, Cambridge, 1970, p. 289. Their source was the *Minerals Yearbook,* 1968.

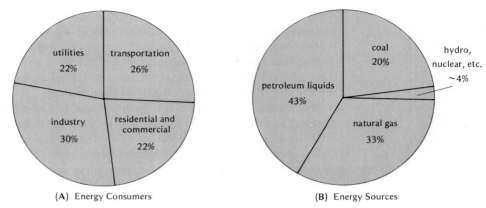

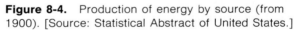

(A) Energy Consumers   (B) Energy Sources

Figure 8-3. Major energy consuming sectors and energy sources in our society (1968).

and the electrical utilities consumed 21.7% of the total fuel energy. As is detailed later, the production of electrical energy is the most rapidly growing energy consumer in our economy. Also particularly important is the fact that our economy is largely based on continuing supplies of petroleum.

Similar results are shown summarized for the period between 1900 and 1970 in Figure 8-4. Whereas "King" coal provided about 87% of our energy

Figure 8-4. Production of energy by source (from 1900). [Source: Statistical Abstract of United States.]

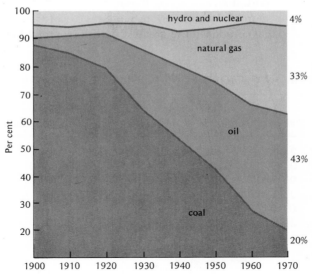

EXAMPLE 8-1

Convert 19×10^{12} kwh to British thermal units using the factor, 3412 Btu[1] $= 1$ kwh.

$$19 \times 10^{12} \text{ kwh} \times \frac{3412}{1} \frac{\text{Btu}}{\text{kwh}} = 64.6 \times 10^5 \text{ Btu}$$

EXAMPLE 8-2

(a) Convert the energy quantity of 19×10^{12} kwh to rate of energy consumption. Note that there are 8760 (8.76×10^3) hr/year. Thus,

$$\frac{19 \times 10^{12} \text{ kwh}}{\text{year}} \times \frac{1 \text{ year}}{8.76 \times 10^3 \text{ hr}} = 2.17 \times 10^9 \text{ kw}$$

(b) If the population of the United States was about 202×10^6 persons in 1969, what was the per capita power requirement?

$$2.17 \times 10^9 \text{ kw} \times \frac{1}{202 \times 10^6 \text{ persons}} = 10.74 \text{ kw per person}$$

in 1900, it provided less than 25% in 1970; the new "Energy King" is petroleum. The gross consumption (energy plus nonenergy) for 1969 was about 19×10^{12} kwh [1] for a United States population of 202×10^6 persons. In the English system, 19×10^{12} kwh is equivalent to 64.6×10^{15} Btu (see Example 8-1).

It is interesting to convert the energy quantity of 19×10^{12} kwh to the rate of consuming energy in 1969, that is, the power requirements, and then determine the United States power requirement per capita (person). The results (from Example 8-2) are 2.17×10^9 kw or 10.74 kw per capita. The latter figure of 10.74 kw per capita is consistent with the figure (10 kw) most often given in the literature.

Again, referring to Figure 8-4, we reemphasize the relative importance of the various fuels since 1900 and the fact that today the economy of the United States is based largely on petroleum (oil and natural gas).

World Energy Consumption

Similar data is available for the world. For 1970, 62×10^{12} kwh [2] of energy was consumed. This converts to a per capita requirement of 1700 w or 1.7 kw, using 3.6×10^9 for the world population (Example 8-3).

---

[1]Remember that 1 Btu is approximately the heat obtained by the complete combustion of one kitchen match. Therefore, 1 kwh = 3412 kitchen matches.

EXAMPLE 8-3

If the world population in 1967 was about 3.6×10^9, calculate the power requirement per person (per capita).

$$62 \times 10^{12} \frac{\text{kwh}}{\text{year}} \times \frac{1 \text{ year}}{8760 \text{ hr}} \times \frac{1}{3.6 \times 10^9 \text{ persons}} = 1.7 \text{ kw per person}$$

Thus, the average American uses power (or energy) six (10.7 kw/1.7 kw) times as fast as the world average. This is even more notable when we consider that the United States has only about 6% [(202×10^6)/ (3.6×10^9) $\times$ 100%] of the world population, and uses 32% of the world's energy [(20×10^{12} kwh)/(62×10^{12} kwh) $\times$ 100%].

Another dramatic way of displaying the affluence enjoyed in the United States is a graph showing the per capita income of various nations against the per capita energy consumption (Figure 8-5). Obviously, the United

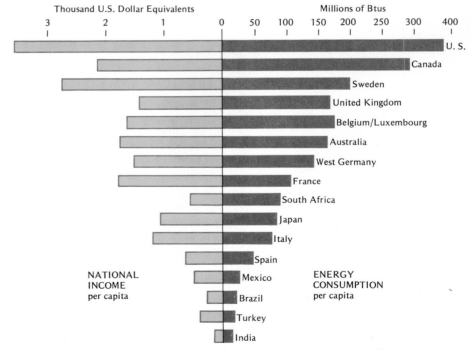

Figure 8-5. Per capita income and per capita energy consumption for selected nations (1969). [Source: U. S. Department of the Interior.]

States has the largest per capita income (also the highest per capita GNP) based on the highest per capita energy consumption. If we may assume that a large per capita income means a high standard of living, then we may conclude that a high standard of living is based on the consumption of large quantities of inexpensive energy. This poses the rather important question about limitations on energy supplies both in the United States and in the world. As we mentioned earlier, growth in GNP and, thus, growth in energy consumption has been taken as a "quality" indicator for life here in the United States. In the next section, we discuss growth in energy consumption and the limitations imposed by finite resources.

Energy Consumption The Future

Since about 1940 (after the depression years), energy consumption in the United States has increased each year. The per cent increase in one year over the previous year is known as the growth rate. Excluding the depression period, 1929 to about 1940, when the consumption of energy actually decreased each year, energy consumption has grown at a rate of about 3.5% per year up to 1965. The average growth rate since 1890, including the depression period, amounts to about 2.8% per year (Chapter 7). A growth rate of 3.5% per year means that every 20 years the energy consumption doubles! Since 1965, the growth rate has been gradually increasing until, in the period between 1968 and 1970, the rate reached 5% per year.

Such a phenomenal growth rate simply means a doubling of consumption every 14 years. For comparison, the growth in energy consumption for the world since 1945 has been 5% per year. Some vital questions with which we must concern ourselves are as follows:

1. How long will this rapid growth continue?
2. Can we produce the fuels (United States and world) necessary to sustain this growth?
3. What environmental impact must we anticipate as a result of the consumption of large quantities of energy?

Granted, these questions may be intricately related to United States and world economics, along with the uncertain political outlook at home and abroad. However, we will take the simplistic approach of assuming these extra pressures will ultimately "iron out" and that we may use past and present trends to project the intermediate and long term outlooks.

Exponential Growth versus Linear Consumption

The type of growth that we have just surveyed is called **exponential growth.** Such growth is a very highly nonlinear type of growth and has

Table 8-2.
Exponential Growth

Growth Rate (per cent per year)	Doubling Time (years)
1	69.8
2	34.9
3	23.4
3.5	20.1
4	17.7
5	14.2
7	10.2
9.25	7.8
10	7.2

the very important characteristic of getting "out of hand" rather quickly. Table 8-2 summarizes the data for various growth rates with their appropriate doubling times. Let us consider annual growth rates in terms of the familiar compound interest on a bank savings account. Suppose that you have invested $1000 in a savings account that pays compound interest of 5% annually and that the interest is paid (added) once a year. From Table 8-2, the appropriate doubling time in value of your savings account is about 14 years. Thus, $1000 invested today will be worth $2000 some 14 years from today.

To see how this exponential growth can be applied to the energy problem, let us assume that we in the United States have consumed E_0 worth of energy to date. (For convenience, we will use 1970 as a reference date.) Just for clarification, E_0 is the energy consumed since the beginning of our industrialization, which has been really significant since about 1900. This is an analogy to our depositing an amount, say $1000, in the bank in 1970. A 5% per annum growth simply means that after 14 years (1984) we will have consumed 2 E_0 of energy (note the doubling). After 28 years (1998), we will have consumed 4 E_0, and so on. When put on a yearly basis, this means that in each succeeding year, we will consume 5% more energy than we did in the previous year. *Or, in each succeeding doubling period, we will consume as much energy as we have in our entire previous history!*

Tables 8-3A and B and Figure 8-6 illustrate the concept of exponential growth. Both illustrations are equally applicable to the United States and world situations because we have estimated that about 5% (10% of the reserve as defined in Chapter 6) of the United States' fossil fuel resource has been withdrawn, and others have made a similar estimate for the world. With these estimates and the assumption of a continuing 5% growth rate, only about 60 years will be required for 100% withdrawal or consumption of our fossil fuel resources (Table 8-3A).

Table 8-3A.
Exponential Growth
(5% per annum
growth = consumption
doubling time of 14
years)

Table 8-3B.
Table A assumes that
5% of the Fossil Fuels
had been used as of
1970, while Table B
assumes that only 0.5%
had been used as of
1970.

Fuel Withdrawn or Consumed	Year
$E_0 = 5\% \times E_{tot}$	1970
$10\% \times E_{tot}$	1984
$20\% \times E_{tot}$	1998
$40\% \times E_{tot}$	2012
$80\% \times E_{tot}$	2026
$100\% \times E_{tot}$	(about) 2030
$E_0 = 0.5\% \times E_{tot}$	1970
$1.0\% \times E_{tot}$	1984
$2.0\% \times E_{tot}$	1998
$4.0\% \times E_{tot}$	2012
$8.0\% \times E_{tot}$	2026
$16.0\% \times E_{tot}$	2040
$32.0\% \times E_{tot}$	2054
$64.0\% \times E_{tot}$	2068
$100.0\% \times E_{tot}$	(about) 2075

Figure 8-6. Energy and exponential growth.

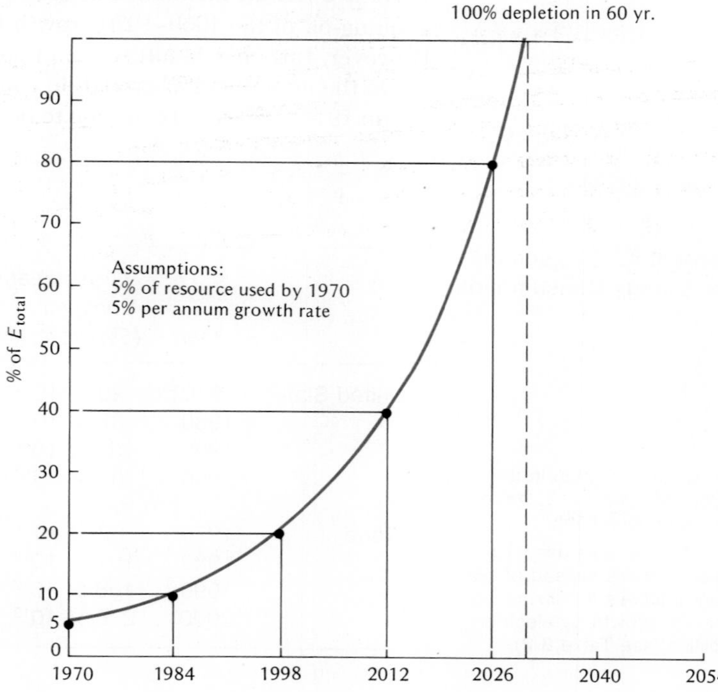

100% depletion in 60 yr.

Assumptions:
5% of resource used by 1970
5% per annum growth rate

% of E_{total}

Let us further assume we have made a mistake and that our resources are ten times as large. This then means that we have only withdrawn to date 0.5% instead of 5%. The 100% depletion date in such a case assuming 5% per annum growth rate in consumption calculates as 2075 A.D., or about 100 years (Table 8-3B). This mere 45 years difference does not substantially change the energy picture so long as exponential growth is the pattern.

We can also perform similar calculations for the yearly rate of energy consumption. In 1970, the United States consumed 20×10^{12} kwh (5% more than in 1969). Again assuming a continuing 5% yearly growth rate, the United States would be consuming at the rate of 40×10^{12} kwh per year by 1984 (note the doubling). Table 8-4 shows projections for the yearly rate of consumption for both United States and the world. We have also included projections for the growth in energy consumed by the electrical utilities, our fastest growing consumer.

Some significant points to note about Table 8-4 are as follows. If the trends continue, the utilities will consume nearly 50% of our fuels in the production of electricity each year by 2000 A.D. If we assume that there are 300×10^6 people in the United States by 2000, the per capita power requirement will be 33 kw, triple the 1969–1970 requirement of 10.7 kw. For a world population of 6×10^9 people (6 billion) by 2000, a similar calculation yields 5.3 kw per capita as compared to 1.7 kw per capita in 1970. We should mention that the National Petroleum Council predicts a falling-off of the 1969–1970 growth to about 4.3% for the United States. However, this change alters our projections very little.

On the other hand, if growth in consumption rate were to stop entirely (say in 1970), we would continue to be consuming at our 1970 rate of about

Table 8-4. Projections for Energy Consumption

	Year	Gross Yearly Consumption* (5%/yr in kwh)	Electrical Utilities (7%/yr in kwh)	Electrical as Per cent of gross (%)
United States	1970†	20×10^{12}	5×10^{12}	25
	1980	32×10^{12}	10×10^{12}	32
	1990	52×10^{12}	20×10^{12}	40
	2000	87×10^{12} (100%)	40×10^{12} (50%)	48
World	1970	62×10^{12}	—	—
	1980	101×10^{12}	—	—
	1990	168×10^{12}	—	—
	2000	278×10^{12}	—	—

*U.S. consumption in the year 2000 will be 31.4% of world consumption.

†Note: We have used 10-year intervals instead of 14-year intervals for illustrating the 7% growth by electrical utilities (see Table 8-2).

20×10^{12} kwh/year. Thus, after 14 years, we would still be consuming only at the rate of 20×10^{12} kwh/year as compared to 40×10^{12} kwh/year with a 5% yearly growth. With "no-growth" and the 1970 use rate, our fossil fuels are calculated to last some 600 years. See Example 8-4. This no-growth situation is called a **linear projection.** Unrealistically, a linear projection requires a pattern of decrease in per capita consumption of energy since a growing population necessarily means a nominal increase in gross consumption. Such an assumption is the opposite extreme to an exponential growth pattern.

In reality, therefore, excluding any catastrophic reduction of world population, the picture of consumption based on the finite, nonrenewable resource of fossil fuels, probably lies somewhere between the two above limits, 60 years and 600 years. We are faced with the prospect of increased populations within the United States and world, so the no-growth projection is clearly not applicable. At the other extreme, we do not know what the per capita energy consumption is going to be—we only project a decrease in the rate of its increase!

The reality of not being able to sustain exponential growth on spaceship earth is based on the fact that fuels become increasingly hard to obtain as we use up the easily accessible supplies and are forced to tap the not so accessible supplies. Such reduced availability in turn increases prices and probably also reduces consumption. Oil exploration is a good example. Earlier we mentioned that a few decades ago substantial oil discoveries required only a few drillings as compared to 70 drillings per discovery in 1970. Furthermore, the average cost per well has gone from $55,000 to $95,000. It also becomes evident that economic considerations (particularly the balance-of-payments problem for oil importation) will become extremely important in considering not only the United States' energy future but that of the world as a whole.

EXAMPLE 8-4

With the assumption of no-growth and the 1970 consumption of 20×10^{12} kwh, estimate the lifetime of the United States' fossil fuel resource.

From Chapter 6, Table 6-7 and 6-9, we obtain the fossil fuel resource as 1.57×10^{12} metric tons of coal equivalent. This, in turn, has the thermal equivalent of 11.6×10^{15} kwh. Then, the lifetime is estimated by the following:

$$\frac{11.6 \times 10^{15} \text{ kwh}}{20 \times 10^{12} \text{ kwh/year}} = 580 \text{ years}$$

Table 8-5. The
Outlook for Fossil
Fuel Production *

* SOURCE: M. K. Hubbert in
Resources and Man. W. H.
Freeman Co., San Francisco,
1969, pp. 157–206.

† The life interval is the time
for depletion between 10%
and 90% of the resource,
that is, time for use of the
middle 80%.

Fuel		Year of Peak Production	Year of 90% Depletion	Life Interval (*middle* 80%)†
Coal	World	2130	2300	200–400 years
	United States	2200	2380	200–400 years
Oil	World	2000	2030	70 years
	United States	1970	2000	70 years
Natural gas	World	no est.	no est.	probably not substantial
	United States	1980	2015	70 years

Something In-Between—A Clouded Crystal Ball

The most significant attempt at projecting the complete cycle of energy production has been made by M. K. Hubbert [3] for the National Academy of Sciences. Table 8-5 summarizes Hubbert's projections and Figure 8-7 illustrates the type of production cycle Hubbert describes with some added detail. Also illustrated is the production necessary for increasing the supply of energy at an exponential rate. A quick perusal of Table 8-5 shows that coal is the only fossil fuel resource of any great long-term significance to the United States and the world. Thus, any oil-based economy is in a very tenuous position, with respect to energy. The time for the complete cycle of consumption of fossil fuels shown in Figure 8-7 further emphasizes the finiteness of the nonrenewable fossil fuel resource and, in particular, the

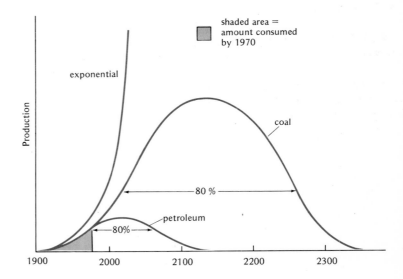

Figure 8-7. Expected production cycle for fossil fuels. [After M. K. Hubbert in *Resources and Man,* W. H. Freeman Co., San Francisco, Calif., 1969.]

Figure 8-8. Fossil fuel production, 2000 B.C. to 4000 A.D. [After M. K. Hubbert in *Resources and Man*. W. H. Freeman Co., San Francisco, Calif., 1969.]

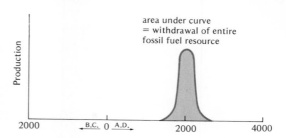

petroleum resource. When compared to the exponential growth curve that characterizes the nature of the present state of our energy habits, one can only conclude that exponential growth cannot be tolerated for any considerable length of time. Finite supplies of energy will be exhausted quickly.

Figure 8-8 compares the entire picture of fossil fuel consumption (excluding wood) with the length of time for man's institutions, starting with the ancient Egyptian societies, −2000 A.D. or 2000 B.C. This figure has been amusingly referred to as "Hubbert's pimple." However, the inference is not very amusing. When viewed in this manner, it is clear that the fossil fuels are at best a temporary resource during man's tenancy on spaceship earth.

The Demand for Electrical Energy (United States)

The pattern of energy consumption has been drastically changing since 1965. Whereas, the portion of the gross energy consumed in 1900 for the production of electricity was negligible, the electrical utilities consumed about 23% of the "gross" in 1969. Why? Our society is more and more demanding of clean heat (locally in the house), convenient power, and clean power. Figure 8-9 charts the growth of electrical energy from 1935 to 1970 and Table 8-6 compares the United States with other major

Table 8-6. World Electric Power (1969)*

Country	Capacity (kw)	Production (kwh)
United States	$310,181 \times 10^3$	$1,436,029 \times 10^6$
U.S.S.R.	$142,504 \times 10^3$	$610,891 \times 10^6$
United Kingdom	$59,628 \times 10^3$	$264,645 \times 10^6$
Japan	$53,187 \times 10^3$	$209,175 \times 10^6$
West Germany	$47,054 \times 10^3$	$189,701 \times 10^6$
Canada	$35,908 \times 10^3$	$176,378 \times 10^6$
France	$34,133 \times 10^3$	$117,925 \times 10^6$
Italy	$30,264 \times 10^3$	$100,249 \times 10^6$
China (mainland)	$15,900 \times 10^3$	$42,000 \times 10^6$
India	$14,314 \times 10^3$	$46,893 \times 10^6$

* SOURCE: Federal Power Commission.

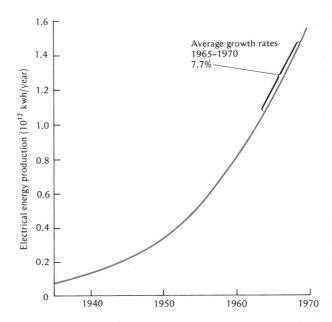

Figure 8-9. Pattern of growth of electrical energy production (United States), 1935–1970. [Source: The Federal Power Commission.]

world powers in the production of electricity. The United States has more capacity and produces more electricity than any other nation. Furthermore, in contrast to the 5% growth rate of gross energy consumption, electrical energy consumption in the United States has been growing at about 7% a year or doubling every 10 years (Table 8-2). Even more startling is the recent data by Earl Cook [4], which indicates that a growth rate as high as 9% is likely.

Interestingly, if the United States were generating at full capacity, the total energy production would be about $2,720,000 \times 10^6$ kwh. By comparing this to the actual production of $1,436,029 \times 10^6$ kwh, we see that the United States was only producing electricity at about 55% of capacity. The reason for this is that, at present, there is no convenient way of storing electrical energy during off-peak hours and then using it during peak hours. Thus, the development of a technology that could provide storage of this energy we could double our actual production without expanding on-line facilities. We have placed further discussion of this idea and of other future technological developments in Chapter 15.

With the changing consumption pattern, the outlook, at least to 2000 A.D., seems to be for a substantial increase in the role of electrical energy for the maintenance of our society. In fact, projections indicate by about 2000 A.D., electrical power production could consume as much as 50% of the total fuel energy consumed (Table 8-4).

As we have already shown (Chapter 7), the use of electrical energy is inherently more expensive, due to lower efficiency. For example, we mentioned that in 1969, 21.7% (4.32×10^{12} kwh) of the gross energy (19×10^{12} kwh) was consumed in electrical power production and, from Table 8-6, the electrical production was 1.44×10^{12} kwh (to consumers). From this data, we obtain an overall efficiency of 33% $[(1.44 \times 10^{12})/ (4.32 \times 10^{12}) \times 100\%]$. Thus, as electrical power generation becomes an increasingly large component of our fossil fuel consumption, the average efficiency of society's machines must gradually decrease from the present overall efficiency of about 50%.

The significance of this observation is that any greater use of electrical energy will represent a larger and larger immediate heat burden on the environment. This additional heat burden will be over and above the increased heat burden, merely due to increased energy consumption alone, based on more people consuming more energy. We will return to the question of environmental impact of heat later in this chapter and in Chapter 13.

Population Growth and Energy

Until the mid-twentieth century, the population of the United States had been growing about 2% annually (ignoring fluctuations during times of depression) which was comparable with the world's rate of growth then and now of 2%. However, in the decade of the 1960s, as shown in Figure 8-10, the rate of increase for the United States slowed to about 1.3%. By 1972, the United States population growth dropped to below 1%. Because our theme emphasizes energy consumption, we must examine the question of how population growth affects energy consumption.

Exponential Growth

Referring to Table 8-2, we see that a population growth of 2% means a doubling time of about 35 years, whereas 1% growth has a doubling time of 70 years. Population growth is then another example of exponential growth. And, you will remember that, although the increase in numbers depends rather sensitively on the rate of growth, any exponential growth rate gets out of hand rather quickly. For example, based on a doubling time of 35 years, the United States would have about 400×10^6 people shortly after 2000 A.D. (2004) (as compared to 200×10^6 in 1969) and the world would have about 7×10^9 people (or double the 3.5×10^9 people inhabiting the world in 1969). We are sure that many of us could not imagine living in most communities of the United States with which we are familiar, or for that matter any place in the world, with twice as many people in these same communities.

With a population growth rate of 2%, a variation of the linear growth

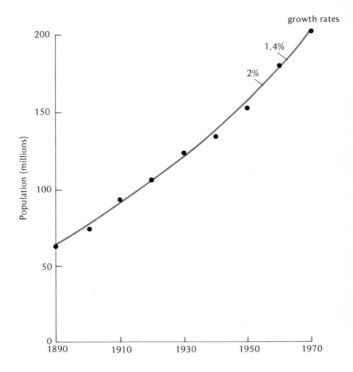

Figure 8-10. Population growth of the United States. [Source: Bureau of the Census.]

concept of energy consumption would be to consider that the same amount of energy *per capita* would be being consumed. Thus, with no change in energy consumption per capita, overall energy consumption would parallel population growth. Thus the growth in energy consumption would be 2% also. However, until recently (more specifically in the period between 1940 and 1965), energy consumption has been growing at a rate of 3.5%, or a doubling time of about 20 years. Since, in fact, energy consumption (at a 5% rate) has been growing considerably faster than the population, there must be other reasons for its phenomenal growth. Furthermore, since 1965 the population growth rate in the United States actually has decreased (to 1.5% or less) while the rate of energy consumption has actually increased to about a 5% growth rate. Clearly, in the United States, the per capita rate of consumption of energy has been increasing, and increasing rather sharply in the late 1960s and early 1970s.

What are the implications of this observation—this increase in per capita rate of energy consumption? Earlier, we noted the fact that beginning in 1965 the energy consumed by our society grew faster even than the GNP. Further, we attributed this change to the increased cost of obtaining energy and to our more expensive way of doing things, the change to electricity being a good example. Not to be excluded, of course, is the greater difficulty in obtaining raw materials necessary for the functioning of society, such as copper and aluminum. This greater difficulty, including using poorer quality ores, transporting materials over greater distances, obtaining materials in difficult climates, manifests itself in more energy being expended to provide the goods demanded by society. These reasons are probably the most obvious ones, but surely are not the only ones in this very complex problem.

Commoner versus Ehrlich

The above discussion leads us directly into the controversy between Barry Commoner [5] and Paul Ehrlich [6]. Briefly, Commoner believes that our pollution and energy plight are due to excessive growth in energy consumption dictated by our more energy consumptive–less efficient ways of doing things. Examples include the highly energy consuming technologies of producing plastics, resins, metals such as aluminum, synthetic fibers, and so on. Specifically, a pound of aluminum requires 30,000 Btu, whereas steel requires only 4600 Btu/lb. The replacement of steel by cement in construction is also highly costly in energy. The shift from rail transportation of goods to truck transportation involves considerably more energy per unit of goods transported. Specifically, transportation of freight by truck requires about 3800 Btu/ton-mi as compared to 600 Btu/ton-mi for rail transportation.

Ehrlich, on the other hand, analyzes world environmental crises as being based on overpopulation and the resulting increasing demands on raw materials by more and more people. In reviewing both sides of the story, we can only conclude that *both* arguments are valid and relevant, even to the point of the controversy being "moot." We simply cannot clearly rule out the possibility of increased population causing increased energy consumption, on the one hand (Ehrlich), and the availability of cheap energy encouraging population growth along with the development of more energy-costly goods and services, on the other (Commoner). Nonetheless a growing population increases the stress on a *finite* supply of raw materials, and thus, more energy is required to obtain the less accessible supplies. We can only conclude that the problem is rather complex and contains many variables, a number of which are not so obvious [7] as those mentioned.

An Exercise for Both Sides: Population versus Energy

We would like to propose two exercises for the reader that have been used many times for emphasis of the impending world population–energy crises. These are as follows:

1. Let the world population double before stabilizing. What would be the lifetime of our fossil fuel resources if at that population level (7 $\times$ 10^9

people) everyone were to consume at the same per capita rate as currently each person does in the United States or 10.74 kw (power) or 9.4×10^4 kwh (energy) per year?

2. How many times could the world population double (at 2% per annum) before there would be one person per square meter of land surface? The area of the earth's land surface is about 1.3×10^{14} m².

The answer (see Example 8-5) to question 1 is 85 years. These levels of consumption and population are considered by experts to be realistic. The answer (see Example 8-6) to question 2 is nine doublings, or a time period of 265 years or by 2235 A.D. (1970 + 265). We should note that many demographers see 15×10^9 people on earth as a likely ultimate population. As an aside to example 8-6, we should mention that the land area includes such uninhabitable regions as the jungles (one third of total land area), the Sahara Desert, Antarctica, and so on. In fact, more than half of the earth's land area is generally considered as uninhabitable. Furthermore, one person per square meter is a bit absurd.

EXAMPLE 8-5

Question 1: For a world population of 7×10^9 people, each consuming 9.4×10^4 kwh per year, we have the gross consumption per year as

$$9.4 \times 10^4 \frac{\text{kwh}}{\text{year} \times \text{person}} \times 7 \times 10^9 \text{ persons} = 6.58 \times 10^{14} \text{ kwh/year}$$

Fossil fuel resources: 56×10^{15} kwh.

$$\text{number of years} = \frac{56 \times 10^{15} \text{ kwh}}{6.18 \times 10^{14} \text{ kwh/year}} = 85 \text{ years}$$

EXAMPLE 8-6

Question 2: If there is to be one person every square meter of land on earth, which has a land area of 1.3×10^{14} m², then the population would be 1.3×10^{14}. The growth rate of 2% means a doubling time of about 35 years, so we have

(0)	3.5×10^9	1970
(1)	7.0×10^9	2005
(2)	14.0×10^9	2040
	. . .	. . .
	. . .	. . .
(9)	1.3×10^{14}	2235

A Question for the Reader: How do you think the assumption made by and the implications deriving from questions (1) and (2) would affect the quality, or, perhaps longevity (quantity), of life on spaceship earth?

Energy Slaves

The term, **energy slaves** (Figure 8-11), used in the title of this text, was originated by Buckminster Fuller. The term very specifically is based on the fact that the 2.8 kwh (2400 kcal) are consumed daily by man in the form of food to support body metabolism. This amount of energy is only about 1% of the daily per capita total energy consumption (per inhabitant) of the United States. The daily energy consumed per capita in the United States is 258 kwh (10.74 kw × 24 hr), which converts to 222×10^3 kcal (258 × 860 kcal/kwh). The balance, 99%, of the total per capita daily energy consumption goes into producing the goods and services that provide both the quality and quantity of life—our standard of living.

Figure 8-11. Our metabolic requirements are only about 1% of the per capita energy consumed in the United States. The remaining 99% goes to run society's machines, and so on.

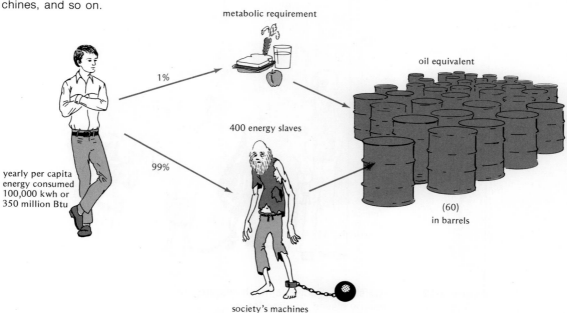

metabolic requirement

oil equivalent

1%

400 energy slaves

yearly per capita energy consumed 100,000 kwh or 350 million Btu

99%

(60) in barrels

society's machines

Thus, each resident of the United States has a large number of "slaves" working for him, day and night. *Today's slaves are society's machines.* If we consider that a human slave can do work at the rate of 75 w or $\frac{1}{10}$ hp (horsepower), and works 8 hr/day, we find that in 1969 every man, woman, and child in the United States had the equivalent of 430 human slaves working for him (see Example 8-7)! Fuller would say that each of us has 430 energy slaves working for us. This is about a threefold increase over the 93 energy slaves man had working for him in 1900. With the prospects of continued increase in per capita energy consumption, we project that each of us will have at least 600 energy slaves working for us in the near future and perhaps even 800 by 2000 A.D. This will certainly prove to be a devastating luxury.

Figure 8-12. Energy slaves. Society's machines are our energy slaves. In 1969, each of us had about 400 energy slaves working for us.

EXAMPLE 8-7

Given that for the United States the per capita power requirement in 1969 was 10.74 kw and that a human slave is worth 75 w or 0.075 kw, find the number of energy slaves working for us 24 hr/day. Assume that a slave can work at the rate of 0.075 kw for 8 hr/day.

Energy from one slave:

$$0.075 \text{ kw} \times 8 \text{ hr/day} = 0.6 \text{ kwh/day}$$

then

$$\frac{258 \text{ kwh/day/person}}{0.6 \text{ kwh/slave/day}} = 430 \text{ slaves/person}$$

Unfortunately, all slaves consume energy, degrade this energy (while producing the goods and services we enjoy), and ultimately discard the degraded energy as heat into the environment. With the phenomenal prospects for growth in total energy consumption worldwide, perhaps we should look back into history to about 1850 and adopt a then popular slogan, "abolish slavery." The following section in this chapter gives an introduction to the burden the environment must bear as a result of energy consumption (Figure 8-12).

Environmental Impact of Energy Consumption

At this point, we have discussed many of the major considerations of energy resources, and the consumption of said resources by society's energy slaves—machines. We now ask the question, "Are there any problems associated with the consumption of energy, or perhaps, more importantly, with the ever increasing rate at which we consume energy?" It was as recent as 1968 that man, from his spacecrafts, looked back at the tiny marbled sphere, since characterized as spaceship earth, and became graphically aware of the finiteness of his tiny ecosphere: finiteness in so far as resources are concerned and finiteness with regard to dissipation of the results of energy consumption.

The purpose of this section is to present an overview of the specific problem of discarding degraded energy as heat from electrical power generation into the nation's stream flow (runoff). We also survey the potential carbon dioxide problem from combustion of fossil fuels, but we want to emphasize that carbon dioxide is not considered as an air pollutant at this time. Discussion of these areas require a more quantitative analysis which, as has been said many times, "adds a concreteness that mere words cannot."

EXAMPLE 8-8

Determine the overall efficiency (η) of society in 1969 if 9.35×10^{12} kwh was discarded as waste heat and 9.65×10^{12} kwh went into useful work. Note the total energy consumed $= 19 \times 10^{12}$ kwh.

$$\eta = \frac{\text{work}_{\text{out}}}{\text{heat}_{\text{input}}} = \frac{W_{\text{out}}}{Q_{\text{in}}} = \frac{9.65 \times 10^{12} \text{ kwh}}{19 \times 10^{12} \text{ kwh}} \times 100\%$$

$$= 0.51 \times 100\% = 51\%$$

Inefficiency of Society

Back in Table 7-1, we listed some representative machines for society and their efficiencies. In 1969, our society consumed 19×10^{12} kwh of energy. Of this, 9.65×10^{12} kwh went into producing work (goods and services) and 9.35×10^{12} kwh was immediately discarded as waste heat. In general, the waste heat is discharged into the hydrosphere, the atmosphere, and the lithosphere. Using the definition of efficiency from Equation 7-1, we obtain an overall efficiency for society of 51%. (See Example 8-8.) Thus, half of the total fuel energy consumed is completely wasted and may eventually represent a severe heat burden on the environment.

Thermal Discharge from Power Plants

Let us turn rather specifically to our energy theme to consider waste heat disposal from electric power generating stations. Because at the present and in the near future, the heat burden due to waste heat disposal from electrical power generating plants must be carried by our rivers, lakes, and estuaries, we must determine the heating (or temperature increase) effect on our nation's runoff. In making our estimate, we will assume initially, for simplicity, that the waste heat is distributed uniformly throughout the runoff.

From Figure 4-4, we obtain the yearly runoff for the United States as being 466×10^{12} gal/year or 30% of the total precipitation of 1552×10^{12} gal/year ($466 \times 10^{12} \times 0.30$). The reader is referred to Chapter 7 for a review of the use of specific heat capacity. The specific heat capacity for water is 1 kcal/kg °C or 1 Btu/lb °F. Hence, to use these conversion factors, the runoff in gallons must be converted either to kilograms or to pounds, depending on our choice of systems, S.I. or English. The literature uses both systems, so we will carry the exercise through with dual units.

In the metric system, the runoff of 466×10^{12} gal occupies a volume of 1.76×10^{12} m^3 (cubic meters) and has a mass of 1.76×10^{15} kg (kilograms). In the English system, the runoff occupies a volume of 62.2×10^{12}

ft^3 (cubic feet) and weighs 3.88×10^{15} lb (pounds). These conversions are shown in Example 8-9a (metric or S.I.) and 8-9b (English), respectively.

The total energy consumed for electrical power generation in 1970 was 4.32×10^{12} kwh. An average of 67% of this (2.89×10^{12} kwh) was discarded as waste heat into the plant cooling water, that is, into the runoff. Remembering that in specific heat problems we commonly use kilocalories (or British thermal units (2.89×10^{12} kwh converts to 2.48×10^{15} kcal (or 9.85×10^{15} Btu). (See Example 8-10 for this conversion.)

If this amount of heat were distributed into all of the runoff uniformly and all at once, the temperature rise of the runoff would be $1.4°C$ ($2.5°F$). However, it should be immediately obvious that we are not going to be able to spread the heat uniformly into the entire runoff because electrical power plants are at specific locations. Thus, let us determine the quantity

---

EXAMPLE 8-9

(a) S.I. (metric) system: Convert 466×10^{12} gal to cubic meters and to kilograms.

$$466 \times 10^{12} \text{ gal} \times \frac{1 \text{ m}^3}{265 \text{ gal}} = 1.76 \times 10^{12} \text{ m}^3$$

The density of water is 1 g/cm^3 or 10^3 kg/m^3, thus we have

$$1.76 \times 10^{12} \text{ m}^3 \times \frac{10^3 \text{ kg}}{\text{m}^3} = 1.76 \times 10^{15} \text{ kg}$$

(b) English system: Convert 466×10^{12} gal to cubic feet and pounds.

$$466 \times 10^{12} \text{ gal} \times \frac{1 \text{ ft}^3}{7.5 \text{ gal}} = 62.2 \times 10^{12} \text{ ft}^3$$

In pounds per cubic foot the density of water is 62.4 lb/ft^3. Thus,

$$62.2 \times 10^{12} \text{ ft}^3 \times \frac{62.4 \text{ lb}}{\text{ft}^3} = 3.88 \times 10^{15} \text{ lb}$$

EXAMPLE 8-10

Convert the waste heat value of 2.89×10^{12} kwh to (a) kilocalories and (b) British thermal units.

(a)
$$2.89 \times 10^{12} \text{ kwh} \times 860 \frac{\text{kcal}}{\text{kwh}} = 2.48 \times 10^{15} \text{ kcal}$$

(b)
$$2.89 \times 10^{12} \text{ kwh} \times 3412 \frac{\text{Btu}}{\text{kwh}} = 9.85 \times 10^{15} \text{ Btu}$$

of water necessary to cool the plant condensers while limiting the temperature of the heated effluent water to a specified temperature rise.

Now, if the temperature of the cooling water is to be limited to an 11°C (or 20°F) increase, all of the power plants in the United States will require 2.25×10^{14} kg (4.82×10^{14} lb) of cooling water. This amount of water is equivalent to about 13% of the total annual runoff in the United States. Obviously, if we were to limit the temperature increase of the cooling water to half or 5.5°C, we would need twice as much water or about 26% of the runoff. These calculations are shown in Example 8-11.

A similar calculation for the year 2000 shows an even more disturbing result. At 7% a yearly growth, electrical power generation will double three times ($2 \times 2 \times 2 = 8$ times). The waste heat correspondingly becomes 19.84×10^{15} kcal, which requires 18×10^{14} kg or 1.8×10^{15} kg of water for cooling. Therefore, on the basis of our projections of a continued 7% per annum growth for electrical power, over 100% of the total runoff would be required for power plant cooling by 2000 A.D.! The practicality of using 100% of the runoff for plant cooling (once-through cooling) is out of the

---

EXAMPLE 8-11

Calculate the percentage of the runoff necessary for power plant cooling, in 1969, while allowing an 11°C (20°F) increase in the cooling water temperature. The waste heat is 2.48×10^{15} kcal (9.85×10^{15} Btu). Therefore, the water required for cooling is

(a) S.I. (metric system):

$$2.48 \times 10^{15} \text{ kcal} \times \frac{1}{\dfrac{1 \text{ kcal}}{\text{kg°C}} \times 11°C} = 2.25 \times 10^{14} \text{ kg}$$

From Example 8-8, the runoff is 1.7×10^{15} kg, so the percentage is

$$\frac{2.25 \times 10^{14} \text{ kg}}{1.76 \times 10^{15} \text{ kg}} \times 100\% = 12.8\% \text{ or about } 13\%$$

(b) English system:

$$9.85 \times 10^{15} \text{ Btu} \times \frac{1}{\dfrac{1 \text{ Btu}}{1 \text{ lb °F}}} \times \frac{1}{20°F} = 4.92 \times 10^{14} \text{ lb}$$

From Example 8-9, the runoff is 3.9×10^{15} lb. So, the percentage is

$$\frac{4.93 \times 10^{14} \text{ lb}}{3.88 \times 10^{15} \text{ lb}} \times 100\% = 12.8\% \text{ or about } 13\%$$

question. Thus, alternate means of cooling must be found. These are discussed in Chapter 13.

Actually the waste heat disposal problem is even more serious than presented above. Because the heat is discarded in the immediate plant vicinity, the discarded heat is very much a *local problem*. That is, the effluent from the power plants is not distributed in anyway uniformly throughout the runoff. Therefore, the immediate volume of water available for cooling is considerably less than our estimate in Example 8-10. Thus, the heating problem will be much more serious locally than estimated in our example! Such a concentration of heat can only have a deleterious effect on the environment. Thermal pollution is the subject of Chapter 13. We will also discuss the potential global heating problem due to energy consumption in Chapter 13.

These calculations have provided us with an overview of the problem of heat and effluent from our maze of power plants. With some additional thought, one can only conclude that the environmental impact in the not too distant future may be enormous.

The Carbon Dioxide Problem

Our earlier discussions of the burning of fossil fuels have noted that besides heat, combustion yields certain substances. Complete combustion of fossil fuels produces carbon dioxide (gaseous) and water (vapor). The minor products of fuel combustion and the products of incomplete combustion of fuel, that is, carbon monoxide and unburned hydrocarbons, will be treated later in our discussion of air pollution (Chapter 14). It is not entirely clear where all this combustion-generated carbon dioxide ultimately ends up or is utilized after burning fossil fuels. However, we know that immediately after combustion the carbon dioxide gas enters our atmosphere.

Let us consider the mass of carbon dioxide released. When 1 metric ton (10^3 kg or 2200 lb) of carbon is burned, as shown in Example 8-12, 3.67 tons of carbon dioxide are produced. However, a typical ton of coal (anthracite) is only 87% carbon, so burning 1 ton of coal would give $0.87 \times 3.67 = 3.2$ tons of carbon dioxide.

Our recoverable fossil fuel resources (reserves) have been given as about 5×10^{12} metric tons of coal equivalent. Let us consider the case of burning all of it. If 1 ton of coal produces about 3.2 tons of carbon dioxide, then 5×10^{12} tons of coal produces 16.0×10^{12} metric tons of carbon dioxide. This is to be compared with the amount of carbon dioxide in the air (0.0318%) which calculates out to have a total of 2.8×10^{12} metric tons. From the results of our example, we see that total burning our fossil fuels will release approximately 6 times the present carbon dioxide content into our atmosphere.

What is the fate of this combustion-generated carbon dioxide? Much

---

EXAMPLE 8-12

If 1 metric ton of carbon (C) is burned, how many tons of carbon dioxide (CO_2) are released?

$$C + O_2 \longrightarrow CO_2$$

Molecular weights: $C = 12$, $O_2 = 32$, $CO_2 = 44$.

$$
\begin{array}{ccccc}
C & + & O_2 & \longrightarrow & CO_2 \\
(12\ g) & & (32\ g) & & (44\ g) \\
(12\ ton) & & (32\ ton) & & (44\ ton) \\
(1\ ton) & & (2.67\ tons) & & (3.67\ tons)
\end{array}
$$

Thus, 1 ton of carbon gives $44/12 = 3.67$ tons, carbon dioxide. How, if coal is 87% carbon, then burning 1 ton of coal will yield

$$0.87 \times 3.67 \text{ tons} = 3.2 \text{ tons } CO_2$$

---

of it would presumably be absorbed by the ocean and by the bio-mass, in both cases by matter capable of photosynthesis (see the following paragraph). Various estimates have been made as to the amount that would be absorbed; two noteworthy ones are 66% [8] and 90% [9]. The former estimate predicts an amount of carbon dioxide in the atmosphere about twice the present, whereas the latter estimate predicts an additional amount which is a little less than present content. Both of these estimates assume that dynamic equilibrium between the atmosphere and the biosphere has been reached. However, due to the slow mixing rate of the various layers of the ocean, the equilibrium (from 66% to 90% absorbed) could take a few hundred years or longer.

Because the fate of atmospheric carbon dioxide is its incorporation into the ocean and into the bio-mass by photosynthesis, let us reexamine the photosynthetic reaction, equation (8-1).

$$6\,CO_2 + 6\,H_2O \longrightarrow 6\,O_2 + C_6H_{12}O_6 \tag{8-1}$$

Equation (8-1) tells us that for each mole of carbon dioxide consumed by plants, 1 mole of oxygen (O_2) is released. In particular locales, man has certainly influenced the capacity of the region to take up carbon dioxide and produce oxygen by man's elimination of plant life. Furthermore, these regions, cities, roadways, industrial parks, and so on, spew forth more than their share of the results of combustion, such as carbon monoxide, carbon dioxide and particulates (dust). Because the capacity of the region to use the carbon dioxide has been decreased, the concentration of it in the air is higher in such regions. Also, the heat put into the environment due simply to the energy consumption in these regions is greater. The higher average temperature of cities and industrialized areas is augmented by the "greenhouse effect" (next section). Worldwide, man's activities apparently have

Table 8-7. Carbon Dioxide Concentration of the Atmosphere*

Year	Carbon Dioxide Content (%)
1880	0.0288
1921	0.0299
1945	0.0314
1968	0.0318

*Data selected in part from G. N. Plass: *Sci. Amer.,* **201**(7):41 (1959)

had the effect of raising the atmospheric carbon dioxide content, see Table 8-7.

From Table 8-7, we observe that there has been about a 10% increase in atmospheric carbon dioxide content between 1880 and 1968. In the next section, we relate the carbon dioxide increase to average global temperature increases. Another aspect of fossil fuel combustion is the fact that there is very little movement (diffusion) of gases produced at the earth's surface above some 10,000 ft in altitude. Many materials do not even rise above 2000 ft. Thus, much of the material is trapped in a layer close to the earth's surface, which probably intensifies the pollution problem (Chapter 14).

Certainly we run the risk that the carbon dioxide concentration of the atmosphere may increase further as we consume our fossil fuel supply. Unfortunately, as we mentioned in Chapter 3, we really do not fully understand all the intricacies of the carbon and other cycles in nature and the magnitude that the above effect might have on them. Nonetheless, we do know that the oxygen needed for total combustion of the fossil fuels will not substantially deplete the earth's atmospheric oxygen supply. The reason for this is that the quantity of oxygen in the atmosphere is more than great enough. Such a change in level of oxygen will probably have little effect on life on earth.

Greenhouse Effect

Probably the major effect of a significant increase in atmospheric carbon dioxide content would be to reflect heat that normally would be radiated from earth into space back to earth, thus tending to increase the average temperature at the surface of the earth[2] (Figure 8-13). This trapping of heat is similar to what happens in a greenhouse. The glass in a greenhouse allows the short wavelength radiation, light, to enter. The plant matter, and so on, absorbs this light and subsequently reradiates energy in the long wavelength, heat region of the electromagnetic spectrum (according to Equation 7-12). The glass is effectively opaque to heat and thus traps much

---

[2]Actually, the carbon dioxide in the air layer close to the surface absorbs the infrared heat radiation and reemits half back towards earth and half to the higher atmospheric layers.

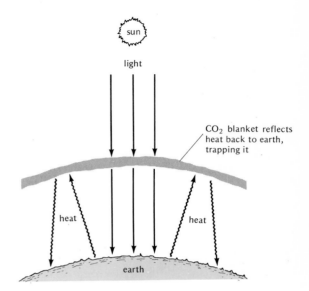

Figure 8-13.
Greenhouse effect.

of the heat inside the greenhouse. For the case of the atmosphere, the carbon dioxide is transparent to visible radiation (light), but, because the carbon dioxide layer is opaque to some infrared radiation, the earth's heat is partially trapped. This phenomena is known as the greenhouse effect, the name having been appropriately chosen (see Figure 8-13).

An interesting example of the greenhouse effect is the surface temperature of the planet Venus, the data having been obtained by the United States and U.S.S.R. space probes. The atmosphere of Venus is thought to be mostly carbon dioxide. The light penetrates Venus' carbon dioxide atmosphere but the carbon dioxide traps the heat reradiated by the surface of the planet. As a result, the surface temperature of Venus may be as high as 900°F. Further evidence of the greenhouse effect may be the very definite warming trend on earth from 1885 to about 1940. During this period, the average global temperature increased about 0.7°C [10]. It has been suggested that most of the increase is apparently due to increased carbon dioxide concentration (Table 8-7).

However, since 1940, the global temperature has decreased by about 0.3°C so that the overall increase since 1885 is only about 0.4°C, although Europe and North America experienced slight additional increases in this recent period. This decrease in temperature is thought to be due to the increased particulate matter (dust) released into the atmosphere (Chapter 3). The increase in "dustiness" of the atmosphere is apparently due to several reasons. As mentioned earlier in the chapter, the phenomenal growth in energy consumption since 1940 resulted, too, in increased materials discharged into the air. During the last few decades, volcanic activity has increased, spewing forth large quantities of dust.

The effect of water in the atmosphere is best illustrated by the temperature moderating effect of a cloudy day. Condensed water vapor as clouds permits light to penetrate through the atmosphere. Some light is scattered and does not penetrate, whereas heat does not penetrate the clouds at all. The result is that the daily temperature has a lower "high" temperature than does a clear, cloudless day. Similarly, heat radiated from the earth is blocked by the clouds and returns to earth resulting in a higher "low" temperature for the cloudy day as compared to a clear day. However, the inter-relations of the potential causes mentioned above, have not been well substantiated. Figures 3-11 and 3-12 illustrate graphically the effect of atmospheric carbon dioxide and water vapor.

Using the observations that we have presented, let us examine roughly the magnitude of the effect of fuel consumption. Let us assume that all of the 0.7°C increase in average global temperature was due to the 10% increase in atmospheric carbon dioxide content. We determined in the last section that the increase in atmospheric carbon dioxide level could be as low as 100% or as high as about 200%, either estimate depending on the absorption by the oceans and bio-mass. Thus, a 100% increase would mean a 7°C temperature rise, whereas 200% would mean a 14°C rise (assuming a linear relationship). Again, even if we have overestimated, by perhaps a factor of 2 or 3, the temperature rise (between 3 and 7°C) is still significant. Such an increase probably would substantially change the air circulation patterns and hence induce large-scale climatic changes.

Indirectly, a large-scale temperature rise would have other effects. The oceans would correspondingly experience a rise in temperature. Since the solubility of carbon dioxide decreases as water temperature increases, less carbon dioxide would be absorbed by the oceans, which would intensify the greenhouse effect. Also, a higher average water (and air) temperature means that larger quantities of water would be evaporated and remain in the atmosphere. As mentioned earlier, water vapor and small water droplets, as found in clouds, contribute somewhat to a greenhouse effect.

Will there be a catastrophic increase in the magnitude of the greenhouse effect as fossil fuels are consumed? And, if so, will there be a large scale climate change? We do not pretend to know. Besides the carbon dioxide theory (greenhouse) of change, there are two other prominent theories. Sunspot and solar activity changes in the solar constant (Chapter 3) have been suggested as the causes of climate change. Also, atmospheric *turbidity* (dustiness) has been implicated as the cause for climate changes.

Perhaps all three have important effects. For example, increased atmospheric turbidity lowers the surface temperature, whereas increasing carbon dioxide content increases the surface temperature. Which dominates? We simply do not know the intricacies of these interacting processes. Again, we emphasize that much study in these areas is needed because we have so little exact knowledge as to how nature behaves with its complex web of interacting cycles and chains. But, we can safely say that to ignore them (as they are our life sustainers) on a tiny spaceship such as ours only means "tough flying ahead"!

Conclusion

The Outlook

In previous chapters, (since Chapter 3), we have discussed the general aspects of energy, the various forms that it takes, the fact that it is conserved, and the laws that govern its use in running society's machines—our energy slaves. The purpose of this chapter has been to survey the actual "quantities" of energy consumed not only by our American society but by the world society as a whole. Some of our estimates may be rough and our numbers not precise. However, even so, we feel that the observations and conclusions that we have made as we have played our "game of numbers" are valid and will continue to be so as long as the pattern of energy consumption of societies on spaceship earth remains as it is. Even if these should change, the basis for predictions of the effect of change has been placed before you.

The features of energy consumption that we have observed are as follows. Not only has the growth in energy consumption been substantial, about 3% per annum since 1900 (United States), but it continues to increase, 5% during 1969, and will probably continue at 4% per year until 2000 A.D. World prospects for growth in energy consumption (5% in 1969) are even greater as other less well-developed societies strive to reach the United States' standard of living. Our early society, pre-1870, depended substantially on wood as a fuel; since then we have observed a shift in importance of fuels for society, from wood to coal to petroleum.

Unfortunately, the petroleum resources and, hence, the reserves are rather negligible when compared with our projected demand over the next 100 years. The future may find us relying more on other sources such as nuclear fission power for our supply of electrical energy. Nonetheless, the outlook to 1985 appears to indicate a heavily oil-based economy. Finally, our patterns of consumption have changed notably with the increased demand for the cleanliness, convenience of electricity. Unfortunately, this indirect use of fossil fuels for meeting our demands is less efficient. Less efficiency necessarily means a greater environmental impact due to degraded energy or heat.

An Energy Crisis

At this time, the economy of our society is based on nonrenewable resources—fossil fuels. A nonrenewable resource means a finite resource, one which is in danger of being depleted because of our ever-increasing appetite for energy. Thus, an economy (ours in the United States) based on a limited supply of energy sooner or later *must* experience a real **energy crisis!** The *only* question that can be raised is "**when** are we to experience severe energy shortages in the United States and worldwide?" The "mild" shortages experienced in 1973-1974 were only a prelude of things to come. It would appear that a major energy crisis looms on the horizon for the United States (within the next 30 years) with the world not very far behind (perhaps 100 years depending on patterns of growth).

Between now and 30 years hence for the United States and the world, there will be energy shortages or crunches that are related to the fact of our limited petroleum supply and that the greatest supplies are located in the Middle East and in East Africa. Accompanying this fact is the realization that **energy means power,** political power. A nation largely dependent on oil importation is in grave danger of being manipulated by oil producing countries. This is the situation in which we could find ourselves here in the United States in the near to intermediate future (1985–1990).

Among the various economic factors not to be ignored with the prospect of an energy crunch is the profit motive. In order to increase profits, individuals and energy industries may withhold fuel supplies in hopes to rake-in "windfall" profits. All economic factors related to "energy crunches" are based on the fact that our present energy supplies are limited.

Such appears to be the heritage of future generations. The extent to which we on spaceship earth may experience shortages of crisis proportions depends on how we use the energy resources we have and on what alternate energy supplies are made available by advancing technology. Our immediate energy future appears to be in electrical power from petroleum and nuclear fission. Nuclear fusion could provide an unlimited source of energy for the future, if and only if scientists are able to prove feasibility and if and only if technologists are able to develop a fusion power plant. Nuclear power is the topic of Chapters 10, 11, and 12.

The ideal, for societies on spaceship earth, of course, would be a stable economy based on renewable nonpolluting energy resources. A survey of the potential future power technologies is made in Chapter 15. Regardless of society's choice of future power technology or technologies, eventually value judgments based on a careful analysis of benefits versus costs must be made: costs in the form of health hazards, economic costs, and costs due to environmental impact. Environmental costs (pollution) are the subjects of Chapters 12, 13, and 14.

In Chapter 9 we turn to a somewhat more philosophical discussion of the limitations imposed by nature on energy conversion and consumption, both with respect to availability of energy for doing work and the consequence, pollution.

Questions

1. How does the unit of kilowatt-hr (kwh) differ from the unit of kilowatt (kw) in energy considerations?

2. Your electrical appliances are rated in kilowatts. What does this mean?

3. When you pay for the use of your electrical appliances at the end of the month, on what units is your "use" based?

4. If the growth in GNP is greater than the growth in energy consumption, what does this mean in terms of prices for units of goods and services?

5. Give two reasons for the growth in GNP now being less rapid than the growth in energy consumption.

6. List the various consumer "blocks" of energy (in the United States) and their percentages of the total. Which of them are expected to be the larger consumers in the next 30 years, and why? What do you think?

7. Consider the different fuels and their contributions to our gross energy budget. How would you expect the contributions to change in the next 30 years?

8. Suggest at least two ways in which you might expect your per capita energy consumption to increase.

9. How does your metabolic energy needs compare with your per capita consumption?

10. How does the United States' population and its energy consumption compare with the world? What evidence can you give for your answer?

11. In what way has the growth in GNP been an indicator of standard of living in the past?

12. Is per capita fuel consumption a good indicator of quality of life? Why?

13. What is meant by the phrase "exponential growth"?

14. What are the implications to us (society) when we are consuming a product exponentially?

15. Using the concept of exponential growth, compare doubling times for consumption with per annum growth rates of 5% and 10%.

16. What is meant by no-growth or linear consumption?

17. (a) Give two reasons why linear consumption is not a reality but is used just as an ideal lower limit.
    (b) Under what conditions might it be a reality?

18. Compare growth in electrical energy consumption with gross energy consumption.

19. Why is electrical energy consumption growing faster than the gross energy consumption?

20. Give three major features of the history of energy consumption and its growth in the United States.

21. Why is M. K. Hubbert's estimate of fuel production probably more realistic than our two extreme cases?

22. Why does the United States actually produce electrical energy only at 50% of rated capacity?

23. (a) Suggest two ways in which population growth contributes to a strain on energy resources.
    (b) Suggest two ways in which plentiful supplies of energy encourage population growth.

24. Suggest reasons why you think that the world population might never reach 15×10^9 persons.

25. The "cry" of the closing years of Millenia Two has been "affluence for all the peoples of the world"! Is this a feasible world goal?

26. What do we mean by energy slaves?

27. What is the environmental impact of energy consumption?

28. What is the greenhouse effect and how might it affect the earth's energy balance?

Numerical Exercises

1. (a) Ten 100-w light bulbs, all in use, have what total power rating?
   (b) If the bulbs in part a are left burning an average of 12 hr/day, how much electrical energy do they consume per day? per month?
   (c) If the cost is $0.04 per kwh, what is the daily cost of burning the bulbs? The monthly cost?

2. If a person consumes energy at a rate of 10.74 kw, calculate the number of kilowatt-hours of energy consumed by this person per month and per year. (*Note:* The answer is used in Example 8-4.)

3. A self-cleaning oven, during the cleaning cycle, requires 9 kw. It takes about 2 hr to do the job.
   (a) How much energy (in kwh) did it consume?
   (b) At a cost of $0.04 per kwh, what is the cost?

4. If the world population were presently consuming energy at the United States' per capita rate, what would be the energy consumed per year on spaceship earth? (*Note:* World population $= 3.6 \times 10^9$.)

5. Show that the United States consumes about 35% of the world's energy per year but has only about 6% of the world's population.

6. (a) If you have invested $500 at 4%, how long will it take to double? to quadruple? to be multiplied by eight?
   (b) Do part a using the per annum growth of 7%.

7. If 63×10^{12} kwh of energy has been consumed by the world from the beginning of its industrialization to 1970, how long will it take to again consume this amount using a 5% yearly growth?

8. If the world was consuming at the rate of 45×10^{12} kwh per year during 1967 at 5% growth, what will be the energy consumed in 1995? in 2009? (*Note:* Not only does the gross energy consumed double in this period, but the yearly consumption does also.)

9. If the United States consumed 19×10^{12} kwh in 1969, *about* how much will it consume in the year 2000 A.D., assuming a 5% growth rate?

10. If the world population were to stop growing and the consumption were to be linear, how much energy would the world consume in the next 400 years while consuming at the 1970 rate? How does this compare with the fossil fuel resources?

11. (a) What percent of the world's oil supply (resource) does the United States have? (See Chapter 6.)
    (b) Do part a for coal.
    (c) Based on parts a and b, how would you assess our position with regard to energy?

12. Show that the estimated fossil fuel resource of 8.5×10^{12} metric tons is equivalent to 63×10^{15} kwh, 54×10^{18} kcal, and 214×10^{18} Btu. (See Appendix A.)

13. (a) If we have consumed 1% of our fossil fuel resource by 1970 and our consumption is growing at the rate of 5% per annum, what is the estimated lifetime of the resource? (Use resource here, not reserve.)
    (b) Do part a using reserves. (*Hint:* Reserves equals one half resource.)

14. If consumption in the United States were to continue at the 1970 rate (linear—no growth), estimate the lifetime of our petroleum resource.

15. The electrical production for the United States in 1969 was 1.44×10^{12} kwh. The energy consumed by the electrical utilities 4.32×10^{12} kwh. What is the average efficiency for electrical power generation?

16. If the electrical utilities consumed 4.32×10^{12} kwh of energy in 1969, how much energy would you expect to be consumed by the utilities each year by 2000 A.D., assuming a continuing per annum growth of 7%?

17. Earl Cook has reported a recent growth of electrical energy of 9.25%. Based on this growth, what would be the estimated yearly consumption by 2000 A.D. (approximately)?

18. We (the authors) have estimated (conservatively) that electrical utilities up to 1970 have consumed about 25×10^{12} kwh. Assuming a 7% yearly growth, how much energy will have been consumed by 1980? by 2000 A.D.?

19. If the average size electrical utility will have a power output of 500 Mw (megawatts) by 2000 A.D., estimate the number of generating stations needed by 2000 A.D. Use your results from problem 16.

20. If the world population is growing at the rate of 2%, how long will it take to reach the 15×10^9 persons figure that many demographers use?

21. If the United States' population is growing at a net rate of 1%, what will be the United States' population by 2000 A.D.? (*Note:* The growth is over and above the death rate.)

22. (a) If the usable land area of the world is about 5×10^{13} m^2, what would be the population for 1 person/m^2?
    (b) At 2% growth, about how long will it take the world population to reach the estimate in part a?
    (c) Using the land area of 5×10^{13} m^2 and a population of 3.6×10^9 people, find the population density (people/m^2 or people/km^2).

23. If 15×10^9 people in the world were consuming energy at the rate of 10 kw (9.4×10^4 kwh/year), estimate the lifetime of our fossil fuels.

24. If an energy slave is worth 75 w or 0.075 kw, and the per capita rate of energy consumption by 2000 A.D. is projected to be 30 kw, how many slaves will each of us have working for us by 2000 A.D.? (Assume that each 75 w slave only works for 8 hr each day)

25. If the per capita rate of energy consumption of the world is 1.5 kw, and an energy slave is worth 0.075 kw, how many energy slaves does each person in the world have working for him (on the average)? (Assume that each 75 w slave can work at this rate of 8 hr each day.)

26. Determine the percentage of the nation's runoff necessary for electrical power plant cooling in 2000 A.D., allowing a 25°F increase in the temperature of the plant effluent. Assume an average yearly growth of 7% for electrical energy consumption. (*Hint:* See text examples.)

27. If 1 ton of coal has 87% carbon, how many tons of carbon dioxide will be produced on total combustion of 1 ton of coal? If 7.5×10^8 tons of coal are burned in the United States in a year, how many tons of carbon dioxide are emitted to the atmosphere?

28. Your local power plant uses 13 tons of coal or (337×10^6 Btu) per capita per year. If the plant is 40% efficient
   (a) Calculate the heat in Btu's per year that is converted into useful heat (or work) per person.
   (b) Calculate the heat per capita per year that is discharged into the environment as waste heat *for each of you!*
   (c) Calculate the number of tons of coal consumed by this plant in providing electricity for a region of 100,000 people.
   (d) Calculate the amount of carbon dioxide discharged into the air for each of you.

References

[1] Bureau of Mines: *Minerals Yearbook, 1970.*
[2] *The National Energy Outlook.* Shell Oil Co., March 1973.
[3] For a more complete discussion and figures, the reader is referred to M. K. Hubbert in *Resources and Man.* W. H. Freeman & Company, San Francisco, 1969, pp. 157–206.
[4] Earl Cook: The flow of energy in an industrial society. *Sci. Amer.,* **224**(3):135 (1971).
[5] B. Commoner: The environmental cost of economic growth. *Chem. Brit.,* **8**:52 (1972) (Chemical Society Publication).
[6] P. Ehrlich and A. Ehrlich: *Population, Resources, Environment.* W. H. Freeman & Co., San Francisco, 1970.
[7] The reader is referred to an enlightening discussion on the topic of population growth and energy consumption by J. Holdren and P. Herrera: *Energy. A Sierra Club Battlebook,* The Sierra Club, New York, 1971, p. 124.
[8] B. Bolin: The carbon cycle. *Sci. Amer.,* **222**:54 (1970).
[9] G. Shirrow: *Chemical Oceanography.* Academic Press, New York, 1965, p. 284.
[10] J. M. Mitchell: *Ann. N.Y. Acad. Sci.,* **95**:248 (1961).

Entropy and Disorder

In the previous two chapters, we discussed extensively the concept of energy flow in our complex machine called "society." In discussing our society and its energy demands, we have likened it to a thermodynamic engine operating between two heat reservoirs at different temperatures. We found that heat flows spontaneously only from a reservoir or system at a higher temperature to one of a lower temperature, never in the reverse direction. We also found that there is an inherent inefficiency in operating any heat engine between two reservoirs at different temperatures; that is, the heat flowing from the reservoir at higher temperature *cannot* be converted entirely (100%) into work. Some of the heat has to be discarded, and we call the discarded heat, "waste heat."

By virtue of the first law of thermodynamics, we have seen that the waste heat or energy has not been destroyed; it still exists and can be accounted for quantitatively. However, there is something peculiar about such waste heat: it is no longer available for producing useful, work. There are various terms that describe this unavailable energy or degraded energy. When we consider the example of the power plant, and particularly the waste heat that results from the generation of electrical energy from heat energy and is then discarded into rivers, streams, and other bodies of water, we can see that the waste heat has become rather "widely distributed" and, thus, surely not available. We can also consider the degraded energy as being highly "disordered" as compared to its more ordered or concentrated initial state, either as the very hot steam in the boiler or as the molecules of a chunk of coal or a barrel of oil. All of the terms or phrases, "degrading heat," "concentration into wide distribution," "order into disorder," are the equivalent to the scientists' or engineers' term, **entropy** or **entropy production.**

In 1865, Rudolf Clausius, a German physicist, introduced the term entropy in his statements of the laws of thermodynamics:

Die Energie der Welt ist constant. Die Entropie der Welt strebt einem maximum zu.

The first statement, even in German, is self-explanatory. However, some of you may need help with the second; "the entropy of the world strives towards a maximum." Clausius's statement of the second law of thermodynamics and its philosophical implications pervade the material of this chapter.

Since Clausius's time, scientists have been fascinated with the concept of entropy or disorder which seems to describe the "natural" trend of nature's processes. However, the authors must acknowledge the fact that all scientists do not agree as to the importance of the entropy concept in considering various thermodynamic processes nor do all scientists agree upon the implications of entropy for various philosophies of life and the evolution of the universe. We mention at the outset that a detailed analysis of "available" work from processes, such as steam engine operation, involves the broader thermodynamic concept of free energy (and enthalpy), but the analysis is too complex for us here and is not absolutely necessary for our purposes. It is the authors' philosophy that the concept of increasing entropy or disorder is the overpowering trend in the evolution of the universe. Furthermore, we feel that a discussion of entropy is both a fascinating and useful way of summarizing the trends that are very real or practical and that seem prevalent in our everyday experience.

In this chapter, we first restate the second law of thermodynamics using the entropy concept. We then present a discussion of entropy in power plant operation (albeit incomplete without free energy) and then relate entropy to everyday experience. We then delve even further into the philosophical implications by relating increasing entropy to the march of time (according to the old cliché) and to the ultimate fate of the universe—its heat death, where heat death means the unavailability of significant temperature differences and does not mean a "burning up."

A Restatement of the Second Law

Before we analyze examples specific to heat engines or power plant operation, let us consider some of the more fundamental aspects of the principle of increasing entropy or entropy production. For convenience, we simply call the concept **entropy.** From the Kelvin statement of the second law, we have found that no heat engine is 100% efficient. The inefficiency manifests itself in waste energy or energy unavailable to do work, which we now call entropy production. In contrast to energy, which is conserved, entropy is not conserved.

A piece of coal is considered to be in a low entropy state. As we use this "high quality" energy, which is released in combustion, we find that as the energy is degraded, the entropy is increased. When the energy has become maximally degraded or used as completely as possible, the entropy becomes a maximum. Thus, for entropy, "we started with none and ended up with some," indicating that entropy is not conserved. A formal restatement of the second law is given as follows: *In an isolated system, the entropy, during processes involving heat flow, either remains the same or increases.* Any process in which the entropy remains the same is a reversible one, such as the Carnot engine (Chapter 7) and, as such, is not a practical reality. Thus, in an isolated system, entropy must increase.

This brings us to the question of what we mean by an isolated system. An **isolated thermodynamic system** is one that exchanges neither energy nor mass with its surroundings. Although earth (with its societies built upon nonrenewable energy resources) has been likened to an isolated thermodynamic system, it really is not because it receives energy from the sun and radiates energy back into space. Actually, earth is described as a **closed system,** which exchanges energy with its surroundings but does not exchange mass with its surroundings (except for a few rockets and meteorites). A truly isolated system is hard to imagine. In fact, you can consider the universe as the only real isolated system (we think). Hence, Clausius's statement becomes "the entropy of the world [the universe] strives towards a maximum [always increases]." Perhaps a more useful way of stating it is that

The entropy of a system and its surroundings always increases.

Furthermore, irreversible processes involve a spontaneous flow of heat, such as radiative heat losses and work against friction. You recall from Chapter 7 that all natural processes are irreversible. Irreversible losses degrade energy, so processes with spontaneous flows of heat are also entropy increasing.

A schematic diagram illustrating spontaneous heat flow is shown in Figure 9-1A. Two heat reservoirs are shown. If these reservoirs are connected, heat will flow spontaneously from the one at the high temperature to the one at the low temperature. Initially, consider both reservoirs as having infinite amounts of heat at their respective temperatures. The change in entropy, ΔS, for this simple situation, is given by

$$\Delta S = \frac{\Delta Q}{T} \tag{9-1}$$

where ΔQ is negative for heat flow out of a reservoir at temperature T (in degrees Kelvin), and correspondingly, ΔS is negative, a decrease in entropy. When heat flows into a reservoir, ΔQ is positive, which corresponds to an increase in entropy, ΔS positive. The more commonly used units for ΔS are kilocalories per degree Kelvin or joules per degree Kelvin.

Study of Example 9-1 reveals that the decrease in the entropy of the "hot" reservoir, which loses heat, is less than the increase in the entropy

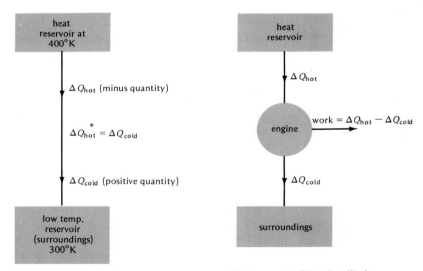

(A) Spontaneous Heat Flow   (B) Conversion of Heat into Work

Figure 9-1. **(A)** Schematic diagram illustrating spontaneous heat flow. According to the second law, heat flows spontaneously from the "hot" reservoir to the "cold" reservoir, never conversely (see also Chapter 7). **(B)** Schematic diagram of a thermodynamic engine.

*ΔQ indicates the absolute quantity, that is, there is no algebraic sign. The inclusion of signed heat and work quantities is beyond our scope.

EXAMPLE 9-1

The diagram in Figure 9-1A represents a process involving the spontaneous flow of a quantity of heat, ΔQ_{hot}, from the high temperature reservoir at $T°_{hot}$. Let us remove 10,000 kcal ($- \Delta Q$) of heat from the hot reservoir. This same amount of heat flows into the cold reservoir ($+ \Delta Q$). From equation (9-1), the entropy lost by the hot reservoir is

$$\Delta S_{hot} = \frac{\Delta Q_{hot}}{T_{hot}} = \frac{-10,000 \text{ kcal}}{400°\text{K}} = -25 \text{ kcal}/°\text{K}$$

Similarly, the entropy gained by the cold reservoir is

$$\Delta S_{cold} = \frac{\Delta Q_{cold}}{T_{cold}} = \frac{+10,000 \text{ kcal}}{300°\text{K}} = +33.3 \text{ kcal}/°\text{K}$$

Therefore, $\Delta S_{total} = \Delta S_{hot} + \Delta S_{cold}$

$$\Delta S_{total} = -25 \text{ kcal}/°\text{K} + 33.3 \text{ kcal}/°\text{K} = +8 \text{ kcal}/°\text{K}$$

of the cold reservoir, which gains heat. Thus, the net change in entropy of the system (hot reservoir and the cold reservoir) is positive and therefore represents an overall increase in entropy of this system. We extend or extrapolate this idea to any thermal change involving spontaneously flowing heat, such as the sun-earth system. We then say that the entropy of the universe has increased, consistent with our restatement of the second law.

What happens if we insert an engine between the reservoirs, as in Figure 9-1B? In so doing, we direct the flow of heat to perform some useful work. As we mentioned earlier, the mathematical complexity of this situation is too great for us here, but we know already that all natural processes are entropy increasing, so our observations for the simple case in Figure 9-1A can be applied in the situation as shown in Figure 9-1B. Thus, as energy is degraded in obtaining useful work, the overall entropy increases.

You will notice that we have neglected to discuss the measurement of an absolute entropy which requires that we know the state of the system or universe for which the entropy is zero. This question is purely academic, as we generally define the entropy of an ideal crystal at $0°\text{K}$ as being zero. This last statement is usually considered the third law of thermodynamics (mentioned in Chapter 7). However, for our purposes, the discussion of entropy changes does not require the definition of a state in which $S = 0$. The entropy changes themselves permit us to see clearly the physical limitations on our lives imposed by nature.

Now let us consider that the heat contained in the hot reservoir in Figure 9-1 (A or B) is finite, or limited, rather than infinite, as we assumed earlier. As heat gradually flows out of the hot reservoir, the temperature of the hot reservoir decreases and, correspondingly, the temperature of the cold reservoir increases. During the heat flow, as we found in our example, the net entropy of the system gradually increases. Finally, the two reservoirs will come to the same temperature, temperature equilibrium, and according to the Clausius statement of the second law, heat flow will cease.

In this state of thermodynamic equilibrium, energy has become **maximally degraded,** and the entropy has become a maximum. Thus, no further work processes can be accomplished when the reservoirs are at the same temperature. The condition for which entropy is a maximum is called **heat death.** However, the term heat death is somewhat misleading because it seems to imply too much heat, or being very hot; in reality, heat death probably is just a lukewarm death!

This game with semantics ignores the real implication of heat death to us here on earth. Degradation of our finite, high quality energy resources will lead us ultimately to the plight of heat death. Whether or not heat death will become a reality on earth probably depends on the ability of scientists and technologists to harness renewable, nonpolluting energy resources. In order that society may plan for the future (Chapter 15), we believe that it is important for you to know the realities along with scientific and technical limitations.

Entropy and Power Plant Operation

Every stage of our power plant operates according to the guidelines set forth by the second law of thermodynamics in one or more of its several guises. The importance of the second law is particularly evident where energy transformations are taking place. The entropy form of the second law is evident in every stage of the physical and chemical processes pertinent to power plant operation. Right in the very first stage of our model power plant we are transforming the chemical energy of the coal, oil, or natural gas into heat. Figure 7-4 records that 12% of the available energy is lost through hot gases and other by-products of the combustion going up the stack and through heat radiated into the boiler room. This loss of heat is necessarily entropy increasing. Figure 9-2 illustrates the various heat energy losses during the irreversible operating cycle of a power plant. The results of our observations about entropy and power plant operation are sequentially summarized as follows.

1. The highly organized, complex, molecular structure of the coal is a low entropy state with a large amount of available energy (in chemical bonds) for doing work. When the coal is burned, much energy is given off as heat. The carbon dioxide, water vapor and other by-products are given off as small molecules in gaseous form. A gas has insufficient structure or order to provide much available energy. Moreover, no

Figure 9-2. Entropy and power plant operation. Each point of heat loss represents a point of increased entropy.

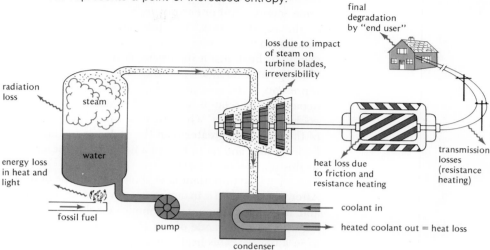

overall heat loss in production (67%) = degraded energy = immediate entropy increase.
Ultimately, 100% of fuel = degraded energy

further heat of reaction is available from the reaction products. The motion of the small molecules is random, and, after leaving the smoke stack, the molecules quickly become widely distributed in the atmosphere. In addition, some heat is radiated into the boiler room and is not used for heating water. It, too, now becomes evenly or widely distributed. Both these losses on combustion of fossil fuels are irreversible and entropy increasing.

2. The useful heat energy is transferred to the water in the boiler. The temperature of each kilogram of water is raised by 1°C for each kilocalorie of heat flowing into it. This heat energy of the combustion reaction increases the average kinetic energy of the water molecules. An increase in the average kinetic energy of the molecules increases randomness of motion and is thus, in itself, entropy increasing.

3. Additional heat from the combustion provides the heat of vaporization, causes boiling of the water, and converts the liquid water to steam or water vapor. The water vapor has much less structure and far more randomness of motion than the liquid and so is in a higher entropy state. However, because the steam is restricted (against expansion) by the boiler walls, and thus is both at a high temperature and at a high pressure, it has energy available for doing work. Fortunately this last is far greater than the portion that has become unavailable due to the increased randomness of molecular motion.

4. The steam is released (at high speeds) onto the turbine blades. The steam does work on blades. The work done is irreversible,[1] which means that some energy is lost or entropy increased merely due to the irreversibility of the rapid process. The mechanical energy of the blades is then transferred to the electrical generator, which loses about 1% of the energy as heat, again increasing entropy.

5. Subsequent condensation of the "cool" steam in the condenser actually represents a decrease in entropy, but this is accompanied by a spontaneous flow of a large quantity of residual heat into a large amount of coolant (river water) which then becomes heated. The increased entropy of the coolant is greater than the decrease in entropy of the condensed boiler water, and there is then a net entropy increase (see Example 9-1) in this portion of the cycle.

6. There is a final important note about the end-use of the electrical energy. After we have used it in our homes, for heating, lighting, running motors, and, in general, running our complex machine called society, the energy regardless of the mode of use ultimately ends up as heat, for example, as in a toaster. This heat, for all practical purposes, has become *maximally* degraded because in a short time period it becomes widely distributed throughout the environment (the home, and so on). Hence, we conclude that all energy use ultimately appears as a heat burden on the environment, consistent with the principle of increasing entropy. As we discuss later, one of the ultimate limitations of our existence on spaceship earth, assuming sane policies on population growth and energy consumption, may be dictated by the heat burden the environment must bear.

The last item, number 6, refers to the global heating problem due to world energy consumption, which appears to be a long range problem (see Chapter 13). The problem of immediate concern is the heat burden due to the heat discharge in the vicinity of the power plant generating electricity.

There are two major points from the preceding discussion for you to keep in mind. In the operation of any machine or engine, there may be entropy decreases, such as in the cooling and condensing that takes place in the condensers and in the use of the work output in providing goods and services for society. When considering the aforementioned decreases alone, there seems to be a contradiction of the entropy form of the second law. However, when all entropy increases are properly included, the total entropy change, system–engine and surroundings–environment, is always positive. All real processes are entropy increasing, which is consistent with the second law. This step by step study of the principle of increasing entropy and its implications in power plant operation has provided us with

---

[1] Refer to Chapter 7 for a discussion of reversible versus irreversible work.

examples that may be generalized to other natural processes and an insight into many equivalent statements for the second law of thermodynamics— some of which follow.

Some Additional Statements of the Second Law

The ways in which the second law touches our life and apparently rules the evolution of the universe are many, but there are a few that are of particular interest to us as we carry out our everyday activities. Some of these are as follows:

1. Concentrations of anything tend to disperse.
2. Structure tends to disappear.
3. The greatest probability for any spontaneous configuration is that one which displays the greatest disorder.
4. "If you think things are mixed up now, just you wait 'enry 'iggins!" [1]
5. Bodies tend to move from regions of higher potential to regions of lower potential energy. That is, bodies tend to fall.
6. In general, "order tends toward disorder" [2].

We could surely make a game of imagining various (spontaneous) happenings and categorizing them according to each of the above statements. We consider just a few and list them in the following examples. We include in our examples any process or action that represents a loss of energy available to do work. We also include examples that represent a loss of order.[2]

[2] We would like to reemphasize that some of the "happenings" may involve more than just entropy changes and, because of this, our choices may be somewhat imprecise. However, our purpose here is not be particularly precise in our discription but to show that the entropy-disorder concept can be understood by students at this level by using a few amusing examples of everyday experiences.

EXAMPLE 9-2

Statement 1. Concentrations tend to disperse.
  (a) A drop of ink in a glass of water very quickly disperses so as to taint (albeit ever so slightly) the entire contents.
  (b) After opening a bottle of ammonia in the corner of the room, the odor quickly spreads throughout the room.
  (c) A pile of leaves left untended will soon need raking again.
  (d) A small grass fire tends to spread.
  (e) A virus for one person in a crowded room tends to be a virus for many.
  (f) Crowds disperse.
  (g) Beaches erode.
  (h) Smoke from a stack disperses rather quickly.
  (i) Sludge (wastes) dumped into waters disperse and seem to disappear.

EXAMPLE 9-3

Statement 2. Structure tends to disappear.

  (a) Ice tends to melt (unless kept in a freezer), and many solids tend to dissolve.
  (b) Liquids left open to the air evaporate and disappear.
  (c) Solid materials rot (for example, the oxidation of metal cans).
  (d) Buildings tend to fall apart.
  (e) Polyatomic molecules tend to break down into the simple molecules from which they were formed. (For example, deterioration of rubber and other hydrocarbons—a child's balloon.)
  (f) Marching formations disperse.
  (g) Mountains crumble (Appalachian Mountains as compared to the Rockies).
  (h) Corpses decay (a natural process).
  (i) Automobiles deteriorate (whether or not in use).

EXAMPLE 9-4

Statement 3. The greatest probability for any particular configuration is that of greater disorder.

  (a) If we take a deck of 52 cards and drop it, what do you think would be the probability of the cards landing by suit and in numerical order? Obviously, the answer is that the probability is vanishingly small. The greatest probability is for having a mess, "52 pick-up!"
  (b) If you were dealt four cards, what is the probability of getting all aces? The answer is about one in 300,000, again remote. The greatest probability is that of getting nothing useful—a disgusting hand!
  (c) If you were to take a big pile of lumber, mortar, nails, and so on, and push it off a cliff, what is the probability that it would land in the configuration of a house? Obviously, the probability is again small, *but finite*. And again, the greatest probability is a mess of building materials at the bottom.
  (d) If you had a box of air molecules, the probability of finding them all moving in the same direction and hence piling up at one side is very small. The greatest probability is that for which their motion is completely random and the various constituents are maximally mixed up. Randomness of motion may be associated with being mixed up, which is a high entropy configuration. Also, the higher the temperature of the gas the greater the probability of random motion—the higher the entropy.

It should be clear from these examples that each of our categories or statements are saying in a little different way that "order tends towards disorder." Perhaps we have extended the principle a little further than originally intended, for example, how do you define the entropy of a garden? But all the statements do seem to be consistent with the concept of a natural tendency towards disorder. All of our examples represent a loss of "availability" and are entropy increasing. Therefore, the terms entropy and disorder are synonomous. Thus, all of our examples of ran-

Figure 9-3. A child's organized toy box becomes very much disorganized by the end of the day—a natural event!

Figure 9-4. A crowded Los Angeles freeway. Before rush hour the travel is bad enough, but during rush hour, travel is "impossible"! [Los Angeles County Air Pollution Control, courtesy of Environmental Protection Agency.]

EXAMPLE 9-5

Statement 4. "If you think things are mixed up now, just you wait, 'enry 'iggins!'"
  (a) A child's well organized toy box in the course of the day becomes disorganized (Figure 9-3).
  (b) The disorder that results when the fog rolls in in London.
  (c) A Los Angeles freeway just before rush hour quickly becomes a mess just after rush hour begins (Figure 9-4).
  (d) The authors' desks before the new semester begins as compared to sometime later in the semester.
  (e) A cocktail party or beer party at the beginning of the evening as compared to later in the evening.

EXAMPLE 9-6

Statement 5. Bodies tend to move from regions of high potential energy to regions of low potential energy. Falling is irreversible.
  (a) Apples tend to fall.
  (b) Rocks and children tend to roll down hill. Did you ever see the reverse? See Figure 9-5.
  (c) Batteries tend to discharge. (A charged battery, which has electrical potential energy, will gradually discharge, losing its potential to do work.)
  (d) Water tends to move from higher elevations to lower elevations.
  (e) Rain falls.
  (f) Your automobile tire goes flat!

---

domness of motion, high temperature, mixed-up-ness and disorder are ways of describing a high entropy state or, conversely, entropy is just a way of saying disorder, randomization, and so on. However, for all the examples exhibiting increasing entropy or disorder we have taken, we can suggest many other examples that represent decreases in entropy or increases in

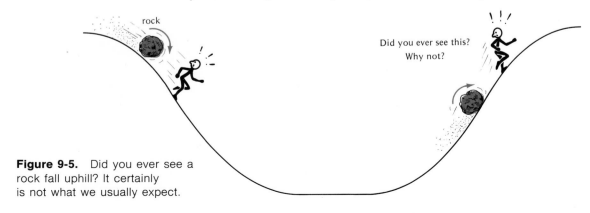

Figure 9-5. Did you ever see a rock fall uphill? It certainly is not what we usually expect.

EXAMPLE 9-7

Statement 6. In general, order tends toward disorder.
  (a) Take a box of black marbles and white marbles, the black ones being on one side and the white ones being on the other to start. As time passes, we know that the natural tendency is toward a mixed up mess. If we shake the box, we can speed up the mixing operation, and, the faster we shake the box, the greater is the probability of finding maximum disorder. This can be compared to gas molecules in a container: The faster they move, the greater is their average kinetic energy and the greater is the probability of being mixed up. Thus, the higher temperature means greater disorder and higher entropy.
  (b) Students in a classroom are "ordered" as compared to the students after class.
  (c) Ice has a more highly ordered structure before melting than after melting.
  (d) The molecules of a liquid before evaporation are more ordered.
  (e) The conversion of highly ordered (or structured) carbohydrates into water and carbon dioxide by combustion represents a change from an ordered state to a disordered one (Figure 9-6). A specific example is that of burning sugar by plants and animals in the process of respiration (the inverse of photosynthesis).

$$C_6H_{12}O_6 + 6\,O_2 \longrightarrow 6\,H_2O + 6\,CO_2 + \text{heat}$$

  (f) Stars like our sun burn their fuel and emit energy.
  (g) A lawn or garden left unattended for the summer becomes overgrown, weedy, and, in general, highly disordered.
  (h) A newly constructed urban housing area soon becomes disordered.

order. Some examples are as follows: taking wood, steel, mortar, and so on, and building a house, a child cleaning his room on Saturday, a plant receiving sunlight and producing sugar and carbohydrates (cellulose) in the process of photosynthesis, or the organizing of the information and presenting it in this text.

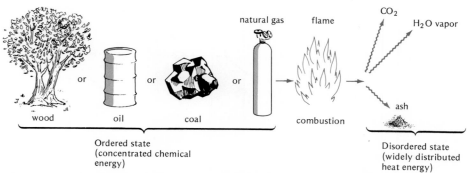

Figure 9-6. Burning. Combustion is a change from order to disorder.

A Question for the Reader: Is life a violation of the second law of thermodynamics?

Do our other examples of entropy decreases (or entropy reversals) represent violations of the second law of thermodynamics? We will try to answer this question with our favorite examples in the following sections.

History's Famous Entropy Reversers

The first example, Maxwell's Demon, is probably one of the best known (at least to scientists and philosophers) attempts of man's imagination to conceive of an entropy reverser that costs nothing in terms of work and energy. The second, Humpty Dumpty, characterizes the second law so well that one can only conclude that the name of Mother Goose should be added to those of Clausius, Kelvin, and Carnot.

Maxwell's Demon

Maxwell's Demon was conceived by James Clark Maxwell, a brilliant Scottish physicist, in 1871. He imagined a box containing a partition with hole and frictionless trap door, see Figure 9-7. The gas in the box on both sides of the partition is at some temperature (the same) which is indicative

Figure 9-7.   Maxwell's demon. Maxwell's demon has the ability to select molecules of different velocities and sort them (see Figure 9-8).

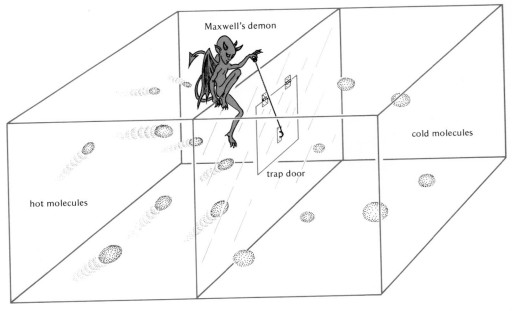

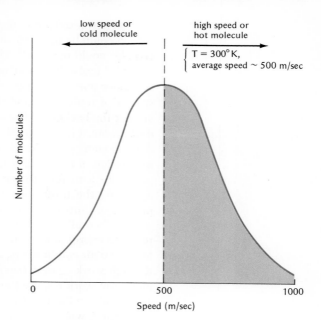

low speed or cold molecule

high speed or hot molecule

$T = 300°\,K$, average speed ~ 500 m/sec

Number of molecules

0   500   1000

Speed (m/sec)

Figure 9-8. Speed distribution of molecules at a given temperature.

of an average kinetic energy and an average speed of the molecules. However, as we mentioned earlier, some of the molecules will be traveling slower than the average (cold molecules) and some will be traveling faster (hot molecules) (Figure 9-8). The clever little demon sits above the trap door and watches the molecules which are moving around like angry bees. When a hot molecule approaches the door from one side, he opens the door and lets it through to the other. When cold molecules approach the door from the other (hot) side of the box, he lets them through. After a while, he has most of the hot molecules in one side of the box and the cold ones in the other.

Because temperature is a measure of the average kinetic energy of the gas molecules, the side of the box with the faster moving molecules is at a higher temperature than the side with the slower moving molecules. There is now a difference in temperature, so we can, in principle, build a heat engine to operate between the two heat reservoirs the demon has produced. Hence, we can say that the demon has taken a disordered system of molecules with unavailable energy and converted it into an ordered system with energy available for doing work, without, in turn, requiring energy "to run" the demon. Maxwell's Demon if he is for real would also do very well as an air conditioner for our homes.

The paradox (the apparent violation of the second law) of Maxwell's

Demon perplexed physicists for about 100 years. Aside from frictional losses (which could be made small) there seemed to be no reason why the demon concept could not work. The paradox was resolved finally about 1950 by two physicists, Brillouin and Demers. They observed that the isolated container and the demon would be in temperature equilibrium with the system of molecules, and the molecules would be impossible to observe against the background black-body radiation. That is, to see objects in the dark we must illuminate them with a light which represents a spontaneous flow of energy and hence is entropy increasing. Have you ever tried to play tennis in the dark? Thus, the little demon would have to light a torch (or strike a match) to see the molecules. And, the entropy increase in operating the torch would be greater than the entropy decrease in the process of "sorting" molecules to create the temperature difference.

The "moral" of the story is that any such entropy reversing process requires energy and work to run the demon (the motor, compressor and fan of our air conditioners, and so on) and the trap door always loses energy through work against frictional forces (a natural irreversibility). Furthermore, the work required is more than the advantage gained. Many of you are familiar with a sorting job. First you must have light to see what you are sorting, which incidentally costs money, and then work is required to do the sorting. Alas, the demon is nothing more than man's dream of getting something for nothing!

Humpty Dumpty

The Mother Goose nursery rhyme of Humpty Dumpty [3] contains much wisdom about the often catastrophic increases in disorder of natural processes.

> Humpty Dumpty sat upon a wall;
>   (a stable but precarious situation)
> Humpty Dumpty had a great fall.
>   (a natural event, gravity being what it is)
> All the King's horses,
> And all the King's men;
>   (not a Maxwell's Demon among them)
> Couldn't put Humpty Dumpty together again.
>   (the second law of thermodynamics still holds!)

Before his fall, Humpty Dumpty represented a well ordered or structured supply of energy. After his fall, he was obviously a disordered mess and a great deal of tedious work would be required to reassemble his shell, if possible. Kieffer [3] suggests using a chicken, "a hen to be sure," for the task of reassembling Humpty Dumpty and, hence, for the role of entropy reverser (see Figure 9-9). Have we found another violation of the second law?

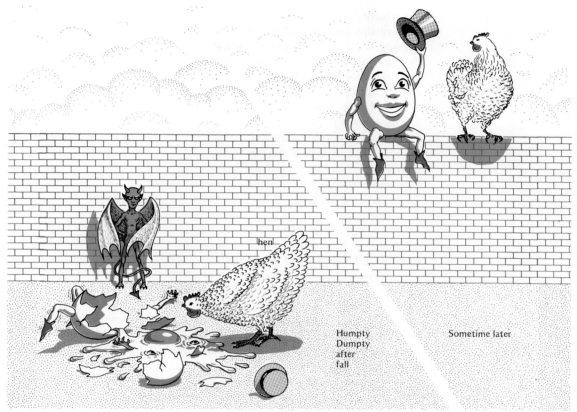

Figure 9-9. Is a chicken, a hen to be sure, a Maxwell's demon?

"Entropy is Time's Arrow"

The suggestion that "entropy is time's arrow" was made by Sir Arthur Eddington, eminent British astronomer and cosmologist [4]. The clichés, "time marches on" and "entropy always increases" seem to imply a definite connection between entropy and time. Time is apparently irreversible, at least to our usual way of thinking. That is, when time is reversed, we get younger, perhaps, eventually returning to our parent's body, and so on. You recall that all natural processes are irreversible and, hence, entropy increasing. Thus, entropy, too, except in small regions of the universe, cannot be reversed, according to the second law of thermodynamics. Further, the suggestion that *entropy is time's arrow* implies that the second law of thermodynamics tells the way in which processes will proceed in time.

In an isolated system (the universe), processes proceed in a direction for which the entropy or disorder tends towards a maximum. That is to say, entropy increases as time unfolds or increases. However, significantly, *entropy*, as the arrow, points forward in time but *does not specify how fast* the processes will proceed. Or, what the entropy arrow does not tell us is the *rate* at which processes will go. For example, the spontaneous energy released from a chemical reaction may take hours, seconds, or less, or it may take thousands of years or even longer. It took the earth about 5 billion years to evolve to its present state, life took about 2 or 3 billion years to evolve to modern man. Buildings take many years to decay, and the Rockies may take hundreds of thousands of years to erode to the extent displayed by the Appalachian Mountains. All of these processes, including their surroundings, have one thing in common, they proceed in the direction for which the entropy of the universe will be a maximum—maximum disorder.[3]

We can think of many personal examples that add credibility to the relationship. The old cliché "you can never go home" is apropos here. We have all experienced that eagerly awaited and triumphant return home from service or from college, our return to the old hangouts. Perhaps you have returned to an old sweetheart or loved one. With few exceptions, we conclude that it is not the same, things are different. Time has marched on and entropy has taken its toll! The song writer (about 1950) who wrote the lyrics. "the Rockies may crumble, Gibralter may tumble, but our love is here to stay," was being a little oversentimental or probably knew nothing about the second law.

But wait, what about the small entropy reversals we mentioned earlier? Do these represent reversals in time too? Such a concept of time is a reality and these reversals can be thought of as small fluctuations or reversals in time. This returns us to our third statement of the second law relating entropy and probability. That is, the greatest probability for a natural process is where the entropy will be the greatest, but there is still a link through probability and fluctuations that, on a small or microscopic basis (when comparing our ecosystem to the universe), an entropy reversal (and a time reversal) can occur. We can liken such a probability to our example of pushing a pile of wood, mortar, and nails over the cliff and asking what chance it has of landing in the form of a house; the probability is small but finite. If we do it enough times, we would obtain a house—only a lifetime may not be long enough.

Because spaceship earth receives such a small quantity of our sun's total energy emission (about 10^{-10} of the total energy emitted), and surely a

---

[3] The authors would like to acknowledge that there may be some differences in opinion, particularly with regard to the evolution of earth. In support of our view, as the earth cools, its entropy is actually decreasing. However, the entropy increase of the surroundings, the universe, is greater, thus giving a net increase which is consistent with the second law.

microscopic quantity compared to the emissions of all the stars in the universe, we conceivably can consider life on our spaceship as a minute flucuation in the universal march of time and entropy. Harold Blum, in his treatise on evolution and the origin of life (via evolution of molecules), compares the living mass on earth, about 10^{19} g, with the earth's mass, about 6×10^{27} g, and suggests that the bio-mass is so minute a portion of the earth's mass and is even more minute when compared to the mass of the universe, that

> we might think of living organisms as a minute quantity of by-products from a complex chemical reaction carried out on a gigantic scale, a tiny mote in a great retort. [5]

Such a thought is a humbling one. W. F. Kieffer, in turn, expresses his retort well when he says ". . .yet this part of the world, humanity, is the only part which is capable of knowing it is alive in the fullest sense." [3] Finally, we know that in order to remain consistent with the second law, the human body eventually must reach the most probable state, the last entropy increase, death! This sobering thought is not new, but was given to us by the writers of *Genesis*,

> . . . dust thou art and unto dust thou shalt return.

Thus perhaps life on earth is a mere fluctuation in entropy and in time but life on earth is in complete accordance with nature's laws, the laws of thermodynamics.

Entropy and the Ecosphere

In Chapter 3, we saw that only a small fraction, 0.02%, of the sun's energy falling on the surface of the earth (that in the visible region of the spectrum) is fixed as sugar and plant fiber (carbohydrate or cellulose) by photosynthesis. The photosynthetic process itself is entropy reversing; that is, it takes light energy along with carbon dioxide and water (all of which may be considered as being in a state of considerable disorder), and transforms them into ordered molecules containing an energy reserve—chemical energy.

This energy then flows, by metabolism of the carbohydrates (controlled combustion within living beings), through the ecosphere via the food chain. However, each step or level of consumer is less than 100% efficient. Our rule of thumb is that the food energy available for each succeeding level of the food pyramid is about 10% of the previous level, that is, 100 lb of forage is necessary to support 10 lb of steer. The 90% loss at each level is mainly in the form of a spontaneous flow of this waste as heat to the surroundings. The process is entropy increasing! The order "fixed" by photosynthesis, which in itself is miniscule, disappears rather quickly through a few consumer levels (see Figure 9-10).

As the final consumers in each chain die, the microorganisms of decay finish the consumption and combustion of animal or plant tissue, giving up most of the energy as heat to the surroundings and, of course, giving carbon dioxide and water, the final combustion (metabolism) products, back to the surroundings. This essentially is the last stage of entropy increase for any particular food chain.

At first glance, the order produced by photosynthesis appears to be in contradiction to the second law. But the universe is truly the only isolated system, so local reversals, as in our earth system, are allowed by the second law so long as the entropy of the universe increases. Earth as a closed thermodynamic system actually receives energy from the sun. This energy is then used to produce the order we see in life. Other than this statement, the remarkable chemical synthesis by plants is far beyond the scope of this book to explain. Moreover, it is the place of science to tell how this process

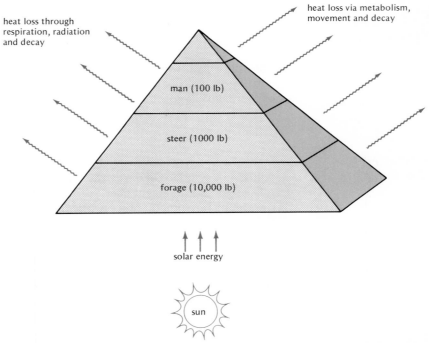

Figure 9-10. A food pyramid. The loss of energy at each level represents an increase in entropy. The order affixed by photosynthesis quickly disappears after a few consumer levels.

happens but not why. Man has been looking for the "why" for centuries and continues to look in hopes that in finding the why, he will be able to find a way to reverse natural trends.

The body itself is an entropy reverser, which is apparent through the healing of various afflictions; healing is an entropy decreasing process, at least when considered locally. However the body degrades large quantities of energy to accomplish the healing. So the total entropy increases as we know it must. Most of us have read of Ponce de Leon's fruitless search for the fountain of youth. There are even those who feel that by consuming the right foods and vitamins in the correct proportions the aging process can be reversed. However, no one has found the key to reverse aging and, hence, overall aging must continue. Maxwell's tiny demon epitomizes the age-old desire of man to find entropy reversers and, equally as important, those that cost nothing to run. However, in the final analysis, the second law always triumphs; the entropy of the universe always increases. The old cliché says, "that is the way it goes." We cannot overcome that compelling principle; entropy and time march on, seemingly together.

Entropy and the Evolution of the Universe

The Big Bang theory of the universe is consistent with the second law of thermodynamics which says that disorder or chaos is the natural trend. The theory proposes that in the beginning, mass was concentrated, a low entropy state. There was then a big bang! And, as a result, according to the Big Bang theory, the mass will expand forever.[4] We see clearly the pattern of a concentration of energy becoming widely distributed energy as per the second law. As the universe expands, the probability or chance of matter interacting by collision to form stars, form planets, and to form life, decreases. Ultimately, one can visualize a state of maximum expansion for the universe which means maximum entropy.

We have characterized this situation by the term *heat death*. However, the fact is that the temperature of empty space is about 3°K now, which would mean a rather "cold" heat death. The real meaning, however, is that at heat death, energy is no longer available for supporting life. Thus, in order to remain consistent with the second law, we conclude that while the clock ticks on, the universe is running down! The rather subtle but important question, though, is how fast is the universe running down? Estimates would lead us to believe that the universe has at least 10 billion more years before it reaches eternal heat death. With the present longevity

[4]The authors would like to acknowledge that there are two other prominent theories of the universe, however, with considerable less supporting evidence than the Big Bang theory, that would appear not to be consistent with the second law. The reader is referred to any basic astronomy textbook for a complete discussion of them.

of man's institutions, this might as well be forever. As you have seen, and will further see in the following chapters, man has more immediate problems that are likely to be critical for his (yours and mine) future, although the ultimate end result at whatever point in time will be the same.

Conclusion   At first, Chapter 9 may seem a deviation from our overall theme of energy. However, its purpose is to summarize, descriptively, the rather involved quantitative discussion of trends in energy production and consumption in Chapter 8. Recall, that Chapter 7 presented the laws by which we must abide on spaceship earth and, in particular, we have found that inefficiency poses the important question: Are we doomed to a continuing decrease in efficiency of our machines or will innovative science and technology bring forth new, novel ways of using and converting energy to the goods and services we enjoy? There is no conclusive answer to this question, but we will peer into our clouded, crystal ball in Chapter 15 with hope of finding some encouragement. Further, you recall that the result of inefficiency is wasted heat energy or energy becoming unavailable to do work. Clausius, in about 1865, used the word "entropy" to describe this unavailable energy.

We have further used entropy or increasing entropy to describe the plight of consumption of our natural resources, specifically our energy resources. The arguments that we have made with regard to energy are equally applicable to mineral resources (one very important one to electrical power generation and its transmission is copper). As we use our resources, such as our fossil fuels, we are gradually increasing the entropy of our ecological system and the universe. At our present rate of use, we will have maximally degraded this store of fossil-fuel energy in a brief period of 100 to 400 years, as compared to the time estimated for the formation of this fossil fuel—600,000 years. Unfortunately, the rate at which the sun is able to replenish the supply of fossil fuels via photosynthesis and subsequent partial decay is too slow. So long as the world society on spaceship earth is based on very finite nonrenewable resources, a crisis looms ahead.

A second important question is how soon must we anticipate this crisis? The concept of entropy only points the direction that things must proceed, and says nothing about how fast we must proceed to that equilibrium condition, termed heat death. From our vantage point, it would appear that we must return immediately to a more sane policy on energy consumption lest we fall prey to the "entropy spectre"!

Questions

1. Explain the implications of Clausius' statements, "Die Energie der Welt ist constant. Die Entropie der Welt strebt einem maximum zu."

2. In your words, what is the meaning of the terms entropy and entropy production?

3. Discuss the implications of the concepts of entropy and disorder to society and its energy consumption habits.

4. Discuss the following terms:
   (a) high quality energy   (d) waste heat
   (b) low entropy state   (e) high entropy state
   (c) degraded energy

5. (a) In the statement of the second law, "the entropy of a system and its surroundings always increases," describe what is meant by the system and what is meant by the surroundings. (b) Give three examples when the entropy of the system decreases. Then, how does the entropy change of the surroundings compare with that of the system?

6. For the case of spontaneous heat flow from infinite reservoirs, we have described the entropy change by equation (9-1), $\Delta S = \Delta Q/T$. In which way does the entropy change as heat flows from the hot reservoir? From the cold reservoir? What is the total change in entropy?

7. Explain any change in significance of equation (9-1) if an engine is placed between the two heat reservoirs in Figure 9-1.

8. How does the discussion of entropy changes in finite heat reservoirs pertain to our energy resources?

9. (a) What is heat death? (b) Does heat death imply that the earth will eventually burn up?

10. How does the irreversibility in operation of a steam cycle relate to entropy?

11. In what way does randomness of molecular motion suggest a high entropy state?

12. Explain how entropy manifests itself in the operation of the following devices:
    (a) household toaster   (d) washing machine and dryer
    (b) electric egg beater   (e) vacuum cleaner
    (c) electric light bulb

13. It has been said that "heat is degraded energy." Explain.

14. An automobile uses chemical energy (source?) to climb a hill.
    (a) Where is this energy after the automobile has reached the summit?
    (b) Where does the principle of increasing entropy manifest itself in this example?

15. A person uses chemical energy (food) to climb a tree and then jumps back to ground. Where has the chemical energy gone? What changes in entropy have taken place?

16. In Finland, people get steamed up in saunas. The steam is commonly produced by dropping very hot stones into vats of water. Is the entropy increasing or decreasing for, (a) the water, (b) the stones, (c) the water plus stones systems?

17. It has been suggested that waste heat from power plants be used to melt icebergs that have been towed from the polar regions in order to utilize the waste heat for producing fresh water from the ice. It was also suggested that in so doing we would be getting something for nothing. What do you think? Be sure to consider all items mentioned.

18. Since our sun is losing energy at a phenomenal rate, its entropy must be decreasing according to equation (9-1). But, we know from the second law that the entropy of the universe must be increasing. How would you reconcile this situation, particularly considering that the sun is one of many stars in the universe?

19. Give two examples for each of the six different statements of the second law (other than those in the text).

20. Explain how each of the six restatements of the second law are related through the concept of "order tends to disorder."

21. Why is the collection of information for this text a decrease in entropy? Explain how this is or is not consistent with the second law.

22. Consider the two processes, photosynthesis and respiration. Write the chemical equation for each and explain how each is an increase or a decrease in entropy.

23. Is life a violation of the second law of thermodynamics?

24. Why is Maxwell's Demon interesting to scientists?

25. (a) Relate entropy to the various stages of the Humpty Dumpty nursery rhyme. (b) What part does a chicken (hen) play? Is the chicken a Maxwell's Demon?

26. What do we mean by a hot molecule? By a cold molecule?

27. Why are the hot molecules in a higher entropy state than the cold molecules?

28. What does entropy mean to a biologist? to a physicist? to a chemist?

29. What does the concept of entropy and disorder imply about spaceship earth and about the universe as a whole?

30. Is star formation an entropy increasing or decreasing situation?

31. How does the concept of heat death relate to a star, for example, the sun?

32. If we flip 10^4 pennies into the air, what is the most probable distribution between heads and tails?

Numerical Exercises

1. Referring to Figure 9-1, the hot reservoir is at 100°C and the cold is at 0°C. The two are placed in thermal contact and 10^3 J (joules) flow from hot to cold. What is the total entropy change?

2. A nozzle on a steam boiler at 1000°F (811°K) is opened, allowing a small amount of steam with a heat content of 10^5 kcal to escape. The surroundings are at room temperature, 77°F (298°K). Determine the total entropy change. What is the entropy change of the universe?

3. To melt 10 g of ice at 0°C (273°K), about 0.8 kcal of heat are needed. What is the increase in entropy of the ice-water system? (*Hint:* Melting ice at 0°C is a reversible process.)

4. For example, let us assume that a typical star emits energy at a rate of 10^{10} J/sec.

(a) If the temperature of the photosphere is $6000°K$, what is the rate of decrease in entropy of the star?

(b) If outer space is at a temperature of $3°K$, what is the rate of entropy increase of outer space (the surroundings)?

(c) What is the rate of entropy change of the universe? Which of the two factors (from a or b) is more significant in this calculation? Why?

References

[1] After a dialogue in *My Fair Lady*.

[2] Other authors have used similar forms of these much used statements. For example, see G. Tyler Miller, Jr.: *Energetics, Kinetics and Life*. Wadsworth Publishing Company, Inc., Belmont, Calif., 1971, p. 45; or W. H. Kieffer: *Chemistry, A Cultural Approach*. Harper & Row, New York, 1971, p. 386.

[3] W. H. Kieffer: *Chemistry, A Cultural Approach*. Harper & Row, New York, 1971, p. 388; 390.

[4] A. Eddington: *The Nature of the Physical World*. The Macmillan Company, New York, 1929.

[5] H. F. Blum: *Time's Arrow and Evolution*. Princeton University Press, Princeton, N.J., 1968, p. 153. The authors have found that this book presents a fascinating view of life and the second law of thermodynamics.

Our
Nuclear
Age

In Chapter 8, we observed that, in its need for more energy, our economy is beginning to shift significantly toward energy from nuclear sources. In the previous five chapters, we examined the generation of electrical energy from the combustion of fossil fuels. The harnessing of thermal energy from the chemical reactions of combustion has been central to both the Industrial Revolution and to the modern way of life in the western world and Japan. These industralized countries—these developed countries—consume vast quantities of inexpensive energy which is used to provide goods and services. The demand for increasingly large amounts of energy motivates the search for energy sources alternate to the nonrenewable fossil fuels—coal, petroleum, and natural gas.

We have already described the chemical energy of the fossil fuels as stored energy of the sun in that the sun's energy powers photosynthetic processes in plants (equation 10-1).

$$6\,CO_2 + 6\,H_2O \longrightarrow 6\;O_2 + C_6H_{12}O_6 \qquad (10\text{-}1)$$

In Chapter 3 we noted that the sun's energy arises from certain transformations of atomic nuclei. In the following three chapters (10, 11, and 12), we present the story of nuclear science and technology which stands as one of the most startling and complex examples of a technology based on scientific endeavor. Chapter 10 summarizes basic nuclear chemistry and physics. Chapter 11 discusses the generation of electricity, using nuclear energy. Chapter 12 focuses on the hazards and benefits associated with the generation of nuclear power.

In order to understand the technology of nuclear power plants and the environmental impact of these plants, we examine in this chapter an area of science called nuclear physics and chemistry. We consider this area of science to be interdisciplinary, particularly because both chemists and physicists have made major contributions to the science of the nucleus.

The energy with which we shall be concerned is often incorrectly called "atomic energy." Atomic structure and energy is primarily concerned with the electrons surrounding the nucleus and the chemical bonds formed through these electrons. During all chemical reactions, such as combustion, the nucleus of the atom plays a passive role. As we see in this chapter, and contrary to the original Dalton view of atoms, the nucleus of an atom can undergo certain transformations. These nuclear reactions, what they are and what they can do, are the subject of this chapter.

More specifically, in this chapter we review our concept of atoms, sketch briefly the discovery of unstable nuclei that are called radioactive nuclei, and then look at these nuclear transformations quantitatively with the assistance of the symbolism of nuclear equations. The instability of radioactive substances is measured by the half-life for radioactive emission, which serves as the basis for our study in Chapter 12 of the handling of spent fuels and radioactive wastes from nuclear power plants. We present a brief story of neutron-initiated chain reactions (fission) as the highly energy releasing process that has been used in nuclear explosions and is increasingly being used as a source of heat energy for electric power generation (Chapter 11). The transformation of nuclear mass into energy is shown to be the source of this vast quantity of available heat energy. Energy, both from nuclear fission and nuclear fusion, is discussed.

The Unstable Atom

A Model

Let us begin our study of nuclear energy by reviewing the modern, simplified model of the atom. The principal particles comprising the atom are the electron, the proton, and the neutron. The nucleus of an atom contains the protons and the neutrons whereas the electron occupies space

Figure 10-1. A helium atom. Most of the mass of the atom is concentrated in the nucleus which contains two protons and two neutrons. The two electrons, located outside the nucleus, balance the electrical charge of the nucleus, thus maintaining electrical neutrality.

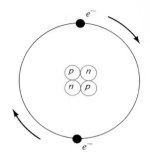

outside the nucleus. Because the electron weighs some 1/2000 as much as a proton or a neutron, the mass of the atoms is almost entirely found in the nucleus. For the hydrogen atom, the nucleus has a radius of 1.0×10^{-13} cm, whereas the hydrogen atom as a whole occupies a sphere of space whose radius is 7×10^{-9} cm. The diameter of the atom then is some 70,000 times larger than the nucleus. If the nucleus were the size of the period at the end of this sentence, then the atom would correspondingly be the size of an average house.

We used the shorthand notation, $^4_2$He or $^{235}_{92}$U, in Chapter 5. The subscript is the atomic number, Z, which is the number of protons in the atom. The superscript designates the mass number, A, of the particular atom. The mass number is the sum of the protons and neutrons of the atom. The reader should note that at times we also will use the popular notation, uranium-235, in the following discussions. Also, you should recall that the neutral atom has the same number of electrons as protons. This simplified view of the atom, is often schematically popularized as in Figure 10-1.

Some History

By the late nineteenth century, many scientists believed that all significant data had been accumulated in the physical sciences and that continued scientific investigations would be confined to more precise determinations of physical constants. However, in 1896, during the course of some studies on phosphorescent materials (substances that, when irradiated with sunlight, give off a visible glow, particularly when viewed in the dark), Henry Becquerel (awarded the Noble Prize in physics jointly with Marie and Pierre Curie in 1903) placed a sample of uranium containing ore, **pitchblende,** next to some photographic film. The film was found to have been exposed, apparently by something emitted from the pitchblende.

The scientists were puzzled as to what could be spontaneously coming out of an ordinary ore to cause such a phenomenon? In July 1898, after months of enthusiastic and dedicated labor, Marie Curie (who also received the Nobel Prize in Chemistry in 1918) obtained a mere one gram of a new element, polonium (named after Marie's native Poland), with atomic number 84, from one ton of pitchblende. Six months later she obtained radium atomic number 88.

In 1903, Marie Curie named the peculiar property of polonium and radium (and also other heavy atoms, such as uranium) **radioactivity.** Within a dozen years, some 30 species of atoms were reported to have radioactive behavior. At present, over 50 naturally occurring radioactive isotopes are known. *Today, we define radioactivity as a phenomenon peculiar to some atoms which spontaneously change by emitting particles and/or rays with tremendous energies.* This radiation comes from the nucleus of the atom.

Radioactive Emissions

Shortly after the discovery of radioactivity, the principal emissions of these heavy atoms, see Table 10-1, were identified (Figure 10-2). Some radioactive atoms, such as radium-226 and uranium-238, break down spontaneously giving off alpha particles, which have been found to be identical to helium nuclei (no electrons), that is $^4_2\text{He}^{2+}$. Other radioactive atoms, such as thorium-234 and polonium-218, lose beta rays or beta particles, which are identical to electrons, symbolized by $_{-1}^{0}e$. The third type of radioactive emission often accompanies alpha and beta particle emissions but involves no particle and no charge. This neutral ray is called a gamma ray, which is now known to be a high energy x-ray. These names, alpha (α), beta (β), and gamma (γ), were assigned before the rays were characterized. In addition, such processes as these are often called radioactive events.

You should note carefully that the alpha particle is a helium nucleus which consists of two protons and two neutrons. Remember that the neutral helium atom has two electrons in the immense space outside the nucleus. Even though $^4_2\text{He}^{2+}$ is a better representation for the alpha particle, we simply write ^4_2He (or α) because this symbol is more convenient and because the high energy alpha particle on collision with matter, for example, the walls of the container, takes on two electrons very quickly, thus becoming a neutral helium atom.

Table 10-1.
Radioactive Emissions

Emission	Mass Number	Charge	Symbol
alpha (α)	4	+2 (+ +)	^4_2He
beta (β)	0	−1 (−)	$_{-1}^{0}e$
gamma (γ)	0	0	γ

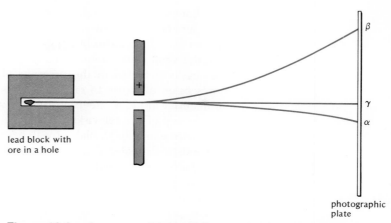

Figure 10-2. An ore containing radium and polonium emits rays that an electrostatic field separates into alpha (α) particles, beta (β) particles, and gamma (γ) rays, all of which expose photographic film.

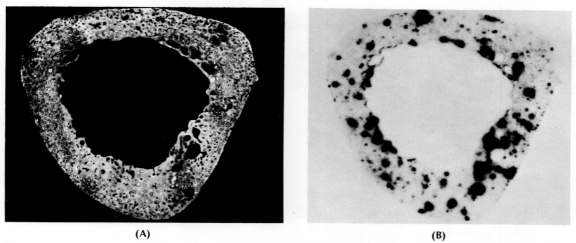

(A)   (B)

Figure 10-3. A section of bone from the body of a former radium watch-dial painter. **(A)** Darkened areas indicate damaged bone. **(B)** An autoradiograph, in which film was merely held against the bone and was exposed by the radium's alpha (α) particle emission. Note that the areas of high alpha activity correspond to the areas of maximum bone damage in (A). [Reproduced from N. Frigerio: *Your Body and Radiation.* U. S. Atomic Energy Commission, Washington, D. C., 1967. Illustrations from Argonne National Laboratory.]

Radium salts, for example, radium bromide ($RaBr_2$) are observed to emit a blue luminescence. This intense luminescence is easily visible in the dark and can even be seen in broad daylight if you have more than 0.10 g of the radium salt. That radium salts emit a high energy particle can be even more easily detected in the presence of zinc sulfide (ZnS). Zinc sulfide fluoresces on exposure to radiation. This is analogous to the behavior of the material inside a television tube as it responds to rays emitted from the rear end of the television tube.

Up until about 1930, dials of watches were made luminous by painting the hands with material containing both zinc sulfide and a radium salt. Regrettably, the dial painter, in order to maintain a very fine point on his brush, frequently touched the $RaSO_4$-AuS coated brush to his mouth. The gradual ingestion of these materials caused serious oral cancers in many of these people (Figure 10-3). We will discuss further some of the health hazards of radioactivity in Chapter 12.

Nuclear Equations

Uranium-238 ($^{238}_{92}U$) spontaneously loses an alpha particle to form an isotope of thorium, $^{234}_{90}Th$, as shown in equation (10-2) (see Example 10-1). You recall that isotopes of an element have the same atomic number, same Z, but have a different mass number, different A.

$$^{238}_{92}U \longrightarrow {}^4_2He + {}^{234}_{90}Th \tag{10-2}$$

In nuclear equations, the sum of the atomic numbers (subscripts) on each side of the equation must be identical. Furthermore, the sum of the mass numbers (superscripts) on each side of the equation also must be identical.

Interestingly, thorium-234 is radioactive and emits beta particles as in equation (10-3) (see Example 10-2).

$$^{234}_{90}Th \longrightarrow {}^0_{-1}e + {}^{234}_{91}Pa \tag{10-3}$$

EXAMPLE 10-1

If ^{238}U emits an alpha particle, what else is formed?

$$^{238}_{92}U \longrightarrow {}^4_2He + {}^A_ZX$$

(a) The atom formed has two fewer protons ($92 - 2 = 90 = Z$) than uranium. From the periodic table, thorium has an atomic number of 90.

(b) The mass number of the thorium is four mass units ($238 - 4 = 234 = A$) less than the ^{238}U because the alpha particle lost has a mass number of 4. Thus, the product is

$$^{238-4}_{92-2}X \equiv {}^{234}_{90}Th$$

or

$$^{238}_{92}U \longrightarrow {}^4_2He + {}^{234}_{90}Th$$

EXAMPLE 10-2

If ^{234}Th emits a beta particle, what else is formed?

$$^{234}_{90}\text{Th} \longrightarrow {}^{0}_{-1}e + {}^{A}_{Z}\text{Y}$$

(a) $Z - 1 = 90$, therefore $Z = 90 + 1 = 91$. Element of atomic number 91 = protactinium (Pa).
(b) $234 = 0 + A$, therefore $A = 234$.

Thus, ${}^{A}_{Z}\text{Y} = {}^{234}_{91}\text{Pa}$, and

$$^{234}_{90}\text{Th} \longrightarrow {}^{0}_{-1}e + {}^{234}_{91}\text{Pa}$$

We consider that the emission of an electron (beta ray) from the nucleus of an atom arises from the decay (breaking down) of a neutron. After an electron has been emitted in beta decay, what is left behind in the nucleus if the electron has negligible mass and negative charge? The particle left must have about the same mass as the neutron and must be positively charged; thus, the particle left in the nucleus must be a proton. Overall then, a neutron has changed into a proton and an electron ($^{1}_{0}n \longrightarrow {}^{1}_{1}p + {}^{0}_{-1}e$).

Because it has been established experimentally that an electron and a proton can combine outside the nucleus to form a neutron, it has been proposed that, in the nucleus, a neutron forms a proton with ejection of an electron. The number of protons has thus increased by 1, so the atomic number must have increased by 1 (while A remains the same) thus producing an atom of atomic number 91 (protactinium). Essentially no mass has been lost; thus, the mass number does not change when a beta ray is emitted.[1]

Protactinium-234 has been found to emit a beta particle giving uranium-234, and uranium-234 in turn emits an alpha particle to give thorium-230. (*Hint:* Write nuclear equations for these two reactions.) These spontaneous nuclear reactions have changed an atom of one element into an atom of another element. These are examples of natural transmutations of elements. Throughout the Middle Ages, transmutation of elements had been sought by alchemists. The alchemists specifically sought to change readily available elements, such as iron, lead, or tin, into gold. This many centuries long search was discarded with the acceptance of the concept of Dalton's indestructible atom. Nonetheless, radioactive disintegrations had been taking place naturally, albeit undiscovered, all the while; however, these natural transmutations were not giving gold! Most, ultimately, give lead.

---

[1] The authors acknowledge that during beta decay a small bundle of energy called a neutrino is also given off, but is of no practical interest to us here.

Radioactive Disintegration Series

The four nuclear reactions indicated above are the first four steps in the sequence by which we now know that uranium-238 decays. In addition to these above four reactions, there are ten more that follow in succession. All of these are indicated in Figure 10-4. The final product of this decay

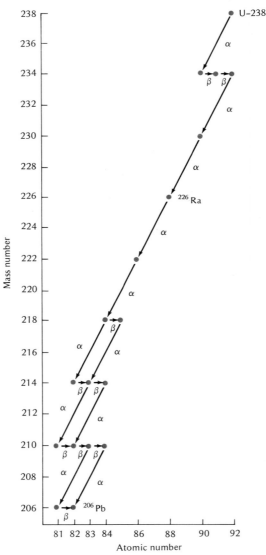

Figure 10-4. The uranium (^{238}U) radioactive decay series. The particle ejected is indicated on the arrow between atoms formed.

sequence is the stable (not radioactive) isotope of lead, lead-206. This decay series is known as the uranium-238 series. Two other series have been found in nature which start with different isotopes. One starts with thorium-232 and ends with lead-208, whereas the other starts with uranium-235 and ends with lead-207.

During radioactive decay, throughout the series, gamma emission often accompanies the alpha and beta particle emissions. Significantly, most of the naturally occurring isotopes of atoms with atomic numbers 84 and higher are members of one of these three decay series. All these atoms of atomic numbers greater than 83 are radioactive (see Zone of Stability below).

In the overall decay of one unstable uranium-238 atom leading ultimately to the stable lead-206, how many alpha particles and how many beta particles have been emitted? Obviously, if you have all the steps written out, you can merely count the steps. There is a simpler way that involves some algebra, as shown in Example 10-3.

EXAMPLE 10-3

How many alpha particles, and how many beta particles are emitted when one ^{238}U atom decays to ^{206}Pb?

The change is

$$^{238}_{92}\text{U} \longrightarrow {}^{206}_{82}\text{Pb} + a\ {}^{4}_{2}\text{He} + b\ {}^{0}_{-1}e$$

where a and b are unknowns. Because no mass is lost as beta particles, all the mass is lost in the form of alpha particles.

$$238 = 206 + 4a$$
$$4a = 238 - 206 = 32$$
$$a = 8$$

Resulting in

$$^{238}_{92}\text{U} \longrightarrow {}^{206}_{82}\text{Pb} + 8\ {}^{4}_{2}\text{He} + b\ {}^{0}_{-1}e$$

The sum of the subscripts on each side of the equation must be the same. Thus

$$92 = 82 + (8 \times 2) + (b \times -1)$$
$$92 = 82 + 16 - b$$
$$b = 82 + 16 - 92 = 98 - 92$$
$$b = 6$$

Conclusion

$$^{238}_{92}\text{U} \longrightarrow {}^{206}_{82}\text{Pb} + 8\ {}^{4}_{2}\text{He} + 6\ {}^{0}_{-1}e$$

This complete disintegration series is summarized by equation (10-4).

$$^{238}_{92}U \longrightarrow 8\,^{4}_{2}He + 6\,^{0}_{-1}e + \,^{206}_{82}Pb \qquad (10\text{-}4)$$

Again, let us repeat, the subscripts on the left of the equation exactly equal the sum of the subscripts on the right. Also, the sum of the superscripts in the left must be exactly equal to the sum of the superscripts on the right.

Concept of Half-Lives

One of the intriguing facts general to the behavior of unstable or radioactive atoms is that the rate of release of particles is completely unaffected by physical treatment (heating, cooling) or by chemical reaction (changing the ore to metal or to some other chemical compound). Another interesting feature is that each unstable isotope has its own characteristic **decay rate.** As a convenient way to compare the unstable atoms, we record **half-lives,** such as the half-lives of the atoms in the uranium-238 decay series in Table 10-2.

The half-life of a radioactive isotope represents the time period needed for one half of the atoms originally present to be transformed to atoms of another element by the appropriate mode of decay of that radioactive isotope.

Thus, at the end of the time period designated as the half-life (or one half-life), only one half of the original number of atoms will remain unchanged, whereas one-half have transformed into some other element.

Table 10-2. Half-Lives of the Isotopes of the Uranium-238 Decay Series

Isotopes	Particle Emitted	Half-Life
^{238}U	alpha	4.51×10^9 years (4.51 billion years)
^{234}Th	beta	24.1 days
^{234}Pa	beta	1.17 min
^{234}U	alpha	247,000 years
^{230}Th	alpha	80,000 years
^{226}Ra	alpha	1,622 years
^{222}Ru	alpha	3.823 days
^{218}Po	alpha or beta	3.05 min
^{214}Pb	beta	26.8 min
^{214}Bi	beta or alpha	19.7 min
^{210}Tl	beta	1.32 min
^{210}Pb	beta	21 years
^{210}Bi	beta or alpha	5.01 days
^{210}Po	alpha	138 days
^{206}Tl	beta	4.19 min
^{206}Pb	none	stable

You will recall that in Chapter 8 we considered growth rates and doubling times and referred to them as being exponential. Radioactive decay is also exponential in behavior. It is just the converse case to doubling; that is, in each given interval of time, half of the atoms decay. This is shown in Figure 10-5.[2] Some half-lives are extremely long, for example, uranium-238 is 4.51 billion years, (the second longest known), whereas others are extremely short, measured in a few days, hours, minutes, or seconds (see Table 10-2). The isotope with the longest half-life (14.1×10^9 years) is thorium-232.

Let us consider the behavior of a sample of thorium-234 whose half-life is essentially 24 days. If your original sample of material contains 128 atoms of thorium-234 then after 24 days, 64 atoms of thorium-234 will have been transformed to protactinium-234. The sample now contains only 64 atoms of thorium-234, and in 24 days one half of these 64, or 32 atoms, will be transformed leaving only 32 atoms of thorium-234 unchanged in the sample. Remember, the half-life is completely independent of how much material is present. The half-life behavior of unstable isotopes all are patterned after the graph of Figure 10-5.

[2]It is interesting to note that, if you draw a curve on a transparent sheet of paper similar to Figure 10-5 and turn the paper over, the curve has the same shape as the growth curve discussed in Chapter 8.

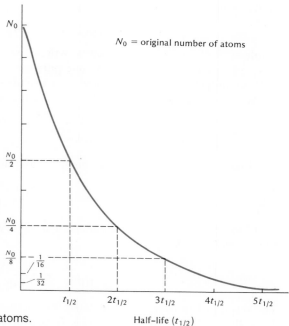

Figure 10-5.   General half-life curve for unstable atoms.

It is important, particularly when we consider the problem of handling radioactive waste materials from nuclear-fueled power plants (Chapter 11) to note that unstable atoms, with a short half-life, decay more rapidly than do unstable atoms with long half-lives. If we compare thorium-234 with uranium-238, we find that, during any given interval of time, thorium gives off 7×10^{10} (70 billion) times as many alpha particles as does the same amount of uranium. (See Example 10-4.)

When each atom of thorium loses an alpha particle, this is a radioactive event. Radioactive events can be followed by devices such as a Geiger counter (Chapter 12). The concept of radioactive events applies to any decay reaction. It should be clear by now that the number of radioactive events in a substance with a short half-life is very much larger than in a sample with a long half-life.

EXAMPLE 10-4

From the half-life data for thorium-234 and uranium-238, calculate how many more particles are given off by the thorium than by the uranium in a given time period.

For convenience, let us choose for our time period 24 days, which is the $t_{1/2}$ for thorium-234. Furthermore, for convenience, let us take 234 g of thorium, which is 1 mole of thorium atoms or about 6×10^{23} thorium atoms. During one half-life, exactly one half of the atoms present decay. Thus,

$$\tfrac{1}{2} \times 6 \times 10^{23} \text{ atoms} = 3 \times 10^{23} \text{ Th atoms decayed in 24 days}$$

Similarly, 3×10^{23} atoms of uranium will have decayed in 4.5 billion years. (*Note:* We have taken 238 g of uranium-238 which, for our purpose, we consider essentially the same mass as 234 g of thorium-234.) Our problem now is to calculate how many uranium-238 atoms decay in only 24 days. Using ev for events and rad for radioactive

$$\frac{3 \times 10^{23} \text{ rad ev}}{4.5 \times 10^9 \text{ years}} \times \frac{1 \text{ year}}{365 \text{ days}} = 1.83 \times 10^{11} \frac{\text{rad ev}}{1 \text{ day}}$$

In 24 days, 24 times as many atoms will have decayed, therefore

$$24 \text{ days} \times 1.83 \times 10^{11} \frac{\text{rad ev}}{1 \text{ day}} = 4.4 \times 10^{12} \text{ rad ev}$$

As a final step, we compare the number of observed events for those two substances in the same time period (24 days).

$$\frac{3 \times 10^{23} \text{ Th ev/24 days}}{4.4 \times 10^{12} \text{ U ev/24 days}} = \frac{7 \times 10^{10} \text{ Th ev}}{1 \text{ U ev}}$$

The result of this example, calculated for a 24-day time period, is the same for any time period. Thus, the rate of decay of thorium is 70 billion times faster than is the rate of decay of uranium.

Of related interest is the use of the uranium decay series to determine the age of rocks and, hopefully, an estimate of the age of the earth. Uranium-238 gives a particularly good estimate because it is one of the longest lived isotopes of a heavy atom known. The relative amounts of uranium-238 and lead-206 present together in a rock permit the calculation of the quantity of uranium-238 that was originally present and at what point in time this rock came into existence. From such analyses, estimates for the age of the earth range from 3 billion to 3.5 billion years; however, such results are considered to approximate the age of the earth usually taken as 4.5×10^9 years, as estimated from other geological studies of rocks.

Artificial Transformations

In the early part of the twentieth century, isotopes of large atomic numbers were intensively studied. All atoms of atomic number 84 and larger have been found to be radioactive. In the uranium series we observe some radioactive isotopes of atomic number lower than 84. In nature, we usually find only the stable isotopes of atoms with atomic number 83 (bismuth) and smaller. Nonetheless, radioactive isotopes of low and very low atomic number elements have also been found.

Shortly before 1920, Ernest Rutherford and his associates examined the effect of alpha particles emitted from radium, and from other alpha emitters, on various kinds of atoms. Such alpha particles have velocities of about 10,000 mi/sec, that is 20,000 times faster than the velocity of a rifle bullet. Bombardment of nitrogen atoms leads to the **artificial transformation** shown in equation (10-5).

$$\ce{^4_2He + ^{14}_7N \longrightarrow [^{18}_9F] \longrightarrow ^1_1H + ^{17}_8O} \tag{10-5}^3$$

Artificially induced transformations have been observed with boron, fluorine, sodium, aluminum, phosphorus, and many other rather light atoms. These reactions are also known as artificial transmutations of one element into another element.

Some of these induced transformations produce unstable nuclei of low atomic number. In the upper atmosphere, neutrons from cosmic radiation penetrate the nucleus of an ordinary nitrogen atom producing an unstable, radioactive isotope of carbon (equation 10-6).

$$\ce{^{14}_7N + ^1_0n \longrightarrow [^{15}_7N] \longrightarrow ^{14}_6C + ^1_1H} \tag{10-6}^3$$

The carbon-14 formed has a $t_{1/2}$ of 5760 years, and emits a beta particle forming ordinary nitrogen, equation (10-7).

$$\ce{^{14}_6C \longrightarrow ^0_{-1}e + ^{14}_7N} \tag{10-7}$$

[3] The square bracket around $^{15}_7N$ and $^{18}_9F$ is used because no physical evidence for the existence of these atoms in this reaction has been obtained. Yet it seems reasonable to believe that, at the point of collision, such a species may have a transitory existence.

The carbon-14 ($^{14}_{6}$C) produced in the upper atmosphere rapidly appears in carbon dioxide molecules which, on contact with the earth's surface, may be incorporated via the photosynthetic reaction into plant matter. The carbon dioxide of the atmosphere apparently contains a steady state, constant concentration of one atom of carbon-14 for every 10^{12} (trillion) atoms of carbon-12. This steady state is maintained because there is a balance between the rate of carbon-14 formation (equation 10-6) and the rate of carbon-14 disintegration or decay (equation 10-7).

All living things contain carbon obtained from atmospheric carbon dioxide and so also take in both forms of carbon in the same proportion as these exist in the atmosphere. When death comes to any plant or animal, the intake ceases, but the radioactive decay of carbon-14 to nitrogen-14 goes on. Thus, old, dead organic matter contains less carbon-14 than newly formed (live) organic matter. Dead plant matter, particularly as paper and wood, may be many centuries old. The age can be determined by the amount of radioactive carbon in a sample. This technique is known as **radio carbon dating** and has been used to authenticate paintings, wooden articles, and the Dead Sea Scrolls. By this same means, it has been determined that the retreat of the most recent great ice sheet in North America began only some 10,000 years ago.

Over 1000 radioactive isotopes have been prepared by various bombardment reactions. Unstable isotopes of most elements have been made. When you plot the mass number of all stable isotopes known against the respective atomic number of each, you obtain a graph, as in Figure 10-6. Atoms falling within the shaded area of the graph are all stable nuclei. This area is referred to as the **zone of stability.** Atoms falling outside the shaded area are unstable nuclei.

Three groups of atoms are unstable. First, all atoms of atomic number greater than 83. These atoms emit alpha, beta, and/or gamma rays. Second, atoms falling above the zone of stability have too many neutrons for the number of protons. Many of these radioactive atoms are observed to be beta emitters. Third, atoms falling below the zone of stability have too few neutrons for the number of protons. The modes of change of this last group to give stable atoms are beyond the concern of our study.

Rutherford used various heavier, radioactive elements as sources of alpha particles to bombard various atomic nuclei. Of necessity, he was limited to alpha particles having just the energy supplied by the decaying atom, such as radium. As a result of Rutherford's artificial transmutation experiment (such as equation 10-5), there was considerable interest in bombarding stable nuclei with high energy particles to discover new nuclear reactions. Scientists recognized that, in order to induce nuclear transmutations, the bombarding particles must have sufficient energy to penetrate the atomic nuclei.

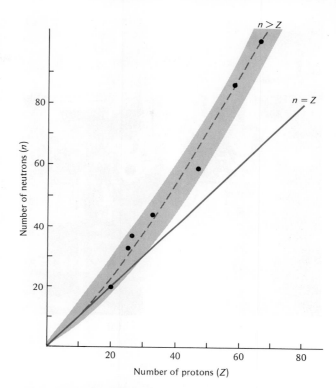

Figure 10-6. Zone of stability. On both sides outside the band encompassing the stable isotopes, radioactive isotopes are found.

The positively charged alpha particle is repelled electrically by the positively charged nuclei of atoms (like charges repel). Clearly it is difficult to get naturally occurring alpha particles with energies high enough to do many interesting transmutations. In 1931, E. O. Lawrence (University of California at Berkeley, Nobel Prize 1939) developed a large electrical device, the cyclotron (Figure 10-7) which is capable of increasing the velocity of positive particles. The particles are accelerated as a result of electrical and magnetic forces exerted on the charged particles. The invention of the cyclotron (and other types of particle accelerators such as the linear accelerator and bevatron) to obtain high energy particles made it possible to bombard heavy nuclei (large positive charges) with the possibility of obtaining even heavier nuclei.

Prior to 1940, the heaviest known element was uranium. Using a cyclotron, McMillan and Abelson (1940) prepared element 93, neptunium, the first synthetic element. They directed a stream of deuterons, $_1^2H$ (a particle containing 1 proton + 1 neutron), onto uranium-238. The results are summarized in equation (10-8).

$$_{92}^{238}U + _1^2H \longrightarrow [_{93}^{240}Np] \longrightarrow _{93}^{238}Np + 2\,_0^1n \qquad (10\text{-}8)$$

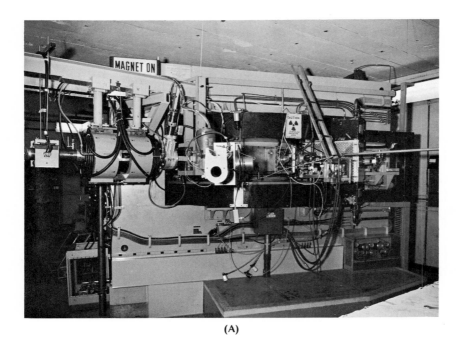

(A)

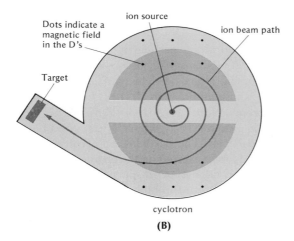

ion source

Dots indicate a
magnetic field
in the D's

ion beam path

Target

cyclotron

(B)

Figure 10-7. A cyclotron. **(A)** The magnet poles are shown located above
and below the particle chamber. The target area is located outside the
magnet area and to the left. [Photograph courtesy of the Nuclear Physics
Laboratory, University of Colorado.] **(B)** Sketch showing the interior of the
cyclotron.

Most nuclear bombardment studies have used protons ($_1^1$H, the hydrogen nucleus), deuterons, and alpha particles. These positive particles must have extremely high speeds in order to bombard and penetrate the highly positive nuclei of the heavier elements. Many of the syntheses of the transuranium elements (atomic number greater than 92) have involved the positive particles mentioned. Studies of the interactions of these highly energetic particles with nuclei have added much to our knowledge of the structure and properties of the nucleus.

Neutron Bombardment of Nuclei

A different kind of weapon for attack on the nucleus is the neutron. The absence of a positive charge would clearly seem to allow the neutron to penetrate a highly positive nucleus far more easily than can alpha particles or some other positive particle. However, the absence of charge on the neutron means that cyclotrons or accelerators are useless in changing the speeds of neutrons. Thus, those nuclear reactions that produce neutrons are the way of "obtaining" neutrons for bombarding various nuclei, such as the reaction (equation 10-9) in which neutrons were first identified (Chadwick, 1932). The neutron obtained from this reaction

$$_4^9\text{Be} + _2^4\text{He} \longrightarrow _6^{12}\text{C} + _0^1 n \tag{10-9}$$

and other neutron producing reactions has great penetrating power. One of the problems with using free neutrons for such reactions is that neutrons are short-lived, $t_{1/2}$ of about 13 min; thus, a free neutron quickly decays into an electron and a proton unless the neutron does something else before it decays.

In Rome, Italy, Enrico Fermi (1901–1954) was probably the first to recognize the significance of the penetrating power of neutrons. During the early 1930s, he examined all the elements from nitrogen to uranium and found that, when neutrons struck all these atoms, radioactive products were produced. Because some of the radioactive products are beta emitters, such as $_6^{14}$C (as in equation 10-7), Fermi (awarded the Nobel Prize in 1938) in the middle 1930s conceived the idea that perhaps elements of greater atomic number than uranium could be produced.

Fermi surmised that if ^{238}U captured a neutron and became ^{239}U, and if ^{239}U were to be a beta emitter, then the unknown element number 93 could be produced, as in equations (10-10) and (10-11). Perhaps then,

$$_{92}^{238}\text{U} + _0^1\text{n} \longrightarrow _{92}^{239}\text{U} \tag{10-10}$$

$$_{92}^{239}\text{U} \longrightarrow _{-1}^{0}\text{e} + _{93}^{239}\text{X} \tag{10-11}$$

by extension of this process, he might be able to develop a whole series of transuranium elements, elements of atomic number greater than 92. This idea was not to be brought to fruition by Fermi. In the neutron bombardment of ^{238}U, Fermi at first believed he had made element 93 because beta rays were emitted from the products of neutron bombardment, but the element produced was shown not to be an isotope of uranium nor was it an isotope of any element near uranium.

This result was most confusing. Irene Curie (daughter of Marie and Pierre) repeated Fermi's experiment in 1938 and concluded that the element produced behaved like an alkaline earth metal, such as magnesium and calcium (group II in the second vertical column of the periodic table). The German scientists Otto Hahn and Fritz Strassman also repeated the experiment. Among the products they specifically identified barium (atomic number 56) which belongs to group II. Such a great change in atomic number during a nuclear transformation had never before been observed!

You recall that in 1938, the clouds of World War II were gathering. The political climate often has profound implications for other activities, including scientific endeavor. Associated with Professor Hahn was Lise Meitner, an able physicist and an Austrian-born Jew. When Hitler's army invaded Austria and had it declared part of the German Reich, Lise Meitner immediately became subject to the German racial laws. She quickly severed her collaboration with Dr. Hahn and managed to escape to Sweden with the assistance of the Danish physicist, Niels Bohr. Here she learned of and grasped the significance of Hahn's identification of barium and conceived the idea that, under neutron bombardment, the uranium nucleus had split into two lighter atoms of approximately equal size.

With her nephew, Otto Frisch, Lise Meitner calculated the energy associated with such a **fissioning** (breaking apart) of uranium. Frisch repeated the now familiar experiment and looked at the energy release attending the process. The energy release was large. The energy release and, as we discuss in the next section, the production of more neutrons caused great excitement among scientists. We will consider both of these phenomena in more detail and will return to a discussion of the energies involved after pursuing the neutron story more fully.

Chain Reactions

Well, what about the neutrons fission produced? Uranium has 92 protons. Fission of uranium forms barium with 56 protons, leaving 36 protons (krypton) for the other fragment. Furthermore, uranium-238 has 146 neutrons. Now, if you look in an atomic weight table for the approximate masses of barium (137) and krypton (84), you see that a stable barium atom has 81 neutrons (137 − 56) and a stable krypton atom has 48 neutrons (84 − 36), which would account for a total of 129 neutrons. Arithmetically, this leaves 17 neutrons of the uranium-238 still to be accounted for. Where are they?

When one neutron is absorbed by uranium, the uranium atom apparently breaks apart to give barium and krypton, among other things, with the release of energy and of neutrons. These released neutrons might then be absorbed by other uranium atoms, each of which would split into barium and krypton, releasing energy and releasing more neutrons. Thus, a nuclear chain reaction, sketched in Figure 10-8, appeared possible. We should mention that krypton and barium are just two possible fragments of the fission reaction. Many others are also possible.

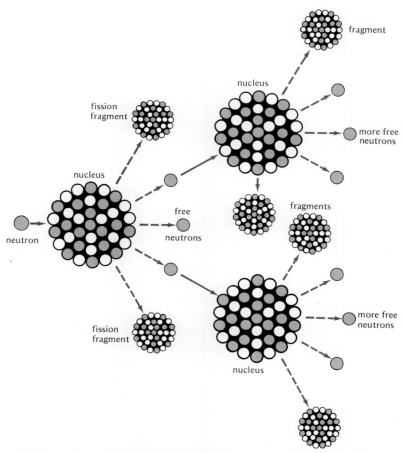

Figure 10-8. Representation of nuclear fission chain reaction in $^{235}_{92}$U. Note that two or three neutrons are released when a $^{235}_{92}$U atom splits into smaller fragments. These additional neutrons in turn trigger additional $^{235}_{92}$U fission reactions, and so on.

About this time, it was discovered that the fission reaction taking place was not a fission of ^{238}U but of ^{235}U ($t_{1/2} = 7.1 \times 10^8$ years). Samples of uranium typically contain 99.3% ^{238}U, 0.70% ^{235}U, and a trace (0.0006%) of ^{234}U. Of these, only ^{235}U is a **fissile atom,** that is, an atom that undergoes fission after capture of a neutron. In terms of the energy release, ^{238}U plays no active part. However, later it will be shown that, when ^{238}U captures a neutron, it eventually transmutates to plutonium-239 (^{239}Pu) which is a fissile atom. Because ^{238}U "breeds," it is called a **fertile atom.** Equation (10-12) lists some of the many known fission fragments from the fissioning of ^{235}U. Some hundred or so have been identified.

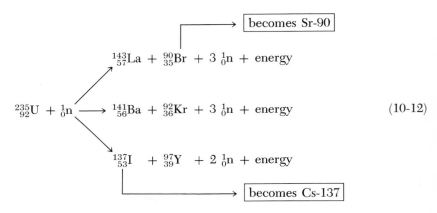

$$\boxed{\text{becomes Sr-90}}$$

$$^{143}_{57}\text{La} + {}^{90}_{35}\text{Br} + 3\,{}^{1}_{0}\text{n} + \text{energy}$$

$$^{235}_{92}\text{U} + {}^{1}_{0}\text{n} \longrightarrow {}^{141}_{56}\text{Ba} + {}^{92}_{36}\text{Kr} + 3\,{}^{1}_{0}\text{n} + \text{energy} \qquad (10\text{-}12)$$

$$^{137}_{53}\text{I} + {}^{97}_{39}\text{Y} + 2\,{}^{1}_{0}\text{n} + \text{energy}$$

$$\boxed{\text{becomes Cs-137}}$$

Example 10-5 illustrates how identification of one fission product permits prediction of another fission product.

Figure 10-8 schematically indicates the chain reaction possible with ^{235}U. If you have a pure sample of ^{235}U, and if all fission reactions produce three neutrons, and if no neutrons escape, then the three emitted neutrons of the first reaction could cause fission of three ^{235}U atoms, each of which would produce three neutrons or a total of nine neutrons. These nine neutrons could in turn cause fission of nine ^{235}U atoms leading to 3×9 or 27 neutrons, which in turn with 27 ^{235}U atoms could produce 81 (3×27) neutrons; the 81 produce 243; the 243 produce 729; and so on.

When we consider that an atom is mostly space, it is clear that the chance of a neutron colliding with a nucleus is small. Hence, there must be enough atoms, that is, enough mass, with a geometry that allows a minimum number of neutrons to escape. Some neutrons can escape from a sample of uranium, that is, from atoms near the surface, without producing a fission reaction. If too many escape, a chain reaction cannot occur. To decrease the chance of loss and to increase the chance of neutron capture, the mass of the uranium must be increased to some quantity with minimum surface area. The optimal configuration for minimal neutron loss

EXAMPLE 10-5

If we know fission of ^{235}U yields $^{141}_{56}$Ba and that three neutrons° are produced, what is the other product of this particular fission reaction? The equation can be written as

$$^{235}_{92}U + ^{1}_{0}n \longrightarrow ^{141}_{56}Ba + 3\,^{1}_{0}n + ^{A}_{Z}X$$

From the subscripts

$$92 + 0 = 56 + (3 \times 0) + Z$$
$$92 = 56 + Z$$
$$Z = 36 \text{ (atomic number 36 is krypton)}$$

From the superscripts

$$235 + 1 = 141 + (3 \times 1) + A$$
$$236 = 144 + A$$
$$A = 236 - 144 = 92 \text{ (mass number)}$$

Thus, the complete equation including the other product $^{92}_{36}$Kr is

$$^{235}_{92}U + ^{1}_{0}n \longrightarrow ^{141}_{56}Ba + ^{92}_{36}Kr + 3\,^{1}_{0}n$$

° *Note:* Experimentally, one usually is concerned only with element identification, for which the number of neutrons produced is not needed.

is a sphere that has minimum surface area for a given mass, and a critical mass of 10 kg. The critical mass of ^{235}U is the mass of uranium below which the chain reaction will not sustain itself. For ^{235}U, the critical mass is a sphere 16 cm (about 6 in.) in diameter.

The first successful self-sustaining nuclear chain reaction was achieved on December 2, 1942, in a squash court under the football stadium of the University of Chicago, as a result of a project under the direction of Enrico Fermi. In 1941, most of Fermi's Columbia University research group (Fermi had left fascist Italy several years before for the protection of his wife) moved to the University of Chicago where Arthur H. Compton (Nobel Prize, 1927) was coordinating the secret Manhattan project (1942–1945). During these years, nearly all nuclear scientists in the Allied World were engaged in the Manhattan project to achieve practical utilization of the great energy release from the nucleus. Militarily, the practical goal was an **atomic bomb** (really a nuclear bomb). By July 1945, enough ^{235}U had been obtained to make an explosive device, which was tested at Alamogordo, New Mexico (Figure 10-9). Then, in August of 1945, a uranium fission bomb was dropped on the Japanese city of Hiroshima (Figure 10-10), destroying 60% of the city, killing 78,000 men, women, and children, and injuring 37,000 more. Three days later, a plutonium fission bomb was

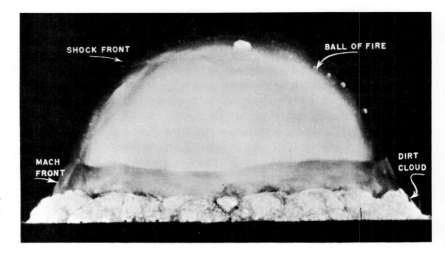

Figure 10-9. The ball of fire touching the ground and the shock front soon after the breakaway in the "trinity" test at Alamogordo, New Mexico, 1945. The mach front due to reflection of the shock wave and the dirt cloud are indicated. [Courtesy of the U. S. Atomic Energy Commission.]

dropped in Nagasaki, producing 100,000 casualties. World War II was concluded shortly thereafter.

In the nuclear bomb, the uranium is kept in separate or several masses, all of which are of a subcritical size or mass. At the time of detonation, all these masses are driven together to make a total assembled mass, which exceeds the critical mass. The liberation of very large quantities of energy heats gases in the vicinity almost instantly to extremely high tempera-

Figure 10-10. Hiroshima—a few days after . . . . The Agricultural Exposition Hall is on the left. The bridge behind is Matoyosa Bridge. [Courtesy of the U. S. Atomic Energy Commission.]

tures—over a million degrees. The very sudden expansion of very hot gases is the explosion that scatters fission reaction fragments and unused material over a wide area.

Most of the fission fragments are unstable, as they typically have a too-high neutron to proton ratio. To correct this imbalance, these radioactive atoms give off beta particles frequently accompanied by gamma rays. These emissions, plus the light and heat, account for the hazards associated with atomic (nuclear) bombs. Interestingly, there is no danger of such an explosion in the mineral deposits of the earth. The richest ores (pitchblende) of uranium typically contain less than 0.2% uranium, any amount of which ore can never constitute a critical mass. Further, of the uranium naturally occurring, only one atom in 140 atoms is a ^{235}U, which is fissile. Remember that ^{238}U is not fissile.

Energy from Nuclear Transformations

Let us return now to the source of energy in nuclear transformations. That particles on emission from the nucleus move away from the nucleus is an indication of kinetic energy, energy of motion. When a collection of particles is moving rapidly, we find that this is manifested as a high temperature reading. A sample of a radium salt maintains a temperature a few degrees above the temperature of the surroundings, which means that an energy releasing process is spontaneously occurring within the sample. More specifically, 1 g of radium metal liberates approximately 580 J/hr from radioactive disintegration. Even after 1622 years, the half-life of radium, the sample still liberates 290 J/hr. Well, why the excitement in the 1930s when fission was proposed?

You recall that we have previously designated the mass numbers of a proton and a neutron as being equal. Earlier we noted that the relative scale of atoms (atomic weights) is based on the arbitrary assignment of $^{12}_{6}C$ as having a mass of exactly 12.000000. Further, you recall that most elements do not have exactly integral atomic weights. Also, we accounted for deviations from exact integral numbers on the basis that most elements in nature are in fact mixtures of isotopes. That isotopic mixtures are most significant is undisputed. However, let us go a step beyond. More careful measurement of the relative masses of isotopes, electrons, protons, and neutrons was made possible at about the same time as the development of nuclear chemistry and physics. Precise measurements of atom and particle masses show that differences from integral numbers occur in nearly all cases! Precise data have been taken to several decimal places (see Table 10-3).

Now you know that we consider the $^{12}_{6}C$ atom to be made up of six protons, six neutrons, and six electrons. If you assume that this is the case,

Table 10-3. Masses of Particles on the Carbon-12 Scale (Atomic Weight Scale)

Particle	Symbol	Relative Mass
carbon-12	$^{12}_{6}C$	12.0000000000000 ... (by definition)
proton	$^{1}_{1}H$	1.00728
neutron	$^{1}_{0}n$	1.00867
electron	$^{0}_{-1}e$	0.00055

you calculate (Example 10-6) the expected mass of $^{12}_{6}C$ to be 12.0990. The actual mass of $^{12}_{6}C$ turns out to be 0.0990 atomic mass units (amu) less than what you calculated. Why?

In general, isotopic masses do not quite add up to the sum of the elementary particles. The missing mass is known as the **mass defect** and is a very small fraction of the total mass. Let us repeat: These differences would not be observable without the masses of isotopes being measured to four and five significant figures. The solution to the meaning of mass defect was actually available when mass defects were discovered. In 1905, Albert Einstein (Nobel Prize, 1921) had discovered, as a consequence of his special theory of relativity, that mass and energy are interchangeable according to the equation

$$E = mc^2$$

wherein E is energy (ergs), m is mass (grams), and c is the speed of light (3×10^{10} cm/sec). Although we will not use this equation for calculations, inasmuch as use of some of the conversion factors take considerable practice, nonetheless, we will use data obtained by use of the Einstein equation.

This mass difference (mass defect) between the sum of the masses of the elementary particles and the actual mass of an atom represents the

---

EXAMPLE 10-6

(a) What is the calculated $^{12}_{6}C$ mass if the atom is made up of free protons, neutrons, and electrons?

Mass of 6 protons   = 6 × 1.00728 = 6.04368
Mass of 6 neutrons = 6 × 1.00867 = 6.05202
Mass of 6 electrons = 6 × 0.00055 = 0.00330
                    Total = 12.09900

(b) How does the calculated mass compare with the measured mass (in this case the defined mass)?

12.09900 − 12.000 = 0.09900 mass units too large

that is, 0.09900 mass unit isn't there!

energy needed to pull the elementary particles apart and completely out of the atom. For lack of a better term, this energy is known as the **binding energy.** The energy released (equivalent to the mass defect) when particles combine in atoms is the same as the energy that must be put in to break them apart again. For every gram of matter converted to energy, 2.15×10^{10} kcal (21,500,000,000 or 21.5 billion kcal) of heat are released; thus, for the formation of 12 g of $^{12}_{6}C$, energy corresponding to 0.09900 g of mass have been released, or 2.13×10^{9} kcal. Actually, in order to compare atoms, we usualy discuss the binding energy per nuclear particle (nucleon).

Both very light atoms and very heavy atoms have significantly smaller binding energies per nuclear particle (nucleon) than do atoms of interme- diate mass (mass numbers $Z = 50$ to $Z = 130$) such as iron (see Figure

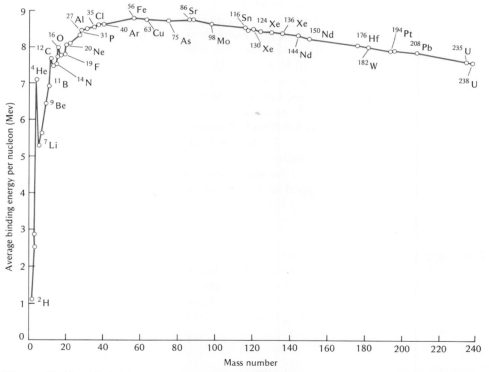

Figure 10-11. Variation of binding energy (mass defect) with mass number. Note that the binding energy per nucleon (not total binding energy) ranges from about 1 Mev for deuterium ("heavy" hydrogen) ^{2}H, with its two nucleons to a bit less than 9 Mev for iron. The elements to the left of the peak can increase their binding energies per nu- cleon only by fusion, those to the right of the peak only by fission.

10-11). Thus, any process that converts very heavy atoms to atoms of intermediate mass will release large quantities of energy. Such a process is fission. Also, any process by which the lightest atoms can be combined (fused together) to produce atoms of intermediate mass will be a highly energy releasing process (see Energy from Fusion). If middle Z atoms were to be either fissioned or fusioned, there would be no release in energy because, the fact is, there is no more stable atom or atoms or collection of particles that can be formed. Atoms of the size of iron are not fissionable and are not fusible; they are stable. Such atoms of intermediate mass are not sources of nuclear energy.

Thus, let us return to ^{235}U fission. A very small amount of mass is lost when 1 g of ^{235}U undergoes fission. The small amount of mass lost [approximately 1×10^{-3} g or 1 mg (milligram)] corresponds to 20,000,000 kcal of energy. For comparison, the heat of combustion of 1 g of coal is a mere 7.8 kcal. In other words, the loss in mass of 1 mg when 1 g of ^{235}U undergoes fission produces as much energy as the combustion of about 2,500,000 g of carbon (2.5 metric tons of carbon) or 5500 lb of carbon. Because coal is not pure carbon, we calculate that when we use an average heating value of coal of 13,000 BTU/lb 2.7 metric tons[4] of coal are equivalent in terms of heat to the mass loss when 1 g of ^{235}U undergoes fission.

Thus, the small decrease in mass occurring during the fission reaction of ^{235}U releases a tremendous amount of energy, mostly as fast moving particles. Remember that thermal or kinetic energy consists of fast moving particles. The energy content in uranium fission is further dramatized when you realize that the fission fragments from 1 kg of ^{235}U weigh 999 g. Only 0.1% of the original mass is actually converted to energy. The fission bomb dropped on Hiroshima contained approximately 1 kg of ^{235}U.

Energy from Fusion [1]

No discussion of energy from nuclear reactions can be complete today without some reference to the other end of the scale of atoms; namely, **fusion** of very light atoms to give heavier atoms. In 1932, J. D. Cockcroft and E. T. S. Walton bombarded $^{7}_{3}$Li with protons that had been accelerated, resulting in reaction (10-13).

$$^{1}_{1}\text{H} + ^{7}_{3}\text{Li} \longrightarrow ^{4}_{2}\text{He} + ^{4}_{2}\text{He}$$
$$^{1}_{1}\text{H} + ^{7}_{3}\text{Li} \longrightarrow 2\,^{4}_{2}\text{He} \tag{10-13}$$

In the study of this reaction, the most startling observation is that the total kinetic energy of the emitted alpha particles had 60 to 70 times the kinetic energy of the bombarding proton. The small decrease in mass in

[4] The metric ton is 1000 kilograms (2204 lb).

the process, when converted to energy according to Einstein's equation, is quite accurately accounted for by the increased kinetic energy observed from the particles produced by the reaction. Please note in this case, as in the fission reactions discussed earlier, that complete annihilation of the atom does *not* take place; only a fractional loss of the mass has occurred (that is, the energy equivalent of the mass defect). The energy released is spectacular. In the above reaction, if 1 g of ^7Li is used, some 516,000,000 kcal (5.16×10^8 kcal) of energy is produced, which translates into 60,000 kwh.

The prospects for successful nuclear fusion recall the myth of Prometheus, who stole fire from the gods and then suffered for his deeds by being chained with a vulture eating at his vital parts. The achievement of nuclear fusion (see Figure 10-12) on earth brings to earth what we believe to be the energy-producing reaction of the sun and other stars. Have we "stolen" the fire of the gods as did Prometheus? As we discuss energy from fusion, both in this chapter and the next two, you should ask yourselves about "suffering" as a consequence of nuclear fusion.

In 1939, Hans A. Bethe proposed that the enormous amounts of energy radiated from the sun and the stars derive from the fusion of hydrogen (^1_1H) into helium (^4_2He). This is consistent with what we know about the composition of the sun from spectral measurements; namely, that hydrogen

Figure 10-12. A thermonuclear (H-bomb) detonation in the Pacific test area on February 24, 1954. [Lookout Mountain Air Force Station, courtesy of the U. S. Atomic Energy Commission.]

and helium are the principal atoms present in the sun. Equation (10-14) summarizes what we believe to be the principal reaction of the sun.

$$4\,{}_1^1H + 2\,{}_{-1}^0e \longrightarrow {}_2^4He + energy \tag{10-14}$$

The key to successful nuclear fusion is an extremely high temperature, for example, 20–100 million °C. Atoms and molecules, as we believe them to exist, are stripped of extra nuclear electrons at such elevated temperatures, and the hot assembly of particles has extremely high kinetic energy. This state of matter is known as the plasma state. Thus, when one says he is working in plasma physics or plasma engineering, he is dealing with this most unusual state of matter. At these extremely high temperatures, the particles have very high kinetic energies—high enough so that positive nuclei can come together and fuse despite their electrical repulsion (recall that like charges repel). Most of the possible combinations of light atoms and their isotopes have been examined as to the energy releasing potential. Some fusion equations are recorded in equations (10-15).

$$\begin{align}
{}_1^2H + {}_1^2H &\longrightarrow {}_2^3He + {}_0^1n + energy \tag{10-15a}\\
{}_1^2H + {}_1^2H &\longrightarrow {}_1^3H + {}_1^1H + energy \tag{10-15b}\\
{}_1^2H + {}_1^3H &\longrightarrow {}_2^4He + {}_0^1n + energy \tag{10-15c}\\
{}_1^2H + {}_3^6Li &\longrightarrow {}_2^4He + {}_2^4He + energy \tag{10-15d}
\end{align}$$

In the deuterium/deuterium (${}_1^2H/{}_1^2H$ or D/D) fusion reaction (equation 10-15a), for each gram of deuterium undergoing reaction some 1.9×10^7 kcal of energy are released, as in Example 10-7. When a similar calculation is performed for the deuterium/tritium (${}_1^2H/{}_1^3H$ or D/T) fusion reaction (equation 10-15c), some 18×10^7 kcal of energy are released per gram of tritium transformed, or more than 9 times as much energy as the ${}_1^2H/{}_1^2H$ fusion reaction produces.

Early fusion work was directed towards the production of an uncontrolled fusion in the H-bomb (Figure 10-12). In order to achieve the very high temperatures needed for the fusion of light atoms, a small fission explosion was set off in which enough heat was generated to trigger a fusion reaction before the bomb blew apart. As we discuss in the next chapters, the key problem yet to be solved by man is the generation and maintenance of a controlled fusion reaction. However, if this problem can be solved, the energy available from fusion (of deuterium) is almost limitless.

Conclusion

Let us summarize so that no mistake can be made about the significant differences between nuclear reactions as discussed in this chapter and the ordinary chemical reactions of Chapters 5 and 6. For convenience, these are organized in Table 10-4.

EXAMPLE 10-7

(a) Calculate the mass loss when 1 g of deuterium (^2_1H) undergoes fusion, equation (10-15a).

For convenience, we will first calculate the mass change for 1 mole of atoms, so we can use precise atomic weights.

$$^2_1H + ^2_1H \longrightarrow ^3_2He + ^1_0n + \text{mass lost as energy}$$

$$\text{mass lost} = \text{mass of "left"} - \text{mass of "right"}$$

$$M_{lost} = \underset{^2_1H}{(2.014102} + \underset{^2_1H}{2.014102)} - \underset{^3_2He}{(3.016030} + \underset{^0_1n}{1.008665)}$$

$$= 4.028204 - 4.024695$$

$$= 0.00351 \text{ g}$$

This is the mass lost from 4.028 g of deuterium. Now we can calculate how much mass is lost when 1.0 g of deuterium reacts.

$$\frac{M_{lost}}{\text{amount } ^2_1H} = \frac{0.00351 \text{ g}}{4.028204} = 0.000874 \ (8.74 \times 10^{-4}) \text{ g lost per gram } ^2_1H$$

(b) Calculate the energy released per gram of deuterium undergoing fusion. (*Note:* 1.0 g of matter completely converted to energy becomes 2.15×10^{10} kcal.)

Thus, for 1 g of ^2_1H undergoing reaction (equation 10-14a)

$$\frac{8.74 \times 10^{-4} \text{ g mass lost}}{1 \text{ g } ^2_1H} \times \frac{2.15 \times 10^{10} \text{ kcal}}{1 \text{ g mass lost}} = 1.9 \times 10^7 \text{ kcal per gram } ^2_1H$$

Table 10-4.
Comparison of Nuclear
Reactions with
Chemical Reactions

	Nuclear Reaction	Chemical Reaction
Atom	May be changed into another atom during a nuclear reaction	Always retains its identity in chemical reactions
Isotope	Different isotopes of an element have different properties as to nuclear reactions.	Different isotopes of an element have practically identical chemical properties
Reactivity	Nuclear reactivity is independent of the state of chemical combination, i.e., free elements or compounds such as oxide, chloride, etc.	Chemical behavior depends on the element's state of combination
Energy	Nuclear reactions involve energy changes that are very large	Chemical reactions involve energy changes that are very much smaller than those in nuclear reactions

The search for the means of obtaining nuclear energy and of putting it to practical use was motivated by many more forces than just a weapon for use in World War II. The building of bombs was a conclusion to a sequence of events that began with the discoveries of Becquerel, the Curies, Einstein, Rutherford, and others, around the turn of the century. The intensive study of nuclear transformations laid the ground work which wartime research was largely built on, and built on to an enormous scale. Actually, the war brought about rather little new fundamental research, that is, few additions to scientific knowledge were made. Nuclear scientists were much too busy in the wartime period putting to practical use the knowledge they already had.

Put simply, nuclear scientists and technologists were at one time intimately involved with warfare. Mere saying that "scientists are guilty" of involvement in weapons development would ignore the subsequent harnessing of nuclear energy for peaceful purposes. The very properties of nature that resulted in bombs and make radiation a significant hazard (Chapter 12) are the same properties that lead to the use of radioactive isotopes in medicine and to a means of generating electric power for society's benefit.

Questions

1. Describe the mass and charge of the particles important to atomic structure; namely, the electron, the proton, the neutron, and the alpha particle.

2. Explain what is meant by the following terms:
   (a) isotopes
   (b) beta ray
   (c) half-life of a radioactive isotope
   (d) radioactive series
   (e) bombardment reaction
   (f) fission
   (g) fusion
   (h) mass defect
   (i) binding energy

3. What is a probable explanation for what happens to excess neutrons when an atom decays by beta emission?

4. Why does ^{235}U not undergo self-sustaining nuclear fission in nature?

5. Both fission and fusion can involve release of energy. Explain.

6. When would you say the Nuclear Age began? Why?

7. Would iron make a good source of nuclear energy? Why or why not?

8. Discuss the possibility that some alchemist may really have succeeded in transforming one element into another but has not been given proper credit.

9. What produces the tremendous energy of a fission neutron? In what form is most of this energy?

10. What is meant by a chain reaction?

11. Can the mass of a nucleus be converted completely to energy by either fission or fusion? Explain.

Numerical Exercises

1. Indicate the number of protons, neutrons, and electrons in each of the following neutral atoms:
   (a) $^3_1$H (tritium)   (e) $^{226}_{88}$Ra
   (b) $^{14}_6$C   (f) $^{83}_{36}$Kr
   (c) uranium-235   (g) $^{239}_{94}$Pu
   (d) uranium-238

2. A radon atom has 86 protons and 136 neutrons. What is its atomic number? What is its mass number?

3. Tritium decays to form another atom and a beta particle. Write the equation.

4. An isotope of hydrogen, tritium, is produced in the upper atmosphere by collision of a neutron with a $^{14}_7$N nucleus. The other product of the collision is $^{12}_6$C. Complete the equation and identify the tritium nucleus.

5. Complete the following nuclear equations, which represent radioactive decay processes. Use appropriate notations and symbols.

   (a) $^{211}_{85}$At $\longrightarrow$ $^4_2$He + ?

   (b) $^{208}_{81}$Tl $\longrightarrow$ $^0_{-1}e$ + ?

   (c) $^{228}_{88}$Ra $\longrightarrow$ $^{228}_{89}$Ac + ?

   (d) ? $\longrightarrow$ $^4_2$He + $^{222}_{86}$Rn

   (e) $^{12}_5$B $\longrightarrow$ $^0_{-1}e$ + ?

   (f) beta emission of $^{60}_{27}$Co

6. Complete the following nuclear equations which represent transmutations caused by bombarding a target with neutrons or an accelerated particle.
   (a) $^{25}_{12}$Mg + $^4_2$He $\longrightarrow$ $^1_1$H + ?
   (b) $^{209}_{83}$Bi + $^2_1$H $\longrightarrow$ $^{210}_{84}$Po + ?
   (c) $^1_1$H + $^{35}_{17}$Cl $\longrightarrow$ $^4_2$He + ?
   (d) $^{24}_{12}$Mg + ? $\longrightarrow$ $^{27}_{14}$Si + ^1_0n
   (e) $^{238}_{92}$U + $^{12}_6$C $\longrightarrow$ $4\,^1_0n$ + ?
   (f) $^{238}_{92}$U + 17$\,^1_0n$ $\longrightarrow$ $^{255}_{100}$Fm + ? $^0_{-1}e$
   (g) ? + $^{11}_5$B $\longrightarrow$ $^{257}_{103}$Ln + 4$\,^1_0n$

7. (a) Consider the radioactive series that starts with $^{235}_{92}$U and ends with $^{207}_{82}$Pb. How many alpha particles are given off in this series? How many beta particles?
   (b) $^{232}_{90}$Th becomes $^{208}_{82}$Pb. How many alpha particles are given off? How many beta particles?

8. The ^{14}C/^{12}C ratio in a bit of charcoal from a cave is one fourth that found in a freshly chopped piece of wood. How long ago did the tree that yielded the charcoal die?

9. Radium with a half-life of 1622 years emits alpha particles. What portion of an original mass of radium remains after 3244 years? After 6488 years?

10. The half-life of $^{70}_{31}$Ga is 20 min. How long will it take for three quarters of a sample of this isotope to decay? seven eighths.

11. A radioactive isotope decays with a certain half-life. A sample is considered to be "radioactively all gone," that is, the radioactivity reduced below a significant amount, after a period of time equivalent to 20 half-lives. Justify the statement.

12. (a) The half-life of $^{218}_{84}$Po is 3 min. How much of a 2-g sample of this isotope remains after 15 min?

    (b) Suppose you wanted to buy some of this isotope and it requires 5 hr for it to reach you. How much should you order if you want to use 0.01 g?

13. 1 mole of ^{235}U undergoing fission releases 4.6×10^9 kcal: 1 mole of carbon (coal) burning releases 94 kcal. There are 454 g in 1 lb, and 2000 lb in 1 ton (short ton).
    (a) How much heat is produced by burning 1 ton of coal? (Assume that coal is 100% carbon.)
    (b) How much heat is produced by the fission of 1 lb of ^{235}U?
    (c) How many tons of coal are equivalent in energy to 1 lb of ^{235}U?

14. The carbon in living systems contains sufficient $^{14}_{6}$C to produce 16 radioactive events per minute per gram of carbon. If the human body is about 5% carbon by weight, show that a 100-lb person experiences over 50 million radioactive events in a day. How many radioactive events occur in a 150-lb person per day?

15. The half-life of ^{238}U is 4.5×10^9 years. Lead is the stable atom formed as a result of the disintegration series of ^{238}U. What will be the atomic ratio of uranium to lead in a rock that is 4.5×10^9 years old? (This assumes that all of the metal, lead, was initially uranium.)

16. In the chapter, note is made that $^{90}_{35}$Br decays to $^{90}_{38}$Sr by beta emission. Write the equations for the transformation. Similarly, write equations for $^{137}_{53}$I decaying to $^{137}_{55}$Cs.

Reference

[1] W. C. Gough and B. J. Eastlund: *Sci. Amer.*, **224**(2): 50 (1971).

Peaceful Utilization of Nuclear Energy

11

Harnessing thermal energy deriving from nuclear transformations stands as one of the most significant scientific and technological achievements of the twentieth century. In Chapter 10, we discussed selected crucial aspects of the behavior of unstable or radioactive nuclei. The study of these nuclei has contributed greatly to our present day understanding of the structure of the atom's nucleus. However, the interest in nuclear chemistry and nuclear physics has two bases—one scientific and one technological. Scientifically, there is the desire to discover more about the structure of matter. Technologically, there is the desire to harness the energy of the atomic nucleus for man's use. The latter is the subject of this chapter.

This chapter and the one following are particularly important and relevant for you now. Is nuclear power a godsend or a devil in disguise? Certainly, most of you are aware that the environmental impact of nuclear power has become an area of great concern and is

323

actually becoming a technology within itself. This technology has already had an impact on our lives, and the prospect is that in the future the impact will be even greater. Chapter 10 delineated selected fundamentals of nuclear chemical physics. As we examine in this chapter some of the relevant facts about power from nuclear fission phenomena and then in Chapter 12 the radioactive waste product problem, we believe that you will develop a knowledge on which you can build as you develop your own viewpoint of this large and significant source of energy for our society.

In this chapter, we discuss the peaceful utilization of nuclear energy with particular reference to the use of nuclear processes in the generation of electrical energy. We focus on the principal features of nuclear-fueled power plants including the nature and source of the nuclear fuels, the means whereby the nuclear chain reaction is controlled, and the materials of construction used to protect the environment from the harmful radioactive emissions from nuclear processes. We then summarize the current state of production of electrical power by nuclear fission processes and project probable developments with particular reference to the nuclear breeder reactor. The prospects for controlled nuclear fusion reactions for peaceful purposes, including the generation of electricity, are also presented.

Controlled Nuclear Fission

Soon after World War II, by Act of Congress (the McMahon Act) the Atomic Energy Commission (A.E.C.) was established. This act placed the huge research and development program of the Manhattan Project and all its plants and locations (Hanford, Washington; Oak Ridge, Tennessee; Los Alamos, New Mexico, and so on) under civilian control. Although military work continued, that is, the H-bomb was developed and tested at Bikini atoll in the South Pacific ocean in November of 1952, the focus of nuclear science and technology began to be shifted towards the development of controlled nuclear reactions for the production of power.

A nuclear reactor is a machine for containing and controlling the release of energy from the nuclear chain reaction in contrast to the instantaneous release in an explosion, that is, a bomb. The fuel that is consumed consists of nuclei that undergo fission after capture of slow neutrons. The atoms that are known to be fissile, that is, to undergo fission, are uranium-235 ($^{235}_{92}U$), uranium-233 ($^{233}_{92}U$), and plutonium-239 ($^{239}_{94}Pu$). The energy of the fission reaction appears primarily (80%) as heat energy or kinetic energy of the fission products, whereas the remainder (20%) is associated with the neutrons and other radioactivity. Thus, at the site of the nuclear reaction, a large amount of heat is available as kinetic energy of atoms. The problem then is how to put this heat to practical use; or more specifically, how to transform this heat energy—kinetic energy—into electrical energy.

Figures 11-1A and 11-1B show the essential features of two types of nuclear power plants. The boiling water reactor (BWR) in Figure 11-1A was the only type on-line until the early 1970s. You should recognize from our earlier discussions of fossil-fuel fired power plants (Chapter 7) that, as drawn, everything to the right of the boiler is the same as in a fossil-fuel

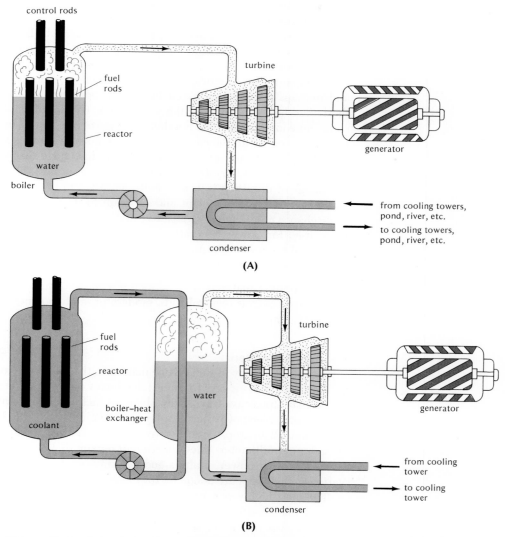

Figure 11-1. Schematic drawing of nuclear power plants. **(A)** Boiling water reactor (BWR). **(B)** Pressurized water reactor (PWR).

fired power plant. The major difference between the two systems is in how the water in the boiler is heated. In the fossil-fuel fired plant the boiler is heated by burning coal, oil, or natural gas in a unit that places most of the heat of combustion at the boiler. In the nuclear power plant, heat from the nuclear chain process, occurring in the reactor core, is used to boil the water in the boiler. The reactor core, in the BWR, is located directly in the boiler. Thereafter, the diagram is really the same as that for the fossil fuel plant.

The pressurized water reactor, PWR, in Figure 11-1B, is essentially a two-stage system. A coolant, for example, water, carries the heat of the reactor core to the boiler, also known as a heat exchanger, and then is recirculated to the reactor core. Future technology will see liquid sodium and perhaps helium used as coolants. The operation of the PWR after the boiler–heat exchanger is the same as the BWR or fossil-fuel plant.

Because in practice there is a slight leakage of radioactive materials from the reactor core into the coolant, one of the major advantages of the PWR is that such materials are contained in the first stage (a closed system). However, the PWR system clearly is more complex. Thus, the major advantage of the BWR is its simplicity; however in the BWR, because the water is used both as coolant and as steam to drive the turbine, there is some chance of leakage of radioactive material (see Chapter 12). The steam cycle used for power generation is the same as for the fossil fuel plant, so we will focus our attention on three key features important for reactor operation and safety. These are as follows:

1. Construction materials.
2. Reactor controls (control rods, moderator, and coolant).
3. Nuclear fuels.

Construction Materials

Early in the study of radioactive materials, the difference in penetrating power of the various rays from unstable heavy nuclei was determined. We found that alpha rays travel only about an inch in air and are stopped by as little as a piece of paper or a layer or two of skin tissue. Beta rays are more difficult to stop. Beta rays travel a few feet in air and are stopped by a 5-mm (millimeter) thick sheet of aluminum or 1 in. of wood. Because the penetrating power of alpha and beta rays is small, the reactor and its various metal parts are adequate to prevent any external exposure hazard.

The most highly penetrating rays, and hence the most difficult to stop, are gamma rays. Gamma rays travel hundreds of feet in air and require several inches of lead or several feet of concrete to be stopped. Because gamma radiation comes from the natural decay of ^{235}U, from the fission reaction of ^{235}U, and from the subsequent decay of the radioactive fission

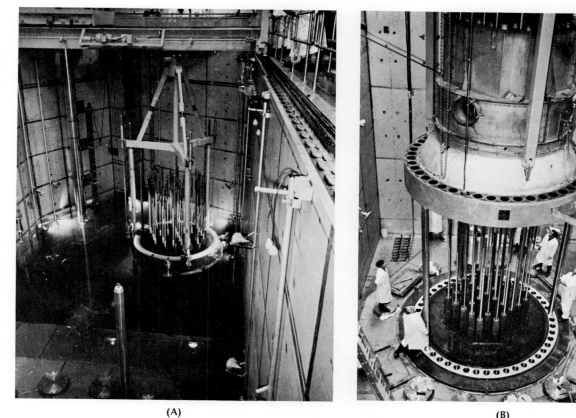

Figure 11-2. **(A)** Reactor vessel with "head" off. During the initial fuel loading at Point Beach Nuclear Plant, Unit 1 (PWR), in October 1970, the fuel manipulator crane (left) moves a fuel assembly from the fuel transfer tube to the reactor vessel. **(B)** The 40-ton reactor vessel head of Unit 1 is shown being lowered over the control rod drive shafts after the core has been loaded full with 121 fuel assemblies. [Courtesy of U. S. Atomic Energy Commission. Photographs, Wisconsin Electric Power Co.]

products (see Chapter 12), there is considerable external exposure hazard. Thus, *massive shielding of nuclear reactors is absolutely necessary.* The gamma rays, which are like x rays, only stronger, can go through most anything and can cause "radioactive damage" to the people, plants, and animals of the biosphere.

The sheer mass of the necessary shielding limits our use of nuclear energy to stationary power generators or such large mobile units as can transport a heavy, massive, shielded nuclear reactor. Safe nuclear flashlight batteries or backpack generators are not possible. The science and technology of shielding is sufficiently well known today that it is possible to have a negligible level of radiation escaping a nuclear reactor during normal

Figure 11-3. Davis-Besse nuclear plant on Lake Erie near Port Clinton. An outside view of the shielding structure which contains a nuclear reactor vessel, (at the right). The cooling tower on the left is one of the largest wet-draft towers in the world.

operation (see later section on nuclear accidents and Chapter 12 on radioactive wastes).

More specifically, in the typical design of the nuclear reactor, the reactor vessel is a massively constructed steel tank (Figures 11-2A and B) that contains the fuel assembly (reactor core). The shield, sometimes known as the reactor pit, that completely encloses the reactor is made of plastics, several different metals, and many feet of concrete. Then, surrounding the entire shielded reactor installation is a large spherical or cylindrical shell made of steel or of reinforced or prestressed concrete lined with steel. This last structure is what you observe as you look at a nuclear power plant (see Figure 11-3).

Reactor Controls

In the previous section we addressed ourselves to one aspect of the control of the nuclear reaction system as we discussed the massive shielding necessary to protect the biosphere from radiation. Put another way, the shielding controls radiation, particularly gamma radiation, emanating from the fuel, the reaction, and the fission reaction products. Let us now consider three other aspects of control: the coolant, the moderator, and the control rods.

The **coolant** provides the means whereby the heat or kinetic energy produced in the nuclear chain reaction is removed from the reactor core. In most nuclear power plants, the coolant is water. Along with the control rods (see below), the rate of flow of coolant controls the operating temperature of the reactor core. The nuclear fuel, in the form of small pellets, is placed in long, stainless steel or zirconium clad cylinders (Figures 11-4 and 11-5). The cylinder walls prevent the coolant from making direct contact with the fuel and, of course, any fission reaction products. Typically, in the BWR, the operating high temperature is about 285°C (545°F),

Figure 11-4. Uranium dioxide pellets ready for loading into fuel rods. [Courtesy of U. S. Atomic Energy Commission. Photograph Batelle-Northwest.]

Figure 11-5. Uranium fuel rods before insertion in a power reactor. [Courtesy of U. S. Atomic Energy Commission. Photograph, Samuel A. Musgrave, Atomic Industrial Forum.]

whereas in the PWR, the operating high temperature is about 315°C (600°F), which necessitates about twice as much pressure as in the BWR. Despite the fact that in both systems the coolant is completely recycled, there is always the possibility of leaks. In the BWR, coolant leakage at the turbine can occur. The design of the PWR considerably reduces the leakage problem and, hence, the resulting radiation hazard.

In both the PWR and the BWR where water serves as the coolant, water also has a moderating effect on the neutrons essential for the nuclear chain reaction. In the fission reaction, most of the neutrons are expelled with a high velocity. These neutrons, on collision with molecules containing light atoms, such as the hydrogen of water or perhaps of some hydrocarbons, lose energy and are slowed down. Sufficient collisions have the effect of slowing the neutrons until they reach a velocity known as the thermal velocity, which is typical of atomic and molecular motion at that temperature. Such thermal neutrons are more easily captured by uranium-235 than are the fast neutrons initially produced and, hence, are more effective for producing fissioning. Thus, the **moderator** is an essential component of the reactor. The original Chicago reactor built by Fermi had no water and used graphite (carbon) as the moderator. In general, molecules made up of light atoms moderate the speeds of neutrons.

Finally, we turn to the real control or the "on-off" switch of nuclear reactors—the **control rods** (see Figure 11-2). Too many neutrons would result in an uncontrolled chain reaction, whereas too few neutrons would slow down the chain reaction and produce too little power. However, even if the reaction were to "run away," there would be no nuclear explosion as in the atomic bomb, because there would not be a critical mass (since the fuel is in separate rods and is only slightly enriched).

The number of neutrons is rather precisely controlled by the use of a material that absorbs or captures neutrons more efficiently than uranium. Cadmium and boron are such substances. Complete insertion of the control rods amongst the fuel rods quenches the chain reaction or "turns off" the nuclear reactor. Most nuclear reactors are designed so that the control rods are raised to permit the neutron-initiated chain reaction to proceed. This means simply that in the event a power loss occurs, or any malfunction occurs that might cause overheating of the reactor core, the control rods drop back down into place, stopping the reactor—an important safety feature.

Nuclear Fuels

As we move to a discussion of the fuels used in nuclear reactors, we encounter one of the major long term problems associated with our consideration of this method of generation of electrical power. The other major problems that we will consider are the waste heat (Chapter 13) and the formation and handling of radioactive wastes (Chapter 12).

Table 11-1. United States Uranium Resources (1971) (as thousands of tons of U_3O_8)

Price U_3O_8 ($/lb)	Reasonably Assured Deposits	Estimated Additional Deposits
8.00	243	490
10.00	300	680
15.00	470	1040
30.00	640	1660
50.00	4640	3660
100.00	8640	8660

SOURCE: Atomic Energy Commission, *WASH-1201, LMFBR Demonstration Plant Program,* 1972, p. 232.

Of the heavy atoms found in ores, uranium-235 is the only fissile one available in substantial quantities. A uranium-rich sample of pitchblende typically contains 2 to 5 lb of U_3O_8 (uranium oxide) per ton. The richest deposits of uranium ores in the United States have been found on the Colorado Plateau. Figure 11-6 also shows other regions where uranium has been found. Most of the other regions have such a low percentage of uranium that the cost of recovery is quite high (Table 11-1).

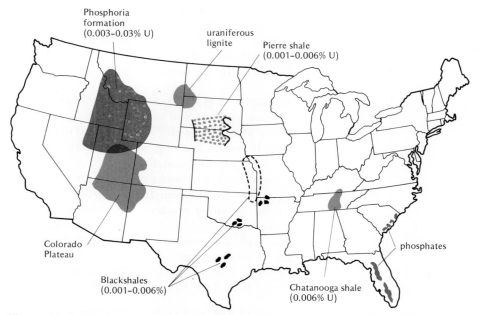

Figure 11-6. The major potential sources of uranium in the United States. [After M. K. Hubbert: *Energy Resources, Pub. 100-D.* Committee on Natural Resources, National Academy of Sciences–National Research Council, Washington, D. C. 1962.]

As a result of the treatment of large amounts of ore to recover the uranium as uranium oxide, much finely divided solid remains, in fact the mass of material is virtually unchanged. Huge piles of these sandy "waste" materials are left in the uranium producing regions. For example, by 1971 some 12 million tons of these processing wastes had been piled up in the Colorado River Basin. There has been little interest in using these materials which, on careful analysis, are known to contain $^{230}_{90}$Th and $^{226}_{88}$Ra. Radium and, to some extent, thorium behave like calcium. Some of the waters in the Colorado River Basin contain twice the stated maximum for radium in water for human consumption. The concern here is simply that, because calcium deposits in bones, materials behaving like calcium may also be deposited in bones. Although relatively few people may be affected, the problem of disposal of these spent or used ores may become a substantial health hazard and is heretofore unsolved.

Uranium is a relatively rare chemical element and ^{235}U actually comprises only 0.7% (one ^{235}U atom for every 140 ^{238}U atoms) of a uranium sample. Under the Manhattan project, huge plants at Oak Ridge, Tennessee; Portsmouth, Ohio; and Paducah, Kentucky, were built to separate these isotopes.[1] Nearly all of the nuclear-fired power plants are light water reactors using ^{235}U. In practice, the nuclear fuel is enriched uranium, which contains about 3% ^{235}U, with the balance being ^{238}U. These reactors are known as **burner reactors** because, in the production of energy from the fission of the ^{235}U, the ^{235}U is lost forever. Furthermore, some of the fission products are more efficient at capturing neutrons than ^{235}U so that long before all or nearly all the ^{235}U placed in the reactor has been used up, the nuclear chain reaction becomes slowed down sufficiently to become useless.

Some nuclear-fuel reprocessing plants have been built to take spent fuels from nuclear reactors and recover the large amounts of unreacted ^{235}U from them. This recovered ^{235}U is then used in new fuel rods, and the cycle repeated. However, the long-term implications are clear. When the ^{235}U supplies have been exhausted in burner reactors, nuclear fission will no longer be available as a source of energy for man's use. Unfortunately, such is the prospect for the near future, 30–50 years, unless breeder reactors (in contrast to burner reactors) are developed. Breeders, to be discussed later, breed more fissile material than they consume.

As we have discussed, the supplies of rather accessible ^{235}U are quite limited. Various projections have been made [1] and these generally agree that, if we in this nation and in the world use only burner reactors and thus consume fissile ^{235}U, the end of the ^{235}U as an energy source will be

---

[1] The method of separation is known as gaseous diffusion. Although this process is of great importance, numerous aspects of physics and chemistry not discussed herein are needed for appropriate understanding—see a text in general chemistry.

Table 11-2.
U_3O_8 Requirements

Year	Cumulative Requirements with 8-year Forward Inventory and Working Reserve Without the Breeder and Without Recycle of Water Reactor Produced Plutonium (10^3 tons of U_3O_8)	Projected Price of U_3O_8 ($/lb)
1970	120	8.00
1975	180	8.00
1980	450	8.00
1985	980	10.00
1990	1740	16.00
1995	2750	over 30.00

SOURCE: Atomic Energy Commission, *WASH-1201, LMFBR Demonstration Plant Program,* 1972, p. 377.

imminent by the year 2000 or 2020 (compare Tables 11-1 and 11-2). This view is in sharp contrast to some which record the vast quantities of uranium found in granite. Granite contains only a trace of uranium, but deposits of granite are very large. However, in order to recover the small amount of uranium the granite must be destroyed as a building material by being pulverized, leaving mountainous quantities of useless material.

A Question for the Reader: Shall we dig up huge regions of the earth, take out the granite, and pulverize vast quantities to get out the small amounts of uranium?

Nuclear Powered Units

Let us survey some of the nuclear powered devices that have been built. In 1954, the United States launched the submarine, Nautilus, and in 1956, the U.S.S.R. completed a nuclear powered ice breaker. In 1951, the U.S.S.R. produced a nuclear powered whaling ship, and in 1959 the United States completed the U.S.S. Savannah. In 1955, electricity from an experimental reactor of the National Reactor Testing Station lighted Arco, Idaho, for one hour. Then in 1956, the first full-scale nuclear power plant went into operation at Calder Hall, England.

A joint venture between the Atomic Energy Commission and the Duquesne Light Company brought into operation in 1957, a 60,000-kw power plant at Shippingport, Pennsylvania, as the first nuclear power plant in the United States. A significant turning point was reached in 1963 when a contract was let to the General Electric Company for the construction of the Oyster Creek plant of the Jersey Central Power and Light Company. This plant was to have a rated capacity of 515 mw (megawatts) and a guaranteed cost of power production equal to or slightly lower than that

(A)

(B)

334

Figure 11-7. Nuclear power plants.

(A) James A. Fitzpatric Power Plant (821 Mw) near completion on the shore of Lake Ontario in the town of Scriba near Oswego, New York. [Courtesy of Power Authority of the State of New York.]

(B) Browns Ferry Nuclear Power Plant near completion at a site on Wheeler Lake 10 mi southwest of Athens, Alabama. It will be the world's largest nuclear plant, with a total generating capacity of nearly 3500 Mw in three units. [Courtesy of Tennessee Valley Authority.]

(C) A nuclear power plant (250 Mw) at Lingen on the Ems in north Germany supplies electric energy for a city with 600,000 inhabitants. [Courtesy of German Information Center.]

(D) Nuclear power plant near Chinon, France. [Courtesy of the French Embassy Press and Information Center.]

(C)

(D)

Table 11-3.
Net United States
Production of Electrical
Energy (units of
10^{12} kwh/yr.)

SOURCE: Adapted from W.
G. Dupree, Jr. and J. A.
West, *United States Energy
Through the Year 2000*,
U.S. Dept. of the Interior,
Wash. D.C.: December 1972.

Year	Coal	Gas	Petroleum	Hydro	Nuclear	Total
1968	0.6 (46%)	0.3 (23%)	0.18 (14%)	0.2 (16%)	0.02 (1%)	1.3 (100%)
1980 (estimated)	1.1 (38%)	0.3 (10%)	0.5 (17%)	0.4 (14%)	0.6 (21%)	2.9 (100%)
1990 (estimated)	1.6 (28%)	0.3 (5%)	0.7 (12%)	0.5 (9%)	2.6 (46%)	5.7 (100%)
2000 (estimated)	2.0 (22%)	0.3 (3%)	0.6 (7%)	0.6 (7%)	5.5 (61%)	9.0 (100%)

of a comparably sized, fossil-fuel fired plant. By the early 1970s over 100 nuclear power plants were in operation at various points in the world, including Belgium, Canada, France, West Germany, India, Italy, Japan, Spain, Sweden, Switzerland, the U.S.S.R., United Kingdom, and the United States (Figure 11-7).

The Atomic Energy Commission [2] anticipates that by the mid-1980s some one third of all new generating capacity in the United States will be nuclear reactors and that, by 1990, over half of the new generating capacity being built will be nuclear. The statistics of Table 11-3 illustrate

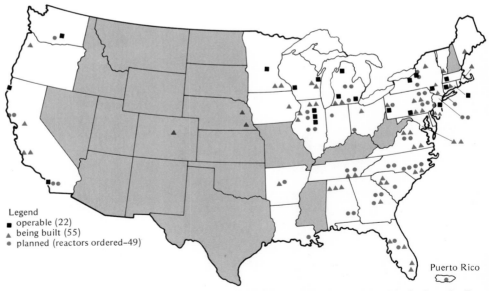

Legend
■ operable (22)
▲ being built (55)
● planned (reactors ordered–49)

Figure 11-8. Location of nuclear power plants in the United States. [U. S. Atomic Energy Commission, September 30, 1971.]

a Bureau of Mines projection that, about the year 2000 in the United States, nuclear power is predicted to become the major fuel source of electrical energy. In the United States, the location of existing and proposed nuclear power plants for electrical power generators is shown in Figure 11-8. You should notice that the locations of these power plants are near regions of high and growing population density. These regions are the ones with the high level of demand for electrical energy.

This extensive building of and planning for water reactors (^{235}U consuming or burner reactors) has necessitated a sharp downward revision of the time availability of reasonably priced ^{235}U. Thus, the possibility of an end to supplies of ^{235}U has raised the priority of the breeder reactors sufficiently so that in early 1972 a contract for a demonstration breeder reactor was let to the Tennessee Valley Authority.

Breeder Reactors

Let us consider the concept of the breeder reactor. Breeder reactors are a reality because the 140-times more available ^{238}U can be converted to a fissile atom by transmutation. You recall Fermi's original idea (Chapter 10) for making heavier elements by neutron bombardment of uranium. Well, in all probability, Fermi may have made some transuranium elements, based on what we now know. However, because of the identification of barium in the reaction products and because of the great release of energy from the ^{235}U fission reaction, he and others pursued the development of fission. The related work on transmutation (overshadowed by research on the fissile atom, ^{235}U) revealed that ^{238}U captures a fast neutron and then releases a beta ray to form the transuranium, element 93, neptunium.

$$^{1}_{0}n + {}^{238}_{92}U \longrightarrow {}^{0}_{-1}e + {}^{239}_{93}Np$$

The formation of neptunium is in accord with Fermi's original prediction. Neptunium-239 has a short half-life, $t_{1/2} = 2.35$ days, and emits a beta particle to give plutonium.[2]

$$^{239}_{93}Np \longrightarrow {}^{0}_{-1}e + {}^{239}_{94}Pu$$

Plutonium-239 (^{239}Pu) has a $t_{1/2}$ of 24,400 years; therefore, among radioactive atoms it is considered relatively stable. Of greatest interest to us, however, is the fact that ^{239}Pu like ^{235}U, is fissile; that is, it undergoes spontaneous fission to atoms of intermediate size on capture of a neutron. The ^{238}U, from which the ^{239}Pu is made, is known as a fertile atom. A **fertile atom** first accepts a neutron and decays to a fissile atom. Similarly,

---

[2]The names of these three heavy atoms, uranium, neptunium, and plutonium, are for the outermost planets of our solar system, Uranus, Neptune, and Pluto.

thorium-232 (^{232}Th, $t_{1/2} = 1.4 \times 10^{10}$ years) has also been found to be fertile. Neutron capture followed by two beta ray emissions gives the fissile ^{233}U. Thorium-232 is the most abundant isotope of thorium in ores.

The relative scarcity of ^{235}U, the knowledge that the more abundant ^{238}U can be transformed into fissile ^{239}Pu, and the fact that about the same amount of thermal energy is released on fission of all three of these fissile atoms have led to the concept of the breeder reactor. In a breeder reactor the initial fuel supply is necessarily ^{235}U. If ^{238}U (or ^{232}Th) is also placed in the reactor core, some of these atoms will absorb neutrons (fast neutrons) and become converted into fissile isotopes. By appropriate control of the neutrons from ^{235}U, one can both sustain the nuclear chain reaction and produce more fissile atoms than are consumed. In principle, then, it is possible to utilize the entire supply of fertile atoms, provided that sufficient ^{235}U is available to start the process. However, it is important to note that a substantial breeding time is required after placing the fuel in the reactor, that is, 5 to 10 years.

The breeder reactor program of the Atomic Energy Commission had a relatively low priority until the mid-1960s, when power companies began intensive planning and growth using nuclear power. A significant turning point was reached with the Oyster Creek plant mentioned above, but several years passed before the 1972 decision to proceed with a demonstration breeder plant. The plutonium breeder reactor under development is called the liquid-metal-fast-breeder reactor (LMFBR). In this system, the term *liquid metal* refers to the fact that liquid sodium will be used to carry heat from the reactor core to heat water in the boiler, schematically like the PWR (Figure 11-1B). As we shall see with even greater urgency below, the degree of success of the breeder reactor program will determine the length of our Nuclear Age.

Although supplies of uranium are not unlimited, readily available uranium-238 at reasonable cost with appropriate nuclear fuel reprocessing techniques can power nuclear breeder reactions for many, many years to come. M. K. Hubbert [1] has thoroughly discussed the potential and necessity of the breeder reactor program with respect to available uranium resources. Hubbert concludes his discussion with the strongest possible plea for development of breeder reactors instead of just continuing to proliferate light water, ^{235}U consuming reactors.

> . . . by the transition to a complete breeder-reactor program before the initial supply of uranium-235 is exhausted, very much larger supplies of energy can be made available than now exist. Failure to make this transition would constitute one of the major disasters in human history.

The concept of a breeder reactor program, however, was not totally new to the 1970s. For example, the breeder technique has been used to produce plutonium for the plutonium fission bomb dropped in Nagasaki, Japan, in August 1945. The technique of breeding plutonium has been developed in three experimental power plants: the first (EBR-I) at Argonne National Laboratory near Chicago, Illinois, the second, EBR-II (20 Mw) developed by Argonne at the National Reactor Testing Station in Idaho, and the third, the Enrico Fermi Plant (166 Mw) 18 miles down river from Detroit, Michigan.

The LMFBR differs from the water cooled reactors (BWR and PWR) in having a more compact core of fuel, which is cooled by rapidly flowing liquid sodium metal. Sodium melts at 99°C (210°F) but does not boil until 890°C (1640°F), which is a great advantage in that high pressures are not required to contain hot sodium. However, the sodium metal is highly reactive to both oxygen gas and water (even as moisture in the air). The sodium metal must be kept away from oxygen and/or water to avoid the hazard of very violent chemical reaction. Furthermore, for a variety of reasons, such as high neutron density, which we will not elaborate in this text, the LFMBR is more prone to reaction runaway (see Nuclear Accidents) than is either the BWR or the PWR. If there is loss of coolant, the more compact core will become very hot in a relatively short period of time.

There is another disadvantage to the LMFBR, which simply is that the form of plutonium (as plutonium oxides) as produced in the reactor is considered to be one of the most toxic substances known. Why? The material tends to form a very fine dust that is hard to control and would rapidly spread out over large areas if released. In contrast to the longer lived uranium atoms ($t_{1/2} = 4.5 \times 10^9$ years), plutonium nuclei are intense emitters of alpha particles, because ^{239}Pu has a $t_{1/2}$ of 24,400 years. Interestingly, most nuclei of atomic number greater than 92 are intense emitters of alpha particles, and, as such, we now realize why these atoms are not found in nature. Even if they had existed, say, when the earth was formed, nuclei of atomic number greater than 92, with their short half-lives compared to ^{235}U, ^{238}U, ^{232}Th, and so on, would have decayed to smaller, more stable nuclei.

Thus, let us summarize these selected disadvantages (there are others) to the LFMBR (breeder reactor) reemphasizing at the same time the fact that abundance of accessible ^{235}U is very limited. The disadvantages of the LFMBR are

1. The hazard of the very reactive sodium metal.
2. The more compact fuel core.
3. The hazard of a runaway accident.
4. The toxicity of plutonium (alpha radiation) especially as a dust.

There is also a disadvantage to the breeder concept which is not directly related to reactor operation. When the plutonium fuel is separated from the original material—^{238}U—it will be in relatively pure form. Once a fissile fuel is relatively pure, it is not overly difficult to make an atomic fission bomb. Storage and surveillance will then present a severe problem.

The question, then, that we all face is how do we weigh these advantages and disadvantages, if we choose to weigh them at all? We elaborate this concept in the next chapter in a section devoted to Tradeoffs: Benefits versus Hazards.

Nuclear Accidents— The China Syndrome

The Enrico Fermi nuclear power plant has been mentioned above both because it was the world's first large breeder power plant and because, after a few months of productive operation, an accident took place in 1966. Numerous subsequent attempts to repair the system were not sufficiently successful, and the plant was closed in 1972.

On October 5, 1966, a blockage in one of the cooling lines caused the fuel elements to overheat sufficiently so that the reactor core partially melted. The safety features built into the plant shut the reactor down automatically, and the radiation was almost entirely held within the outer containment shell. However, at the time of the blockage, the plant was producing power only at about 10% of full power. Considerable controversy has ensued as to whether the carefully engineered safety features are adequate. Some say, "yes" and some say, "no." However, this accident in particular led scientists and technologists to consider not only the maximum credible accident but also the worst case.

The worst situation is *less* than an explosion because the concentration of fissile atoms is far less than needed to achieve the necessary extremely rapid reaction—a critical mass. Nonetheless, complete loss of coolant and failure of the emergency cooling systems could lead a complete "meltdown" of the reactor core in either a burner reactor or a breeder reactor. In this case, a ton of molten fuel, which would be kept hot mainly by the radioactive decay of the fission products, might possibly melt its way down through the bottom of the outer containment shell and sink or penetrate some distance into the earth. Would it reach China? Well, surely not. The phenomenon, although never yet observed, is called the China syndrome.

No one is really sure how far the meltdown would proceed. But, the fear is that it could proceed several hundred feet into the earth and contaminate underground water supplies. It also is a greater worry with the breeder reactor because the reactor core and sodium coolant are run at higher reaction rates and hence, at higher temperatures than the present burner reactors. The plutonium also presents a greater contamination hazard. The radioactive materials would then move into the ecosphere, causing damage and possibly death to many (see Chapter 12). However, in all fairness, such an accident with the present safety system is felt to be very remote. Unfortunately, one can never say with complete assuredness that it will never happen. Does the potential benefit justify the risk or hazard?

Fusion Power

Let us now turn to another nuclear process—nuclear fusion—which scientists and technologists have been attempting to harness for man's use. The quest for fusion power began in the early 1930s with the discovery of nuclear fusion reactions (Chapter 10), followed by the development of the theory of these processes which also explains the source of energy of the sun and other stars.

The motivations for achieving fusion power have remained basically the same from the beginning. Nature has provided us with an available, huge supply of fuel ^2_1H, heavy hydrogen, better known as deuterium, and the energy output per gram of reactant of fusion reactions is higher than that of fission reactions (Chapter 10). Although only 0.1% of the mass involved in fission reactions is convertible into energy, between 0.4% and 0.7% of the mass in fusion reactions is convertible into energy. Thus, on a mass basis, fusion reactions produce 4 to 7 times as much energy as do fission reactions. Unfortunately, even though the motivations for fusion research have been strong, the massive financial support has been and still is lacking, while scientific and technological problems continue to be very great.

In the 1950s, following the development of the thermonuclear bomb, which employs the fusion reaction, the extreme difficulty of controlled, sustained nuclear fusion processes became apparent. In the 1960s, theoretical, experimental, and technological groundwork established the credibility of confining a nuclear fusion process. For example, in 1963, a very brief but controlled fusion of ^2_1H (equation 10-15a) and ^3_1H (tritium) nuclei (equation 10-15c) at about 50 million $(50 \times 10^6)°C$ (it makes no significant difference whether we talk about °C or °K) was achieved. In the 1970s, the main task appears to be to prove the scientific feasibility on confinement and control of a nuclear fusion for time periods sufficient to generate power. Put another way, the science and technology of nuclear fusion has not yet reached the comparable point of nuclear fission work which was reached by Fermi's group on December 2, 1942. That is to say, fusion power is at least 30 years behind fission power.

A successful, power producing, controlled thermonuclear process has three prime requisites. First, the temperature must be high enough for the fast moving nuclei to collide and fuse. Recall that high temperature means high average speed. The $50 \times 10^6°C$ is one of the lowest ignition temperatures of the many possible fusion reactions. At these ignition temperatures, no ordinary substance can exist, or, more specifically, none of the known metals or other construction materials can withstand these reaction conditions. The most stable materials known to man melt below 4000°C and vaporize below 6000°C. These temperatures are far lower than the fusion ignition temperatures of close to 50×10^6°C.

At these very high temperatures, molecules and atoms become charged particles (ions) in the state known as the plasma state. Thus, the second requisite is the confinement of the plasma state long enough for the fusion reaction to occur and, hence, to release a significant net output of energy. Numerous clever and elaborate devices (Tokomak, Russia; Stellarator, United States; and so on) have been developed to contain the plasma state and the fusion reaction. Most containment devices utilize magnetic fields—known as a **magnetic bottle**. Finally, the third requisite is that the energy released must be recoverable in a useful form, for example, in electrical energy. Since feasibility is yet to be established, there is much work to be done!

When such control is developed and perfected, man will have obtained a piece of the sun—an infinite supply of energy—to do with what he chooses. Let us consider the advantages of fusion power, which, very

---

EXAMPLE 11-1

Calculate the number of deuterium atoms in $1 \, m^3$ of water if there is one deuterium atom for 6500 hydrogen atoms.

(a) $1 \, m^3 = 10^6 \, cm^3$. Density of water = $1 \, g/1 \, cm^3$. Then

$$1 \, m^3 \text{ of water} = 10^6 \, cm^3 \times 1 \, g/cm^3 = 10^6 \, g = 10^3 \, kg = 1 \text{ metric ton}$$

(b) Water has molecular weight of 18 and contains 2 mass units of hydrogen. Thus, the hydrogen content of water is $\frac{2}{18}$ or $\frac{1}{9}$ of the mass.

Thus, $\frac{1}{9} \times 10^6 \, g$ is the mass of hydrogen atoms in $1 \, m^3$ of water.

(c) Each ^2_1H atom has twice the mass of an ^1_1H, ordinarily hydrogen.

$$\frac{1 \, ^2_1H \text{ atom}}{6500 \text{ H atoms}} = \frac{2 \text{ mass units}}{6500 \text{ mass units}} = \frac{1}{3250} \text{ mass is } ^2_1H$$

(d) Mass of ^2_1H in 1 metric ton of water.

$$\frac{1}{9} \times 10^6 \, g \times \frac{1}{3250} = 34 \, g \, ^2_1H \text{ in 1 metric ton water}$$

(e) Number of ^2_1H atoms in 1 metric ton of water.

$$34 \, g \, ^2_1H \times \frac{6.02 \times 10^3 \text{ atoms of } ^2_1H}{2 \, g \text{ of } ^2_1H} = 1.02 \times 10^{25} \, ^2_1H \text{ in 1 } m^3 \, H_2O$$

practically motivates the quest. There are five principal advantages of fusion power. First, the fusion process consumes none of the world's oxygen and none of the world's fossil fuels. Second, the fusion process, in general, releases no combustion products to the atmosphere, such as carbon dioxide or sulfur dioxide. Third, the proposed fuel cycles produce at most very small quantities of radioactive materials. The principal radioactive material produced is ^3_1H, tritium, a beta emitter, $t_{1/2} = 12.26$ years, which is itself a valuable fusion fuel and therefore would preferably be captured, retained, and recycled to the reaction system. Fourth, the extreme conditions and the small amounts of material involved in the reaction are such that a runaway accident is essentially impossible. The reaction is very easily "quenched" by cooling, even slight cooling. Fifth, and finally, the fuel supplies are potentially very large compared to the heavy nuclei used in fission, such as uranium and thorium.

Let us consider the fuel supplies for a fusion process. In water, we find that there is one deuterium atom for every 6520 hydrogen atoms. In 1 m³ (1 metric ton) of water there are 1.02×10^{25} deuterium atoms or 34 g as calculated in Example 11-1.

The total volume of the oceans is about 1.5×10^9 km³ and there are 10^9 m³ in every cubic kilometer. An approximate figure for mass energy conversion in fusion is 1.9×10^7 kcal of energy (Chapter 10, Example 10-7) released when the 1 g of deuterium fuses. Even if only 1% of the deuterium in the oceans were to be used, the energy (10^{25} kcal) released would amount to some 2,000,000 times that available from the world's total initial supply of the fossil fuels (5×10^{19} kcal) [1] Example 11-2. Calculated another way the deuterium content of about 8 km³ or less than 2 mi³ (1 km³ equals 0.23 mi³) could provide in theory an amount of energy equal to the energy

---

EXAMPLE 11-2

Calculate the energy released if 1% (0.01) of the deuterium of the oceans were to undergo nuclear fusion.

From Example 11-1d we know there are 34 g of ^2_1H per metric ton of water and 1.9×10^7 kcal/g of ^2_1H fused is estimated as the energy produced. Thus

$$34 \text{ g } ^2_1H/m^3 \times 10^9 \text{ m}^3/\text{km}^3 \times 1.5 \times 10^9 \text{ km}^3/\text{oceans} = 51 \times 10^{18} \text{ g } ^2_1H/\text{oceans}$$

Then

$$51 \times 10^{18} \text{ g } ^2_1H \times 0.01 \times 1.9 \times 10^7 \text{ kcal/g } ^2_1H = 97 \times 10^{23} \text{ kcal}$$

That is, about 10^{25} kcal would be released upon fusion of 1% of the deuterium in the oceans.

obtained from all the earth's initial supply of coal and petroleum. Such potential spurs the quest!

Although the discussion and calculations serve to illustrate dramatically the potential of the deuterium-deuterium fusion reactions, experimentally, a deuterium-tritium reaction seems to be even more attractive. Why? The deuterium-tritium fusion reaction requires only 50×10^6 °C in contrast to the 100×10^6°C required for the deuterium-deuterium reaction. Since tritium, $_1^3$H, has a $t_{1/2}$ of 12.26 years, the waters of the world are not practical sources. Rather, an isotope of lithium, $_3^6$Li, on capture of a neutron is converted to $_1^3$H.

$$_3^6\text{Li} + {}_0^1n \longrightarrow {}_2^4\text{He} + {}_1^3\text{H} + \text{energy}$$

Then

$$_1^3\text{H} + {}_1^2\text{H} \longrightarrow {}_2^4\text{He} + {}_0^1n \tag{11-1}$$

Equation 11-1 is highly energy releasing, much more so than is the deuterium-deuterium fusion. Although development of this thermonuclear reaction sequence appears more attractive from an ignition temperature viewpoint, there is an additional disadvantage. Specifically, lithium is far less abundant than is hydrogen and this isotope comprises only 7.4% of the lithium in nature. Hubbert has calculated that the deuterium-tritium fusion process will be limited by the availability of $_3^6$Li rather than by deuterium (Figure 11-9). Nonetheless, both processes continue to be studied.

Among the disadvantages of fusion power is the fact that large quantities of waste heat are likely to be generated unless some process for transferring the energy of the charged particles in the plasma directly to electrical

Figure 11-9. Schematic drawing of an imagined fusion reactor.

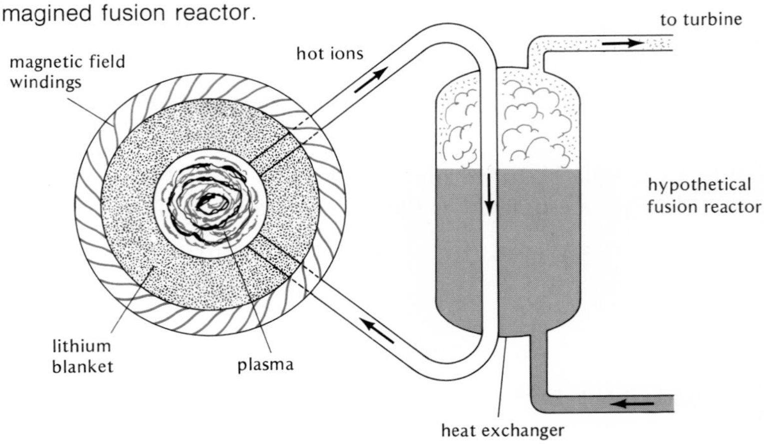

energy can be developed. Another disadvantage is that some radioactive tritium will certainly be a by-product. Containment of tritium, like ordinary hydrogen, is not easy. The principal disadvantage, though, of relying on fusion power centers on the fact that some of the science and nearly all of the technology of fusion power is yet to be developed. Scientists hope to prove feasibility in the next 5 years or so. Whether or not fusion power will be a reality is very much up for speculation.

The annual funding of the United States' fusion program reached a peak in 1960 and for the subsequent 10 years was funded at a constant level of some $17–18 million per year (corrected for inflation). Most people associated with fusion research and development believe that a practical fusion-power reaction can be made and have found that there is no law of nature preventing such a development. The key question for all of us is how much effort (measured in dollars) will our society invest. The dollar factor is considered the most significant factor in determining the time-table to successful fusion power. As NASA (National Astronautics and Space Administration) has shown, major technological objectives can be achieved on a greatly shortened time scale, given the basic scientific information, adequate priority, a clear goal and adequate funds for research and development (some $40 billion was spent by NASA between 1960 and 1970).

Interestingly, the Soviet Union is putting about four times as much effort into fusion research and development as is the United States. How important is the fusion energy source and the necessary technology to our energy hungry society? We face the fact that the technology, when and if successful, will be highly complex, highly technical, and highly sophisticated. In brief, the time-table might be short enough that a prototype (not full-scale commercial) reactor could be produced by the early 1980s. The actual time-table will depend on our priorities. Most experts in the field estimate that some 20 years of development will be needed *once* there has been scientific demonstration of the sufficient control of the fusion reaction before useful thermonuclear reactors could provide energy for society's use.

Conclusion

In this chapter, we have surveyed the present and growing use of the uranium-235 fission reaction in nuclear reactors. The reactors (burner reactors) presently in use are consuming uranium-235 at such an ever-increasing rate that we are already worried about the finiteness of our reasonably accessible uranium-235 supply. We have also examined the two additional nuclear systems that are today receiving the most intensive study—in dollars and in material—in the hopes that these systems will safely provide more energy for an energy hungry world. These two are the LMFBR (breeder reactor) and the fusion or thermonuclear reactor. There are other variations of the breeder concept that have received some study, and, although these have not received the priority of the LFMBR or of fusion, further study seems appropriate. For example, a breeder reactor using fertile thorium-232 has an advantage of producing fissile uranium-233 which is far less toxic (far more stable with its 1.59×10^5 year half-life) than is plutonium-239.

However, beyond these essentials of this incomplete story of nuclear power or the peaceful utilization of nuclear energy as presented in this chapter, we have a major aspect of these systems yet to discuss. Many of the nuclear reactants and most of the products, particularly those from fission processes, are highly radioactive. Radioactive materials are toxic and dangerous to the ecosphere—to you and me. What is the danger? How is it controlled? These are the subjects of the next chapter—Chapter 12.

Questions

1. Which are the most important fissile atoms?

2. How is a nuclear fission reaction controlled?

3. What might be some of the hazards associated with the mining of uranium ores?

4. What conditions limit the generation and use of power from nuclear fission reactions?

5. What is a breeder reactor?

6. Write the equations for the transformation of fertile ^{232}Th to fissile ^{233}U.

7. What is the importance of purity of the fissile matter in a nuclear reactor? The purity of fertile and fissile matter in a breeder reactor?

8. In view of the data of this chapter, discuss M. K. Hubbert's statement, "Failure to make this transition (to a breeder reactor program) would constitute one of the major disasters in human history."

9. Why is it considered unlikely that a nuclear power plant, such as that at Oyster Creek, will explode?

10. By reference to a sketch of the geological strata underlying the regions of the nuclear power plants, which are pictured in Figure 11-7, predict what might happen if the worst possible accident (China syndrome) occurred at each plant.

11. Why cannot glass or metal containers be used to hold thermonuclear plasmas?

12. Summarize some of the advantages and disadvantages of fusion power.

Numerical Exercises

1. In an ordinary nuclear reactor using ^{235}U, if no neutrons are lost, what is the minimum number of neutrons that must be released in a fission reaction in order to maintain the chain reaction?

2. (a) From data in this and the preceding chapters, for a plant with a capacity of 1000 Mw (1,000,000 kw output rating), calculate how many grams of ^{235}U will be needed per day. (*Hint:* 1 kwh corresponds to 8.6×10^2 kcal.) How many kilograms? How many pounds?

   (b) How many metric tons per year of ^{235}U will be needed to operate this plant for 1 year at peak efficiency. (*Note:* Assume the plant is 30% efficient in utilizing the heat from the fission reaction.)

3. The rock formation of Chattanooga Shale found in Tennessee, Kentucky, Ohio, Indiana, and Illinois is about 15 ft (5 m) thick and contains 0.0060% by weight of uranium; 1 m^3 of this rock weighs 2.5 metric tons.

   (a) What is the uranium content in grams per metric ton and in grams per cubic meter?

   (b) If 1×10^6 metric tons of uranium is needed by 1985 and if this rock were the only source of uranium, how many square kilometers of land would need to be mined to provide this uranium? Assume 100% recovery of the uranium. How many square miles?

   (c) Look up the size of a county or two in these states and estimate what fraction of a county would need to be completely mined to provide the quantity of uranium required in part b.

4. In a breeder reactor using fissile ^{235}U for fuel and fertile ^{238}U to produce fissile ^{239}Pu, if no neutrons are lost, what is the minimum number of neutrons that must be released in the fission reaction in order both to maintain the chain reaction and to replace the spent fuel?

5. (a) From the data in the chapter, calculate how many cubic kilometers of water are needed to provide the deuterium (for $^2_1$H-$^2_1$H fusion) which, when fused, would give an amount of energy equivalent to the earth's initial supply of fossil fuels.

   (b) Convert your results from part a to cubic miles.

References

[1] M. K. Hubbert in *Resources and Man.* W. H. Freeman & Co., San Francisco, 1969.

[2] U. S. Atomic Energy Commission: *Forecast of Growth of Nuclear Power WASH.* 1139 (rev 1). Washington, D.C., 1971.

Is Your Geiger Counter Working?—Pollution from Radiation

The previous two chapters have set forth the essentials of the technology of electrical power generation from nuclear reactions and the key scientific aspects for understanding the peculiarities of that technology. This chapter is a bridge chapter. Up to now we have only hinted at the effect of radioactive substances or, more specifically, the emissions themselves on people and the environment. In fact, the effects of producing electrical energy on the environment of our energy-hungry society have only been alluded to occasionally as we developed some essential science on which any real or practical discussions of pollution must be based. Thus, in this chapter and the following two chapters we now turn to pollution of three different types—radioactive wastes (Chapter 12), heat (Chapter 13), and fossil-fuel combustion products (Chapter 14).

By way of general background, in this chapter, we first consider precisely what is meant by pollution. We examine how one detects radioactive substances with gadgets or devices such as the Geiger counter. Detection provides the means for quantitative assessment of radioactivity, which is measured two ways—one purely physical and the other biological—for ultimately we are concerned with the effect of the various forms of radiation on our environment and on us! We then discuss a very few of the most important radioactive substances found in the environment, and more particularly, those which man by his energy-producing activities may be placing into the environment. The maintenance and control of hazardous wastes is surveyed.

We then turn to a central concept, fundamental to man's continued existence, that is, benefits versus hazards or trade-offs. The concept of hazards includes costs, so that another way of phrasing trade-offs is benefits versus costs: environmental, economic, personal, or societal. This topic, although presented as scientifically and dispassionately as possible, necessarily carries with it overtones of emotional reaction. At the heart of this discussion, within this chapter and the next three, lie questions of opinion and judgment with which you, we, and all members of society not only should be familiar but on which we must make decisions for our way (or any way) of life.

We then turn back from the more philosophical view to present the facts as they are best available to us on effects of radioactive substances on the population as a whole. We present data on society's regulation of these materials. These regulations are necessarily both a matter of judgment and a matter for continued vigilance and concern on the part of all of us.

Pollution Defined

What is a pollutant? What is pollution? What does it mean to pollute? Webster's dictionary defines to pollute as "to make or render unclean; to defile; to desecrate; to profane; to contaminate." To pollute has also had a moral overtone in the sense of "to impair the purity of." Certainly air pollution and water pollution clearly refer to the contamination or impairment of the air environment or the water environment, respectively. One of our concerns is to measure and understand what our reference points are. What is a clean environment? What is clean air? What is clean water? For example, we often long for a drink of cool, clear water such as from a bubbling spring. This water contains small and measurable quantities of dissolved substances. These actually enhance the flavor of the water. Pure water, that is, distilled water, containing almost no dissolved substances, is quite tasteless or flat. You would probably spit it out despite its higher purity!

Beginning in the 1960s, new expressions have become common, such as radioactive pollutants, thermal pollution, and air pollution. Although radioactive pollutants can be added to the environment, thus impairing the environment, the damage comes from the radiation emanating from the materials, not directly from the materials themselves. Thermal pollution (waste heat), discussed in Chapter 8 and again in Chapter 13, is the change of the environment—air, water, or land—by raising its temperature. Is this always a desecration of the environment? Pleasantly warmer waters might encourage more swimming for people, but can be deadly for certain species of fish and other aquatic life. What should be our perspective on heat? We suggest you turn back to our earlier discussions of heat and the environment.

Similarly, air pollution has little to do with the temperature of the air: dirty, smoggy air can be cool or hot. Noise pollution has little to do with either the purity or the temperature of the air. Dirty, smoggy air can be very quiet and peaceful, and pure air can be shattered by noise. Briefly, noise pollution is the desecration of the environment by noise. Thus, we ask you to consider very carefully, rationally and unemotionally, what is pollution? We submit that you must consider each situation, case by case, on its merits and demerits (benefits versus hazards).

In this chapter, we will probe into some of the scientific facts and some of the technological data surrounding radioactive pollution of our environment. Of the many factors surrounding man's use of nuclear energy, we have examined already all the key scientific facts and problems, except for the effect of radioactive materials on the environment. Actually, in Chapters 10 and 11, we discussed the key physical concepts, so let us now turn to the detection of radioactive materials, the effect of these materials on the environment, and some necessary controls.

The Geiger Counter

Although we now know that our body is affected by the particles and energy produced in radioactive decay, none of our senses respond directly to radioactivity. How then, can radioactivity be detected? The discovery (1896) of radioactivity followed from an exposure of photographic film placed next to a mineral. Thus, exposure of a photographic film is today one method of detection. In fact, all people whose work causes them to be exposed to alpha, beta, and gamma rays wear film badges (Figure 12-1) that are checked regularly for the amount of exposure.

A second method, discovered shortly thereafter, is based on the fact that radioactivity produces scintillations (bursts of visible light) when the rays strike certain substances such as zinc sulfide. A third method, which is one of the most practical and widely used, is the Geiger counter, which is based on the conversion of gas molecules to charged particles (ions) when struck by radioactivity. As radiation passes through matter, energy is transferred from these high energy rays to the molecules (and the constituent atoms

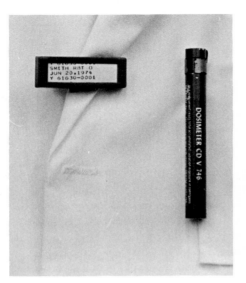

Figure 12-1. Two types of radiation detectors worn by laboratory personnel, x-ray technicians, and so on. The film badge, left, is used for a permanent record while the dosimeter is an instant read-out device used for temporary records.

of molecules) that make up matter. These rays are often collectively called ionizing radiation because the one common result of their passage is the conversion of atoms or molecules (electrically neutral) to charged particles or ions. You recall that in Chapter 5 we discussed the electrical nature of the atom with its proton-containing, positively charged nucleus surrounded by negatively charged electrons. The Geiger counter gives a measure of radioactivity by counting the number of ionizations produced in the device.

The Geiger counter (Figure 12-2), named in honor of its inventor, consists of a metal chamber with a window (W) made of thin mica (a mineral having suitable strength and transparency to radiation) or some other material easily penetrated by radiative emissions. Running through the center of the tube but insulated from the metal case is a wire that is highly charged relative to the metal case. The chamber is filled with an inert gas, such as argon, at about 0.05 atm (atmospheres) pressure. The charge in the wire is set to be almost great enough, yet not quite high enough, to cause an *electrical discharge* through the case from wire to wall. However, if a single, fast particle enters the chamber through the window, the particle will hit an argon atom causing the atom to lose an electron (outside the nucleus) and thus be converted to a positively charged particle (Figure 12-3).

The result is a cascade effect as the positively charged particles rush to the negative wall and the electrons to the positive wire. This movement of particles is observed as a sudden massive discharge in the chamber, all produced by a single high energy particle. As soon as the charged particles

Figure 12-2. The Geiger counter. **(A)** Schematic diagram of the Geiger counter. **(B)** A portable, battery powered, Geiger-Müller counter showing the cylindrical chamber or Geiger tube and a meter or an audio indicator. The Geiger tube at the left, showing the charged rod or anode, is connected to another type of counter, a scaler. The numbers on the instrument actually count radioactive events passing through the Geiger tube. The Geiger counter however gives an integrated countrate in counts per minute or milliroentgens per hour.

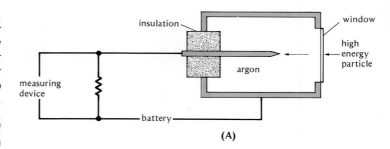

(A)

(B)

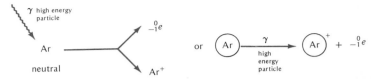

Figure 12-3. Visualization of the gamma ray induced ionization of argon.

have been collected at the wire and wall respectively, the discharge stops. The tube is then recharged and is ready for another event. The number of discharges per second can be counted with appropriate devices, such as flashing lamps, mechanical counters, or even devices that produce audible sounds (speakers). Thus, the Geiger counter detects products of nuclear decay and also actually counts their number as well. The counter is rather rugged and dependable. Portable counters can be made for prospecting for radioactive minerals or for searching for lost radioactive materials. The Geiger counter suffers from one disadvantage, namely, to be detectable the particle must have sufficient energy to penetrate the mica window. To detect the weakest emissions, even more sensitive devices have been developed; however these need not be discussed here as the mode of operation is similar in principle to the Geiger counter.

Measures of Radiation

Since the Geiger counter measures counts per unit of time, a physical unit of radioactivity has come into use that is based on this counting. The unit of activity (rate of disintegration) of a radioactive source is the curie; 1 curie is approximately the rate of disintegration of 1 g of $^{226}_{88}$Ra. Today, the **curie** is defined as 3.7×10^{10} cps (counts per second or disintegrations per second) or as 3.7×10^{10} radioactive events per second.

Smaller units are the millicurie (37×10^6 disintegrations per second) and the microcurie (37,000 disintegrations per second). These smaller units are frequently used in describing the amount of radioactive fallout from nuclear explosions. However, this physical unit, the curie and its derivatives, is not that useful in biological (environmental) studies because the physical unit merely indicates the number of disintegrations per second regardless of the type of radiation *and* regardless of the effect of that radiation on tissue.

Well, what about some of the different kinds of radiation? The three types of radiation discussed in Chapter 10 are not equally dangerous to the environment or to living organisms because they have different penetrating ability. Alpha particles are the least penetrating, for they can be stopped by a thin sheet of paper, a thin film of water, or the outer layer of skin. But, please note that alpha particles, which are heavy and have a doubly positive charge, are very damaging to tissue because, if the source is in the skin or inside the body, many ionizations are produced and all the alpha radiation is absorbed in a local area.

Beta rays penetrate materials more deeply. It takes up to an inch of wood or a few millimeters of aluminum sheet or a few millimeters of skin to stop beta rays. If the source of beta rays is left in contact with the skin, severe burns (Figure 12-4) will develop after some time. Although normal clothing provides protection from both alpha and beta radiation, gamma radiation passes through the clothing, through skin, through organs, tissue, and bone. Gamma rays (Figure 12-5) are highly penetrating and cause cellular damage as they travel through the body. Gamma rays are stopped by several inches of lead and/or several feet of concrete!

All three types of radiation, when they enter skin tissue, cause ionization of atoms in the molecules of cells in roughly the same manner as argon is ionized in the Geiger counter. Such changes disrupt the normal chemical processes within the cell, which may cause that cell to grow abnormally or to die. Let us then consider two biological measures of radiation.

The **rad** (**r**adiation **a**bsorbed **d**ose) refers to the amount of energy actually absorbed by tissue that has been radiated. One rad[1] corresponds to the

[1]In most texts, the rad is defined as the absorption of 100 ergs of energy per gram of tissue. An erg is a very small unit of energy; 1 cal is 41.8×10^6 ergs.

Figure 12-4. **(A)** Beta burns on a Rongelap native (1954). Hyperpigmented raised plaques and bullae on dorsum of feet and toes at 28 days. Feet were painful at this time. **(B)** Beta burns six months after exposure. Foot lesions have almost healed with the exception of the deeper ones. [Courtesy of U. S. Atomic Energy Commission, photographs by Navy Medical Research Institute.]

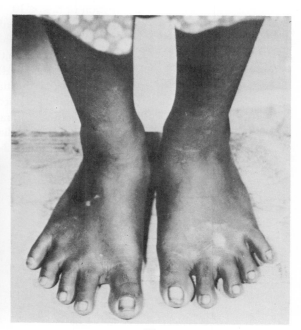

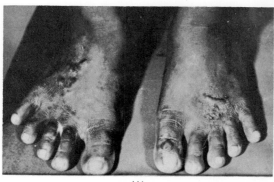

(A)

(B)

Figure 12-5. The effects of nuclear radiation on our environment are studied in a 50-acre forest on Long Island. A gamma ray source is located at the end of the path in the picture, and nearby trees and vegetation have died as a result of exposure to these rays. [Courtesy of Brookhaven National Laboratory.]

absorption of 2.4×10^{-6} cal of energy by 1 g of tissue or 2.4×10^{-6} kcal of energy per kilogram of tissue. The rad is a very small unit of radioactive energy when we think of the purely temperature effect of this amount of heat. However, because the particles of radiation are individually so highly energetic, even this small dose can be devastatingly effective as it disrupts vital molecules or the functioning of these molecules.

The second biological measure of radiation, rem—pronounced as spelled—combines the rad with the **relative biological effectiveness** (RBE) of various kinds of radiation. The RBEs of some types of radiation are summarized in Table 12-1.

Table 12-1. RBEs of Various Forms of Radiation

Type Radiation	RBE
gamma rays and x–rays	1
beta rays	1
thermal (slow) neutrons (as from a reactor)	5
alpha particles	10
fast neutrons (as from fissioning)	10

EXAMPLE 12-1

(a) What dose in rads of alpha particles is equivalent in rems to a 1-rad dose of gamma rays?

Gamma:
$$1 \text{ rad} \times 1 \text{ RBE} = 1 \text{ rem}$$

(*Note:* 1 rad of gamma rays = 1 rem of gamma rays.)

Alpha:
$$? \text{ rad} \times 10 \text{ RBE} = 1 \text{ rem}$$
$$\text{rad} = \frac{1}{10} = 0.10$$

Thus, 0.10 rad of alpha particles has the rem equivalent of 1.0 rad of gamma rays.

(b) Compare the effect in rems of a 1 rad dose each of alpha particles and gamma rays.

Gamma:
$$1 \text{ rad} \times 1 \text{ RBE} = 1 \text{ rem}$$

Alpha:
$$1 \text{ rad} \times 10 \text{ RBE} = 10 \text{ rem}$$

A 1 rad dose of alpha particles is 10 times more damaging than a 1 rad dose of gamma rays.

The **rem,** the **r**ad **e**quivalent to **m**an, is the dose of the radiation in rads multiplied by the RBEs of the type of radiation, see equation 12-1.

$$\text{rem} = \text{rad} \times \text{RBE value} \tag{12-1}$$

Thus, 1 rem is the amount of radiation of any type that, when absorbed by man, has an effect equal to the absorption of 1 rad of gamma rays or x–rays. When the RBE is equal to 1, as for beta and gamma rays, the rem and the rad are indeed equal. However, a 0.1-rad dose of alpha particles has the same biological effect in rems as does a 1.0-rad dose of gamma rays (Example 12-1a). Put another way, a 1-rad dose of alpha particles is 10 times more damaging than a 1 rad dose of gamma rays (Example 12-1b). Thus, the rem, or the millirem (mrem), is a measure of biological damage (see Medical Radiation) and is commonly used in technical (medical and engineering) descriptions of actual and permissible doses of radiation (or exposures to radiation).

Our Radioactive Environment

Life on earth has probably always existed in the presence of radiation. In Chapters 10 and 11 we discussed many kinds of radioactive heavy elements and also a number of radioactive light elements including carbon-14. Some of the radioactivity present on earth prior to the twentieth century came (and has come for many centuries) from radioactive isotopes (radioisotopes) that survived over the history of earth. For example, the naturally occurring radioisotopes found in ores in any significant quantities are those with very long half-lives, ^{235}U, ^{238}U, ^{232}Th, ^{226}Ra, and so on (Chapter 10).

In addition, cosmic radiation is coming to earth and, throughout earth's existence, presumably has been coming to earth from outside our immediate solar system. Cosmic rays enter the upper atmosphere, and their very high energy shatters the nuclei of atoms in the gaseous molecules of the air. Among the various particles produced by collision of cosmic rays with the atmosphere, neutrons start nuclear reactions such as in equation (12-2). Earlier we discussed this process, which results in the formation of carbon-14, a radioactive isotope of carbon (a beta emitter) with a half-life of 5760 years.

$$^{14}_{7}N + ^{1}_{0}n \longrightarrow [^{15}_{7}N] \longrightarrow ^{14}_{6}C + ^{1}_{1}H \tag{12-2}$$

Radiations caused by these reactions compose what is known as the **natural background radiation.**[2]

This natural background radiation has existed, exists, and will continue to exist for the foreseeable future no matter what man does or does not do. Put another way, there is neither any way that we can avoid natural background radiation nor any way to reduce it. Thus, there is no such thing as *no* (or zero) exposure to radiation. In Table 12-2, we show a few of the important contributors to natural background radiation.

[2] For convenience in discussion, we discuss background radiation as being made up of two components—natural and manmade. In practice, we can measure only the total radiation in our environment as background radiation.

Table 12-2. Some Important Natural Radioisotopes

Origin	Radioisotope	Half-Life	Radioactive Emission
Ores	^{238}U	4.5×10^9 years	alpha, gamma
Ores	^{235}U	7.1×10^8 years	alpha, gamma
Ores	^{232}Th	1.4×10^9 years	alpha, gamma
Ores	^{40}K	1.3×10^9 years	beta, gamma
	Derived from above		
Ores	^{226}Ra	1622 years	alpha, gamma
Ores	^{222}Rn	3.8 days	alpha, gamma
Ores	^{210}Po	138 days	alpha, gamma
Ores	^{210}Pb	21 years	beta, gamma
Radioisotopes formed from cosmic radiation	^{3}H (tritium)	12 years	beta
	^{14}C	5760 years	beta

Table 12-3. Sources
of Natural Background
Radiation in New
Jersey

cosmic rays	50 mrem/year
the earth	15 mrem/year
building materials	30 mrem/year
air environment	5 mrem/year
food consumption	25 mrem/year
Total	125 mrem/year

What are we in the United States exposed to as regards natural background radiation? Interestingly, there is some variation from location to location. In the United States, the average per capita dose of natural background radiation from all sources is about 125 mrem/year. Table 12-3 shows a breakdown of the 125 mrem background radiation.

The cosmic component of natural background radiation increases with elevation, that is, Colorado residents are exposed to cosmic radiation alone amounting to some 120 mrem/year, because cosmic ray activity is greatest in the upper atmosphere. In addition to altitude, the level of natural background radiation varies depending on the amount of uranium and other radioisotopes in the local soil. You recall the mention in Chapter 11 that granites contain small amounts of uranium. The gross variation of natural background radiation in the contiguous states of the United States are summarized by region in Table 12-4.

A Question for the Reader: Even though the natural background radiation of Colorado is double that of New Jersey or 2.5 times that of Texas, how many residents will change location for this reason? Would you?

Table 12-4. Regional
Natural Background
Radiation In The
Geographical Regions
of the United States

Region	Average Annual Background (mrem/year)
New England	135
Mid-Atlantic (N.Y., N.J., Pa., Del., Md.)	125
Southeast	135
Midwest	135
North Central (except S. Dak.)	140
S. Dak.	210
South Central (except Tex.)	135
Tex.	100
Rocky Mountain (except Col., Wy.)	170
Col., Wy.	245–250
Pacific	125

Man's Contribution to Environmental Radioactivity

We have examined one of the sources of radiation received by man—natural background radiation—in the previous section and turn now to two other sources, medical radiation and radiation derived from man-controlled nuclear processes.

Medical Radiation

The amount of medical radiation varies with the type and frequency of medical treatment. In the United States, the average per capita dose received in medical and dental diagnostic x–ray examinations has been estimated to be between 60 and 90 mrem/year. We may consider this dosage as a **whole-body dose.** However, x–ray photographs of specific parts of the body may give that body part a very high amount of radiation. Because the damage in this case is localized, the overall effects are substantially less than if the same high dosage were received over the whole body. Table 12-5 indicates the number of millirems received during some types of typical medical treatments.

The physiological effects of radiation are cumulative over a period of time. One of the principal effects of radiation is the shortening of life. Skin cancer, bone cancer, thyroid cancer, leukemia, and other cancers are products of radiation exposure. We have noted earlier the effect of the radium decay on watch-dial painters (Chapter 10). Physicians and radiologists, who work constantly with x–rays, have been found to live an average of 5 years less than other physicians. Even specialists having some exposure to radiation such as urologists and dermatologists show an average of 2 years shorter life expectancy. The effects of single whole-body dosages of radiation are summarized in Table 12-6.

As shown in Table 12-6, a 100-rem dose causes mild radiation sickness in man; a 250-rem single dose causes severe sickness and a few deaths; a 500-rem dose causes about 50% deaths, as illustrated in the curve of Figure 12-6. Let us leave further discussions of this topic to later in the chapter.

Nuclear Processes

In the years since 1945 by his technological activities man has increased the quantity of radioactive materials on earth with the necessary result

Table 12-5. Amount of Radiation Received During Various Types of Medical Treatments

Type of Treatment	Dose (mrem)
chest x–ray, full size	40–50
pregnancy x–ray examination	50–100
dental x–ray, single picture	100–150
dental x–ray, whole mouth	5000
gastrointestinal (GI) series	1000–8000

Table 12-6. Effects of Single Dosages (rems) of Radiation in the Body (whole-body dose)

Dose (rems)		Effect
0–25	0.125 rem/yr background radiation 0.005 rem/yr fallout 0.010 rem/yr radium watch dials 0.05 rem/yr chest x–ray 5.0 rem/yr dental x–ray set	no observable effect
25–50		small decrease in white blood cell count
50–100		marked decrease in white blood cell count; lesions
100–200		nausea; vomiting; loss of hair
200–300		hemorrhaging; ulcers, some deaths
300–500		hemorrhaging; ulcers; up to 50% deaths
greater than 700		fatal

that the total background radioactivity (natural plus manmade) in the environment has increased. This increase derives from two sources: one, the development, testing, and use of nuclear weapons, and, two, the development, construction, and operation of nuclear reactors for the generation of electrical power. The two sources are based on the neutron-initiated fission reaction of fissile ^{235}U.

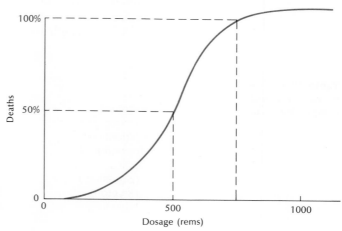

Figure 12-6. Percentage of cases in which death occurs for various radiation doses.

In the fission reaction, a great many radioactive fission products are produced. Many of these materials are intensely radioactive, as they have very short half-lives, that is, of the order of seconds or minutes. Other fission products have rather longer half-lives, that is, years. In either case, these half-lives are short compared to naturally occurring radioisotopes.

Total release of these radioactive materials to the environment occurs in a nuclear explosion. Some of these products are recorded in Table 12-7. These substances are in the "mushroom cloud" of the nuclear explosion, and these attach to dust particles by chemical or intermolecular forces or join water droplets. The contaminated dust and rain that eventually come to earth are known as radioactive fallout. This fallout begins locally a few hours after detonation of the explosion, but some of the radioactive materials circulate globally (Chapter 3), particularly in the stratosphere, for years. Global circulation means that these radioactive materials now contaminate to some degree the entire earth environment.

A 1966 United Nations report has calculated that the probable dose to the average person from various radioisotopes produced by nuclear explosions from 1954 through 1965 amounted to about 200 mrems. Although atmospheric nuclear weapons testing has been discontinued under the 1963 Nuclear Test Ban Treaty, in the early 1970s, the residual fallout still amounted to an average of 5 or 6 mrems per year. Thus, we can consider that an average person in the United States is exposed to a total of 210 mrems of radiation deriving from natural background (130), medical and dental (75), and radioactive fallout (5). Continuation of the ban on nuclear weapons testing is imperative if we are to keep radioactive fallout from increasing man's exposure to radiation. Furthermore, there is an obvious, real danger to the proliferation of nuclear weapons for mankind in the fallout that necessarily will come from the use of such weapons.

An interesting "use" of radioactivity in the atmosphere from nuclear explosions has been to detect forgeries of paintings, whose originals were painted in the nineteenth and early twentieth century. The explosions since

Table 12-7. Yield of Some Radioisotopes from U-235 Fission

Radioisotope	Half-Life	Yield (atoms per 1000 ^{235}U atoms fissioned)
^{90}Sr (strontium)	28.1 years	58
^{137}Cs (cesium)	30.2 years	61
^{131}I (iodine)	8.1 days	31
^{95}Zr (zirconium)	65 days	62
^{84}Kr (krypton)	10.8 years	3
^{140}Ba (barium)	12.8 days	64

1945 have increased the carbon-14 content of the environment, with the result that the carbon-14 content in plant life should be greater in plants alive after 1945 than in plants that died before 1945. Linseed oil, a carbon containing compound, is commonly used as a base for artists' paint, and is prepared from the seeds of flax, an annual plant. The carbon-14 content of the paint (linseed oil based) has been shown to reflect, as predicted, the environment in which the seeds were grown and harvested. For example, the carbon-14 content of the paint of a painting done in 1968 has been found to be double the carbon-14 content of one done in 1933 (prenuclear weapons).

A Question for the Reader: Does carbon-14 decay by one-half in 35 years? Estimate how much of the original carbon-14 might decay in 35 years.

Handling of Hazardous Wastes (Power Plants)

Certain Fission Products

We have spoken in general terms about radioactive substances being placed in the environment as a result of the ^{235}U fission reaction in explosions. Before we discuss the handling of these same radioactive wastes in nuclear power plants, let us consider the behavior of two radioisotopes, strontium-90 (^{90}Sr) and cesium-137 (^{137}Cs), which will help us to understand the mechanisms necessary for environmental protection.

Both ^{90}Sr and ^{137}Cs are found after ^{235}U is fissioned (Table 12-7) and are beta-ray emitters with substantial half-lives, 28.1 years and 30.23 years, respectively. The lifetime of these two radioisotopes is such that significant decay does not occur quickly. When these materials arise from an explosion, they become widely distributed throughout the environment prior to any significant decay. All radioactive materials decay on a time scale characterized by the half-life of the isotope. A general decay curve is shown in Figure 10-5. The Health Physics Division of the United States Atomic Energy Commission (A.E.C.) many years ago established the guideline that a radioisotope must be kept separate from the biological environment for a time period equivalent to at least 20 half-lives. In a 20-half-life time period, the radioactivity of a substance will decay to about one millionth of its original radioactivity (actually 9.5×10^{-7} times). This means that about 600 years (20 × 30) are needed for the isolation or storage of any materials containing ^{90}Sr and/or ^{137}Cs!

However, there is more reason for concern about these two radioisotopes than merely their half-lives. Examination of the periodic table reveals that the element strontium is in the same vertical group, group II, as are magnesium, calcium, and barium. In the discussion of the periodic table (Chapter 5), we pointed out that elements of the same group, in this case the group known as the alkaline earth elements, have great chemical similarities.[3]

More specifically, calcium is one of the principal constituents of bones. Calcium (from milk) is deposited in bones as calcium phosphates, particularly when bones are healing, or more importantly, while bones are growing in young children. Thus, strontium, when present, is deposited along with calcium as bones are formed. In such bones, some radioactive ^{90}Sr, a beta-ray emitter, is located near cells along the bone and near the bone marrow where the body's red blood cells are produced. The bone marrow and red blood cells are damaged or destroyed by beta rays. This can cause leukemia.

Similarly, examination of the periodic table reveals that cesium is in the same vertical group (group I) as are sodium and potassium. Most compounds (salts) containing the group I elements are soluble in water. Moreover salts containing sodium and/or potassium are important in many bodily functions, particularly as components of body fluids. The similarity in chemical behavior of the members of this group results in the easy entry of cesium, including the radioactive ^{137}Cs, into body fluids. As long as any ^{137}Cs exists in these fluids, all parts and cells of the body (the average water content of the human body is 65%) are exposed to beta and gamma radiation. Clearly, neither ^{90}Sr nor ^{137}Cs (nor any other radioactive fission products) should be permitted to enter the environment from any source in any significant quantity (see discussion of standards p. 370).

Put another way, ^{90}Sr and ^{137}Cs are the most troublesome of the ^{235}U fission products or ^{235}U fission-derived products. These two radioisotopes have short enough half-lives (about 30 years) to have a very intense beta-ray decay, but yet a long enough half-life that this decay does not decrease in intensity very much, even after many years or as when compared to a person's lifetime. Fission products with half-lives of only a few days lose their intensity rather quickly, and any with rather long half-lives (for example, technicium-99, $t_{1/2}$ of 10^6 years) undergo decay so gradually that the intensity is very weak, so neither constitute generally a serious personnel hazard.

Wastes in the Power Plant

In Chapter 11, we examined in some detail selected critical features of nuclear power plants. Among these features are the fuel (fissile atoms such as ^{235}U or ^{239}Pu), which is in itself radioactive (alpha emitters) and the massive or bulky shielding that is necessary to protect the environment from the radioactive emissions, particularly gamma rays. In the section

---

[3] Since ^{90}Sr is part of radioactive fallout, cows ingest ^{90}Sr when they forage on materials dusted with ^{90}Sr. In the cow's body, the ^{90}Sr is then concentrated along with calcium in the cow's milk and is then consumed by people.

above, we discussed in some detail two of the materials found amongst the many products of the fission reaction. Let us now utilize this information as we consider how to handle these radioactive materials in the power plant and after use in the power plant.

Initially stainless steel or zirconium metal tubes contain nuclear fuel—pellets of enriched ^{235}U. The tubes are known as fuel rods. In the reactor, many of these fuel rods are grouped together. When the control rods, which move between and amongst the fuel rods, are withdrawn (Chapter 11), the nuclear chain reaction proceeds. As the rods become heated, the heat is carried away by the coolant (water) circulating across the rods.

What goes on in the fuel rods? The neutron-initiated nuclear chain reaction proceeds and generates the various fission products and heat, as in equations (10-12). The fission process in the nuclear reactor leads to the same radioisotopes as are found in nuclear explosions. The fission products are retained within the fuel rods. The fuel in the rods gradually becomes less capable of maintaining the fission reaction (less reaction means less heat), because ^{235}U is consumed and because some of the ^{235}U fission products are very efficient capturers of neutrons. Such capture prevents some of the neutrons from fissioning ^{235}U. At some point, then, a given fuel rod stops producing sufficient heat even with the control rods rather completely withdrawn. Such a fuel rod is now known as a "spent" fuel rod.

What is in a spent fuel rod? A spent fuel rod contains all the materials initially placed in the fuel rod, less about 6% of the ^{235}U. This amount of ^{235}U has been transformed into fission products with the concomitant release of heat energy. Those fission products with very short half-lives (minutes, days, even weeks or months) are very, very intensely radioactive. Other products such as ^{137}Cs or ^{90}Sr are intensely radioactive but less than the first-mentioned materials. A few products of very long half-lives are weakly radioactive as is the unfissioned uranium. A spent fuel rod, just after removal from the reactor, is extremely radioactive (hot), and must be handled with the greatest of care (protection of the environment and all people working at the power plant).

When a spent fuel rod is removed and replaced by a fresh fuel rod, the spent rod is stored at the power plant in a well-shielded area, under water for some months to dissipate the heat energy produced as the very short-lived radioisotopes decay. Although storage at the power plant, but outside the reactor unit, for several months solves the problem of the highly radioactive materials, such a time period of storage does not result in any significant change in the levels of materials such as ^{137}Cs or ^{90}Sr. Furthermore, some 94% of the original uranium still remains in these spent rods, which, if recovered, could be put to good use in a new fuel rod. The reprocessing of spent fuel to separate useful uranium and plutonium from radioactive, nonfuel material is done at reprocessing plants, in isolated locations, away from nuclear power plants. This, of course, brings up the potential transportation problem as nuclear power grows.

Before we discuss transporting, reprocessing, and storage of the final wastes, let us examine briefly another aspect of power plant operation—the formation of radioative gaseous products in the power plant. One of the many fission products is krypton-85, $^{85}_{36}Kr$. Krypton is a gas that does not dissolve in water. This specific radioisotope has a $t_{1/2}$ of 9 years. Because this element is a gas, it may diffuse through the thin casing of the fuel rods. Certainly if there are any faults in the casing, such as pin holes or cracks, the ^{85}Kr will easily pass out and into the coolant water. The chance for escape of ^{85}Kr is real. In addition, wherever boron is used, say in the control rods, another radioactive gas, tritium, $^{3}_{1}H$, can form as shown in equation (12-3).

$$^{10}_{5}B + ^{1}_{0}n \longrightarrow ^{3}_{1}H + ^{8}_{4}Be \qquad (12\text{-}3)$$

Gaseous tritium, a beta emitter, with a $t_{1/2}$ of 12 years, is like its isotope hydrogen (a small molecule), extremely difficult to contain. Such radioactive gases as ^{85}Kr and ^{3}H are found in extremely small quantities. The disposal of these materials is by a method familiar to all of you, "dilution as the solution of pollution." Specifically, most of these materials go up a chimney stack as gas and are diluted in the atmosphere. Full-scale operation of some nuclear power plants has been delayed because of too high a level of these radioactive materials passing out the stack.

In operating practice, the radioactive gases are allowed to move into a separate storage area inside the outermost reactor shield. There, the gases are held until the radioactivity due to any components shorter-lived than krypton-85 or tritium has dropped to a safe level and then the gases, including krypton and tritium, are vented off through the tall stack. These gases are the only source of radiation routinely given off by an operating nuclear power plant.

Storage of Wastes

At the power plant site, then, it is possible to retain spent fuel rods for a time period long enough for the highly radioactive materials with short lives to decay, but the time period necessary for decay of ^{90}Sr and ^{137}Cs to safe levels is just too long, that is, 600 years. Thus, the spent fuel rods after several months storage at the power plant are shipped in casks (Figure 12-7) to reprocessing plants where the unconsumed uranium fuel is chemically separated from the radioactive, useless (nonfuel) materials containing some cesium and strontium. These leftover wastes must then be stored someplace where they will *never* enter the environment. As a result of the fuel reprocessing, these wastes are in the form of a water solution and/or a water slurry containing some solid material. By the early 1970s, the

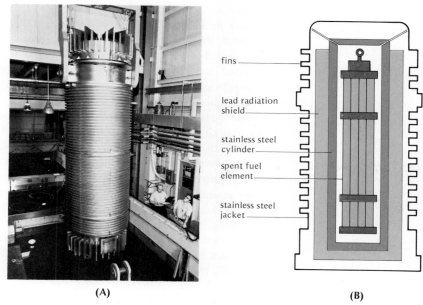

fins

lead radiation
shield

stainless steel
cylinder

spent fuel
element

stainless steel
jacket

(A)   (B)

Figure 12-7. **(A)** Shipping cask for spent fuel rods. [Courtesy of The General Electric Co.] **(B)** Sketch showing the interior of a shipping cask.

Atomic Energy Commission (A.E.C.) had in isolated storage some 93 million gal of these aqueous liquids. A typical storage container is shown in Figure 12-8.

Where are these to be stored? Most of the A.E.C. locations are relatively isolated from major centers of population: along the Savannah River in

Figure 12-8. High-level waste storage tanks for solutions or slurries containing radioactive wastes. The tanks are built of carbon steel, surrounded by concrete encasements 2–3 ft thick, set about 40 ft in the ground and back-covered with dirt. [Courtesy of U. S. Atomic Energy Commission, Photograph, E. I. DuPont DeNemours & Co., S.C.]

South Carolina; Hanford, Washington; and the National Reactor Testing Station in Idaho. Although wastes are taken to these out of the way locations, the storage of liquid wastes presents some problems. For example, there has been some leakage from the containers, particularly at the Hanford storage facility. The containers do deteriorate in a relatively few years. Thus, there is need for constant monitoring and periodic replacement of the containers.

The water slurries are relatively dilute, that is, do not contain large amounts of radioactive materials. Such quantities of water conveniently dissipate the heat released during radioactive decay of the wastes. However, the large volumes of these slurries, projected to increase manyfold as more nuclear power plants are constructed, pose massive storage problems. Many, many containers will dot the landscape and will need monitoring unless alternative methods of storage and methods of reducing liquid volume are developed. Water, if it contains *no* radioactive materials, can in principle be returned to the environment without polluting it. However, the technology for reducing the volume by recovery of water free of radioactive materials, leaving the radioactive material behind as a solid, has not been fully developed.

Predictions have been made that the waste in storage plus all wastes to be accumulated through 1980 can be reduced to a concentrated water slurry whose total volume would be only 5 million gal. Furthermore, this 5 million gal, in theory, could be converted to solids, which would occupy only one-tenth this volume. Obviously, shipping and storage of reduced volumes seems easier, although in the more concentrated forms, the modes of heat dissipation from the radioactive decay processes are not fully developed. Volume reduction continues to be studied intensely at all three A.E.C. locations because increased use of nuclear power necessarily means more and more radioactive wastes requiring long term storage.

Interestingly, in the 1960s, considerable study was made of abandoned salt mines as a place for radioactive waste storage. The A.E.C. has operated a pilot project at Lyons, Kansas, for many years (Figure 12-9). Salt mines are very, very dry places and are expected to remain dry and undisturbed for thousands of years. Salt (sodium chloride) is very soluble in water. In order to have had salt remain deposited in such a place within the earth, there must be no water in the vicinity. Furthermore, salt mines are good for radioactive waste storage because salt is a good radiation shield that will melt around the wastes, locking them in place, and because regions containing underground salt deposits are not known to be involved in earthquakes. The only clear disadvantage of salt mines is that one must guarantee that man does not intervene by boring holes, that is, drilling for oil, and so on. Holes in the salt might allow water to enter the mine, and the integrity of the salt deposit with its stored radioactive wastes could

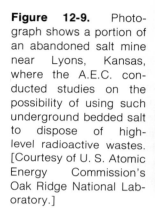

Figure 12-9. Photograph shows a portion of an abandoned salt mine near Lyons, Kansas, where the A.E.C. conducted studies on the possibility of using such underground bedded salt to dispose of high-level radioactive wastes. [Courtesy of U. S. Atomic Energy Commission's Oak Ridge National Laboratory.]

not then be guaranteed. Certainly monitoring these storage sites would be less of a problem than monitoring the tank storage previously described.

However, as attractive as the use of abandoned salt mines may be, the A.E.C. has deferred plans for such a central, radioactive waste repository pending several years of additional planned study. In 1972, the A.E.C. committed itself to engineered surface storage in isolated locations with constant technical surveillance until the 1980s. Of major concern for each and every person must be the continued search for sufficient and satisfactory sites for the storage of fission product wastes, especially ^{90}Sr and ^{137}Cs. Such storage studies are increasingly important as the number of nuclear-fueled power stations increases both in the United States and elsewhere in the world.

Alvin M. Weinberg [1] has stated with penetrating clarity the fact that nuclear energy involves a "Faustian bargain with society." Nuclear reactors, both burner and breeder (whose waste problems to a first approximation are the same as those described above), offer an almost inexhaustible supply of energy that is cheaper than energy from fossil fuels and is almost nonpolluting. "But, the price that we demand of society for this magical energy source is both a vigilance and a longevity of our social institutions to which we are quite unaccustomed, and to a degree that is probably unparalleled in history. We have relatively little problem dealing with radioactive wastes if we can assume *always* that there will be intelligent people around to cope with eventualities . . . as we deal with our wastes indefinitely. Is mankind prepared to exert the eternal vigilance needed to

ensure proper and safe operation of its nuclear energy system?" This is
the Faustian bargain. Are you prepared to make this choice rationally?
The scientific and technological community is not 100% agreed in favor
of controlled nuclear energy development. Before you choose, let us consider standards of exposure to radiation.

Radiation Standards   Earlier in this chapter we discussed various measures of radiation and
noted some difference in biological effect of various kinds of radiation. We
also considered the natural background of our radioactive environment,
which varies from 100 to 250 mrems/year in the contiguous states of the
United States. The official United States standard-setting body was the
Federal Radiation Commission (F.R.C.) until the early 1970s when the
Environmental Protection Agency assumed this role. The F.R.C. shortly
after its formation in the early 1950s, proposed a set of principles that are
valuable for us today and include:

1. Exposure to radiation should always be as low as practicable.
2. No exposure should be allowed without expectation of benefit.
3. It is assumed that all radiation is potentially harmful.
4. It is appropriate to set different standards for different classes of sources.
5. Unless it can be proved otherwise, the biological risk associated with
   low levels of exposure is proportional to the risk that has been estimated
   or demonstrated at higher exposure levels.

Furthermore, the F.R.C. set the maximum allowable average exposure for
the general population at 170 mrem/year above background radiation with
the provision that no individual was to be exposed to any more than
500 mrem/year above background radiation. This figure was based on the
recommendation that in a 30-year period, a person should receive no more
than a total of 5 rems (5000 mrems) of radiation above background radiation.

Pressure from concerned scientists and concerned citizens plus more
recent data of the A.E.C. has led, in 1971, to the proposal of more stringent
standards; namely, that the maximum allowable exposure of an individual
or any of his organs in an unrestricted area shall be no more than
about 17 mrems/year.

How does this standard compare with experience? In the region surrounding the Oyster Creek Nuclear Generating Station, New Jersey, considerable data has been collected. In the 1966–1968 period before the plant
went into operation, the background radiation was 105 mrem/year. In
downwind locations as close as 2 miles to the plant, in operation since 1969,
there has been no observable change in radiation level. Only on-site loca-

tions (restricted areas) have been measured as high as 124 mrem/year (105 as background, 19 from plant operation). Not all nuclear plants record such low levels of radiation, but recall that the Oyster Creek station has served as a model for all stations contracted for subsequently (since 1963). Nonetheless, A.E.C. officials state that even by the year 2000, radioactive levels from all the nuclear power plants projected to be built by then will contribute only about 3 mrem/year of radiation to the environment.

There is, however, another way at looking at the effects of radiation on man. Joseph Liberman [2] of the Environmental Protection Agency summarizes the concept of manrems. The manrem is obtained by multiplying the rems by the number of people. For example, if 1000 people each received 1 mrem of radiation per year, we say that 1000 millimanrems per year or 1 manrem/year would be the radiation exposure of this population group. Radiation exposures quoted in manrems represent a combination of the per capita exposure and the size of the population exposed to the radiation. Demonstration of health risk (in terms of manrems) can be obtained from information in the International Commission on Radiological Protection reports.

As an example, Table 12-8 sets forth some Environmental Protection Agency data. In the first three items of this table, the intensities of radiation are not expected to rise, but the population of the United States will and they will obtain better medical care. However, the figure for nuclear power is based on the A.E.C.'s forecast that, by the year 2000, over half of our electric power will be nuclear fueled. With regard to the hazard, some estimates indicate that for every 7000 manrems, we can expect one additional cancer. It seems probable that, if the data is correct, then a relatively small number of additional cancer patients can be expected on the basis of the radioactivity emitted from the nuclear power industry.

The word of caution here is in respect both to the data (additional check plus recheck of data is necessary) and to a key assumption (not stressed previously), that the nuclear-fueled power facilities will be very carefully monitored and that all possible safety precautions will be observed scrupulously. Vigilance is necessary, particularly when one is dealing with a benefit that carries such risk along with it.

Table 12-8. Radiation Impact on the United States, 1970 and 2000

Source	Estimated Exposure (in manrems, 1970)	Estimated Exposure (in manrems, 2000)
natural background	27,000,000	40,000,000
medical dental x rays	18,000,000	40,000,000
weapons fallout	1,000,000	1,000,000
nuclear power	430	56,000

**Benefits versus
Hazards:
Trade-Offs**

The essential facts are before you. The extremes are easy to define. One position is that no radioactive materials be permitted to escape into the environment. Technologically speaking, if we are to have nuclear-fueled electric power, we must have some radioactive materials produced that may pass into the environment. It is interesting to note that a certain amount of radioactivity is released in a fossil-fuel system. This radioactivity is not a product of the energy releasing reaction, as in a nuclear reactor, but is evolved into the environment because of very, very small (trace) quantities of radioactive radium atoms as impurities in coal or oil. Thus, no radioactive materials in the environment literally must mean no electrical power generation by burning coal or oil or by use of nuclear reactions.

However, the other extreme—building power plants at any location without regard to safety—must not be permitted. You should note that historically there has been wide acceptance that environmental pollution is an inescapable by-product of industrial development. That we are in or are approaching a different period in history necessitates our pausing to evaluate and consider how much disturbing the environment can tolerate and then recover reasonably quickly, if it can recover at all.

To many of you, the position stated above may seem most unsatisfactory, but you and I are frequently faced with a combination of benefits and hazards. Radioactive materials merely represent one of these. Do we tolerate the hazard of radioactive wastes appropriately monitored and controlled in return for the benefit (or convenience) of electric power? A position is stated by Rene Dubos and Lady Barbara Ward in *Only One Earth* [3]. Their view is that the world in general and the American people have already indicated a willingness to take the risk, even if that risk may mean death. Consider that in the early 1970s some 50,000 deaths are directly attributable to the automobile. How many people are prepared to abandon automobiles completely? Are you? In the minds of most, the convenience and benefits of the automobiles seems to out-weigh the risk, even of death. Similarly, in 1971, 741 billion cigarettes were consumed in the United States despite the correlation of lung cancer (since 1950) with smoking and despite the specific labelling of the cigarettes with the statement, "Warning: *The Surgeon General has determined that cigarette smoking is dangerous to your health.*"

So it is with the question of nuclear power. You must ask the question: Is the benefit worth the hazard? This problem of balancing hazard or risk against benefit is perhaps the oldest problem in human experience.

Conclusion

You now have before you the dilemma imposed by the production of electrical power using nuclear reactions. We have examined all the key facets of this source of electrical energy except the thermal effects on the environment in the vicinity of the power stations. These heat effects are discussed next in Chapter 13. At this point, you should be aware that the world's appetite, and America's in particular, for electrical power is voracious. The generation of electricity with hydropower plants changes rivers (by damming) drastically. The generation of electricity in fossil-fuel fired plants pollutes the air (Chapter 14), consumes limited resources (Chapter 6 and 8), and has thermal environmental effects in the vicinity of the power plants (Chapter 13).

The nuclear story, covering three chapters, began in Chapter 10 with a setting forth of the essential chemistry and physics of nuclear transformations. In Chapter 11, we examined the harnessing of nuclear reactions for peaceful purposes, particularly for electrical power generation. The present chapter has focused attention on the detection, measurement, and control of radiation in the environment. We necessarily have defined pollution and have examined some of the effects of radioactive emissions on the ecosphere, and particularly on man.

The fact that the nuclear energy has been as carefully studied as it has, truly places us in the position where we must do more than merely acknowledge the Faustian bargain posed by Alvin M. Weinberg. Consumption of electrical energy is rising at some 7% annually. To begin to meet this demand requires more extensive energy sources (some say inexhaustible) which are minimally polluting. Of the possibilities visualized in the mid-1970s, only one—the nuclear breeder reactor—appears to be technologically and economically realistic. Our energy crisis is essentially made up of the people's energy wants *and* of technology's (science-based) ability or inability to provide energy along with a low enough level of environmental pollution. We will return to this dilemma in Chapter 15 after exploring in greater depth thermal pollution and air pollution.

Questions

1. Define pollution, pollutant, polluter.

2. Define curie, rad, rem.

3. How does a Geiger counter detect and count ionizing particles?

4. What is natural background radiation? From what sources does this radiation arise along the East Coast of the United States? Colorado?

5. What is a spent fuel rod?

6. How are radioactive wastes in spent fuel rods handled and then disposed of?

7. Certain gaseous radioactive fission products of ^{235}U have half-lives sufficiently long as to retain some radioactivity when they leave the plant's stack. What are they?

8. Why are ^{90}Sr and ^{137}Cs of concern?

9. Why is it imperative to reprocess the materials in a spent fuel rod?

10. What use are radiation standards?

11. What problems are associated with the storage of radioactive wastes?

12. Write equations for the following:
    (a) Beta decay of strontium-90.
    (b) Beta decay of cesium-137.
    (c) Beta decay of tritium.

Numerical Exercises

1. (a) How does a 5 rad dose of gamma rays compare in biological effects with a 2 rad dose of alpha particles?
   (b) How does a 5 mrem dose of gamma rays compare in biological effect with a 2 mrem dose of alpha particles?

2. (a) Calculate the manrems of exposure in 1970 of natural background radiation in the United States population of 210×10^6 people. (b) What will be the manrems if the United States' population is 280×10^6? 300×10^6? 350×10^6?

3. The incidence rate of cancer in the population is about 280 cases per year per 100,000 people. This rate of development has occurred and occurs in an average radioactive environment of about 200 mrem/year.
   (a) How many cases of cancer develop each year in a population of 210×10^6 people?
   (b) If we assume that all cancers are caused by radiation, how many additional cases might develop annually if the average radiation level were 5 mrems greater?
   (c) Testimony [4] before the Senate Committee on Public Works indicated that a 1 rem (1000 mrem) exposure, over and above dosage that we normally receive (200 mrem/yr), would increase the cancer incidence rate by 1%. If we are exposed to an additional 5 mrem on the average [as in part (b)], what will be the increase in cancer incidence rate annually for the United States population of 210×10^6 people? Compare your answer to that found in part (b).

References

[1] A. M. Weinberg: *Science*, **177**:27 (1972). For another viewpoint using Weinberg's data, see J. T. Edsall: *Science*, **178**:933 (1972).
[2] J. A. Liberman: *Phys. Today*, **24**(11):32 (Nov. 1971).
[3] R. Dubos and Lady Barbara (Ward) Jackson: *Only One Earth*. Norton, New York, N.Y. (1972).
[4] Testimony, dated Nov. 18, 1969, by J. W. Gofman and A. R. Tomplin (Berkeley and Livermore, Calif.) before the Senate Committee on Public Works.

We May Be in Hot Water— Thermal Pollution

Because biological activity occurs in a rather narrow temperature range, principally between 0 and 100°C, ecologists consider the temperature the primary control of life on earth. If so, why should a "little" extra heat dumped into the environment constitute a problem? As we have discussed in Chapter 3, nature is made up of a large number of interacting cycles and chains (food, carbon, oxygen, and so on) all existing in a dynamic equilibrium, often referred to as the balance of nature.

Temperature is one of the primary controllers of these complex cycles and chains, so that any permanent change in average surface temperature of the entire earth or in any locale will necessarily affect these chains. Unfortunately, not enough is known about the various natural balances to predict with some confidence the extent of the effects of even small temperature changes. However, enough is known about temperature effects on various species of aquatic life to provide some basis for concern

375

that the thermal discharge from power plants may present a *potential hazard* to the balance of nature and perhaps to life in general. It is for this reason that the thermal discharge has been called **thermal pollution.**

In Chapter 8 we concerned ourselves with the overall problem of heated discharges into the runoff during steam electric power generation. We have seen that in regions local to the power plants, the waste heat can cause a rather severe heat burden. Such heat is expected to cause major damage to the biotic communities in the plant vicinity. Is the problem of degraded heat energy even more universal? According to the second law of thermodynamics, the result of the consumption of a quantity of energy is the degradation of some (and ultimately all) into heat. Thus, as we consume energy worldwide, we must ask questions about the possible effects of increases in global temperatures. At some point, that is, when a large enough temperature change has taken place, there will be the interruption of the balance of nature on a global basis, with consequent changes in weather and climate. Unfortunately, again we are faced with inadequate knowledge about the causes of changes in climate and, in particular, of the effects on climate of small temperature variations. In our brief treatment, we will attempt to summarize the current thinking in the area of global heating effects on weather and climate. It should suffice to say that such studies are an entire science within themselves.

In this chapter, we discuss the basis for thermal pollution with regard to a specific example—the Oyster Creek nuclear power plant. We generalize the problem in order to have a rule of thumb for heated discharge versus the operating power of a plant. We then consider various media available for the disposal of waste heat, including pros and cons of present practice and suggestions for alternatives. Because the question of thermal pollution ultimately depends on the extent of impairment of the quality of the environment, a discussion of the biological effects of waste heat on the environment then follows.

Thermal Pollution— A Natural Consequence

From our study of nature's guidelines, the laws of thermodynamics, we have seen that energy could be transferred and transformed but could neither be created nor destroyed in the process. However, you remember that the first law said nothing about the direction of heat flow in various processes nor did it set any guidelines for the conversion of heat or other forms of energy into work. It is the second law that says no machine is 100% efficient, which means that only a portion of the energy being converted becomes useful work, whereas the remainder becomes waste. It is on this waste (heat) that we focus in this chapter.

We have seen the second law in practice during our study of steam turbine operation. Recall the simplified view of the continuous operation of the power plant (Figure 7-4). To complete the operating cycle and to increase the plant's efficiency, heat is removed from the used steam by circulating cooling water (from a lake or river) over the condenser coils, thus condensing the used steam to water. This water is then pumped back to the boiler for reheating and the start of a new cycle.

You also recall that the cooling water is not at a high enough temperature to operate (in principle) a heat engine nor does it contain enough heat for space heating in distant towns or cities. Thus, the heated cooling water is pumped back into the rivers, lakes, or estuarine waters, raising the average temperature of that body of water in the vicinity of the power plant. It has been the concern of the ecologist and is now the concern of society that this thermal waste (or waste heat) may eventually ruin our rivers, lakes, and estuaries.[1] In all cases, the thermal waste, which we have called thermal pollution, is a direct consequence of the second law of thermodynamics.

From our restatement of the second law in Chapter 9, we also know that the entropy of our closed thermodynamic system, that is, the entropy of spaceship earth (and the entropy of the universe), must have increased by virtue of the heated discharge. Furthermore, that law actually states that the entropy in an isolated system (which for this case is the universe) either remains the same or increases. However, the condition of the total entropy remaining the same represents an idealized (reversible) process that never occurs in nature, so we can ignore this part of the statement for our study. You recall that increased entropy means energy becoming unavailable for producing useful work. Energy that has become widely distributed is unavailable energy in this respect. Thus, waste heat from power plants quickly becomes widely distributed in our runoff and is surely unavailable for producing useful work. Therefore, we again conclude that energy consumption for power generation and the corresponding thermal waste increases the entropy of our system—earth and the whole system, the universe—which is consistent with that overpowering trend in nature.

From our preceding discussion, we may equate pollution and, in particular, thermal pollution with increasing entropy. Even though we have taken a specific case, an electrical power generating plant, for studying the "natural trend of things," this example is the epitome of our technological society, a machine that consumes energy and produces goods and services with an overall efficiency of about 50%. The remaining 50% represents thermal waste and an immediate entropy increase, usually right in the vicinity of where the energy is being used. It should be reemphasized that our nonrenewable energy resources (fossil, nuclear, and geothermal fuels)

---

[1] The actual effect of heated discharge is discussed later in this chapter.

represent a low entropy state. As we use them, we increase the entropy of our system. However, the concern of environmentalists, scientists, ecologists, and so on, is the seeming rush toward using our fuels at an ever-increasing rate and the maximizing of the entropy of our system in only a few decades or at most a few hundred years. You recall that maximum entropy is synonomous with the term heat death for our thermodynamic system—spaceship earth.

Now that we have set forth the meaning of thermal pollution, in a qualitative way, let us turn to a specific example for a more quantitative approach to the problem of thermal pollution.

A Specific Case Study— Oyster Creek

We have selected for our study the Oyster Creek Nuclear Generating Station near Toms River, New Jersey. The station is owned by the Jersey Central Power and Light Company (J.C.P.L.) which is a subsidiary of the General Public Utilities Corporation (G.P.U.). The G.P.U. system is part of an Interconnection System operating in New Jersey, Pennsylvania, and Maryland. Physically, the station is located near Barnegat Bay on a 1412 acre site, 35 miles north of Atlantic City and 60 miles east of Philadelphia.

Our interest in this particular example stems from several reasons. The Oyster Creek facility (Figure 13-1) is located on estuarine waters. Some of the more highly publicized controversies about plants have been directed at those located on various rivers throughout the country. Secondly, the Oyster Creek facility is considered to be a "model" plant incorporating some of the latest technological advances (up until the plant going on-line in 1968) for BWR's (boiling water reactors, Chapter 11). Thirdly, it represents the first case for which a significant fish kill has been reported upon the plant shutting down (for repairs). Most of the concern with power plants to date has been with the effects of heated discharge during continuous operation.

The Oyster Creek facility contains a single BWR, turbogenerator (steam turbine and electrical generator), and accessory equipment with an expected maximum electrical capacity of 640 Mw (640 × 10⁶ w). Once-through cooling water is used to remove waste heat from the condensers. Water is taken from Barnegat Bay by way of the south branch of Forked River and a dredged canal. The condenser discharges into a dredged canal flowing into Oyster Creek and eventually into Barnegat Bay. A map of the area, including the coolant flow pattern, is shown in Figure 13-1B.

The station discharges waste heat nominally at a rate of 4.4 × 10⁹ Btu/hr [1], utilizing a flow through the condensers of 460,000 gal/min or 27.6 × 10⁶ gal/hr. Because 1 gal weighs about 8.3 lb, the flow converts to 2.3 × 10⁸ lb/hr. Using the conversion factor (1 Btu

(A)

Figure 13-1. (A) Oyster Creek Nuclear Generating Station. **(B)** An aerial photograph of the Oyster Creek electrical generating facility. The dredged canal is fed by Barnegat Bay and Forked River shown at the top, and the heated effluent is diluted by Oyster Creek, below. [Photographs courtesy of General Public Utilities Service Corp.]

(B)

Flow

EXAMPLE 13-1

Waste heat from the Nuclear Generating Station at Oyster Creek is discharged at the rate of 4.4×10^6 Btu/hr. If the water flow through the condensers is 2.3×10^8 lb/hr, determine the rise in temperature of the coolant.

$$\frac{4.4 \times 10^9 \text{ Btu/hr}}{2.3 \times 10^8 \text{ lb/hr}} \times \frac{1 \text{ lb } °F}{1 \text{ Btu}} = 19°F$$

---

raises the temperature of 1 lb of water $1°F$), we find that the water flowing through the condensers is heated by $19°F$. This calculation is shown in Example 13-1. We note at the outset that we have used the English system of units, mainly because all data and reports published in the literature are in this system for Oyster Creek and other facilities.

Using the rate of the heat discharge, we can find the rate at which heat is obtained from the nuclear pile. From Example 13-2, that value is 6.58×10^9 Btu/hr. If this were a fossil-fuel fired plant, this would be the heat obtained from burning oil, coal, or gas at the rate of about 200 metric tons/hr of coal equivalent. Assuming the plant operates at 33% efficiency,

---

EXAMPLE 13-2

Given that the rate of thermal discharge of the plant is 4.4×10^9 Btu/hr and that it is operating at 33% efficiency, which leaves 67% for the inefficiency, find the rate at which heat must be produced by the nuclear pile. Let x be the heat produced by the nuclear pile, then

$$\text{inefficiency} \times \text{total heat} = \text{waste heat}$$

$$(0.67)x = 4.4 \times 10^9 \text{ Btu/hr}$$
$$x = 6.58 \times 10^9 \text{ Btu/hr}$$

EXAMPLE 13-3

If the rate of heat input to the boiler of a power plant is 6.58×10^9 Btu/hr and if the plant operates at 33% efficiency, what is the power output of the plant?

$$0.33 \times 6.58 \times 10^9 \text{ Btu/hr} = 2.17 \times 10^9 \text{ Btu/hr}$$

$$2.17 \times 10^9 \text{ Btu/hr} \times \frac{1 \text{ kw}}{3412 \text{ Btu/hr}} = 636 \times 10^3 \text{ kw}$$

$$= 636 \text{ Mw}$$

which is consistent with the stated operating level. (See Appendix A for conversions) we find that the plant operates at 636 Mw, which is close to its design objectives. The details of this calculation are shown in Example 13-3.

As we have mentioned, the 460,000 gal/min of coolant is provided by waters from Barnegat Bay and the south branch of Forked River, which are channeled into the facility by a dredged canal (as shown in Figure 13-1). The canal is about 100 ft across and about 10 ft deep. The water flow speed is about 1 ft/sec. For comparison, the canal can be considered as a typical stream or river. If we assume that all of the flow is diverted through the condensers, the above flow which calculates to be 1000 ft³/sec (cubic feet per second) yields the necessary 460,000 gal/min or 2.3×10^8 lb/hr. The details of the calculations are shown in Example 13-4.

Also shown in Figure 13-1 is the dilution channel that diverts water from entering the condensers to the discharge channel. By-passing cool water from the intake point directly to the discharge canal allows the temperature of the estuary and bay waters (during summer months) to be maintained below the 95°F limit set by the New Jersey Environmental Protection Department (E.P.D.). The water temperature is monitored continuously in the bay not far from the discharge channel. The beneficial result of diverting some cooling water is the dilution of the high tempera-ture effluent water coming from the plant. *Dilution, then, is a solution to thermal pollution!* The adverse effect of diverting part of the flow from the condenser is to raise the temperature of the now smaller coolant volume even further, greater than 19°F. In Chapter 7, it was emphasized that a

---

EXAMPLE 13-4

A stream or canal 100 ft wide, 10 ft deep, and with a flow speed of 1 ft/sec supplies cooling water for a power plant. How much water is available for cooling in cubic feet per second, pounds per hour, and gallons per minute.

(a) volume flow = cross-sectional area of stream × average speed of stream
              = width × depth × speed
              = 100 ft × 10 ft × 1 ft/sec = 1000 ft³/sec

(b) If 1 ft³ weighs 62.4 lb and there are 3600 sec/hr, then

$$\text{weight flow} = 1000\,\frac{ft^3}{sec} \times 62.4\,\frac{lb}{ft^3} \times 3600\,\frac{sec}{hr}$$

$$= 2.25 \times 10^8\,\frac{lb}{hr}$$

(c) flow $= 2.25 \times 10^8\,\dfrac{lb}{hr} \times \dfrac{1\,gal}{8.3\,lb} \times \dfrac{1\,hr}{60\,min} = 450{,}000\,\dfrac{gal}{min}$

higher operating temperature for the low temperature reservoir, the coolant, decreases the efficiency. The output power level of the plant correspondingly drops. Unfortunately, the plant power level drops when it is most needed, during hot days because of the extensive power used for air conditioners.

The example that we have considered here is similar to plants operating on various rivers and estuaries around the country, particularly with regard to flow requirements. For example, the Vermont Yankee Nuclear Station on the Connecticut River (in Vermont) at times diverts two thirds of the river flow through the condenser, heating it by 20°F. Similar situations exist for other plants: for example, the nuclear plant near Haddam on the Connecticut River and the Indian Point plants on the Hudson River. The states of Oregon and Washington have set a 68°F limit on the Columbia River, which directly affects the operation of the Hanford nuclear plants.

The public is also becoming increasingly sensitive to the fact that nuclear plants discharge some 35–40% more waste heat into the waterways than does a fossil-fueled plant operating at the same power level. One reason is that about 12% of the total waste heat from a fossil-fueled plant goes up the smokestack with the flue gases, whereas all the waste heat from a nuclear plant enters the waterways. Another is the lower operating efficiency (about 10% lower) of the nuclear plant. Thus, the search for various media for waste heat disposal that minimize the environmental effects is *foremost in our concern about thermal pollution.*

Before discussing the various ways in which waste heat is disposed, we will generalize our above problem for a "rule of thumb." Because many future electrical power generating stations will be 1000-Mw (or so) facilities, our rule of thumb will be for this level of plant operation. A 1000-Mw nuclear plant operating at approximately 33% efficiency consumes 10.3×10^9 Btu/hr of heat energy while discharging 6.9×10^9 Btu/hr as waste heat. Operating under these conditions and limiting the temperature increase of the coolant to about 20°F requires 7.07×10^5 gal/min (a stream 157 ft wide, 10 ft deep, and flowing at 1 ft/sec). In contrast, a similar coal-fired plant (40% efficient) consumes heat energy at the rate of 9.7×10^9 Btu/hr, noting that 12% (1.2×10^9 Btu/hr) of this immediately goes up the stack and 5.1×10^9 Btu/hr is discharged into the cooling water. You should further note that the 6.9×10^9 Btu/hr thermal discharge for the nuclear plant is 35% greater than the 5.1×10^8 Btu/hr thermal discharge (into waters) for the fossil fueled plant.

In either case, the cooling water requirement is equivalent to the entire flow of an average sized river. In conclusion, then, any selection of plant site of necessity requires substantial water flows so that the water available for dilution is adequate for minimizing the elevation of the temperature of the coolant and its environmental impact.

Media for Waste Heat Disposal

The average American is not aware of the severe demands being made on one of our most important resources—**water.** As we have seen, the steam–electric power industry alone requires large quantities of water for cooling the plant condensers. In Chapter 8 we showed that, if the present rate of growth for electrical power continues, the entire runoff for the United States could possibly be needed by 2000 A.D. Assuming *once-through cooling* (use of rivers, and so on), this amount of heat will raise the temperature of this cooling water by 20°F.

However, to even assume that the waste heat can be distributed throughout the runoff is contrary to present practice and future planning. Whereas, 15 to 25 years ago, the maximum size of steam electric generating plants was about 200 Mw, new facilities are single 1000-Mw plants. The waste heat is then much more concentrated and the environmental impact can be expected to be far greater since *the water for dilution immediately available is far less.* This, then, brings us to the question of what natural media are available for cooling, and, in light of a projected severe heat burden on the aquatic environment, what other means are feasible for cooling? We have only three choices: water, air, or land.

Natural Media—Rivers, Lakes and Estuaries

Presently, our rivers are the most widely used media for waste heat disposal and, as many of you may have observed, are also widely used for industrial and municipal disposal. However, with our rapidly increasing rate of consumption and the consequent waste, many of our rivers are becoming unfit for aquatic life and human consumption. Perhaps the most obvious reason for the use of rivers, streams, and so on, as garbage dumps (including waste heat) is their wide geographical distribution across the country. However, this wide distribution must not be taken too literally as any geographer knows. Sections of the United States are water-poor, for example, the Southwest.

Power plant cooling water must be available in sufficient quantity the year round. How many rivers, streams, lakes, and so on are available to meet such quantity demands? Where are they geographically? The whole point here is that there are water limits on power plant siting or where power plants can (if we choose) be located. Many major cities that are large electrical energy consumers are located on these rivers and it is to the utilities' advantage to locate near the consumer to minimize energy losses in transmission. However, a major concern is that electricity consumption by the major population centers is growing, which means more and larger generating stations and more demand for cooling water.

For example, a large generating station or group of stations (Indian Point stations 1, 2, and 3 on the Hudson river with proposed total capacity of 3000 Mw) will need cooling water at the rate of about 40,000 gal/sec

(2.4×10^6 gal/min). This is nearly equivalent to the water requirements for New York City, and this also represents a substantial portion of the Hudson River flow. Once-through cooling, as it is called, is more efficient and more economic. Obviously, to recycle already heated water would reduce plant operation efficiency. The economic aspect is self-evident.

The extent to which a particular power plant may affect the aquatic life in the proximity of the plant and elsewhere is determined mainly by how much of the river's flow is needed for cooling. If the river is not very wide and is shallow, the entire flow of the river in the proximity of the power plant will be heated to a higher temperature than the waters upstream and way downstream. The effects of the heated effluent may be evident for relatively large distances downstream from the plant, depending on how rapidly heat is lost by evaporation, radiation, conduction, and dilution.

In Figures 13-2 and 13-3, are shown two sketches of the effect of the thermal plume on a river. Figure 13-2 shows an aerial view of a section of the river in the vicinity of the power plant. The thermal plume is evident in Figure 13-2 and the extent to which it covers the surface of the river naturally depends on the width of the river, the turbulence, and the speed of the flow. Figure 13-3 shows a projected temperature profile (a cross section) of the river. The lines shown are suggestions as to how the water temperature varies with the river's depth. Such an effect is called **stratification**. The density of water, a maximum of 1 g/cm^3 at $4°C$, depends on temperature. The higher the temperature, the lower the density; water at $15°C$ has a density of 0.99 g/cm^3, 1.0% lower than the density at $4°C$. Such a small difference in density is sufficient to cause the cooler, heavier

Figure 13-2. An aerial infrared photograph of the Connecticut River at the Yankee ''nuke'' plant at Haddam shows how the surface plume of warm condenser water spreads across the river. The white region of the plume is at $93°F$ whereas the darkest regions are at about $75°F$. [Courtesy of U. S. Geological Survey, Photograph by HRB-Singer, Inc.]

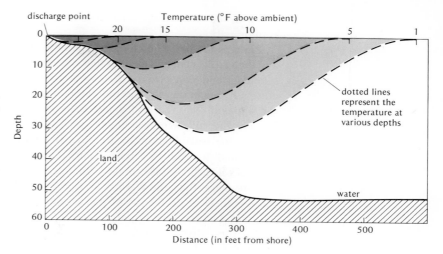

Figure 13-3. A temperature profile of a river showing stratification. The hot water discharged at the surface remains at or near the surface due to its lower density. The temperature scale at the top and the color density indicate water temperature at various depths.

(for a given volume) water to stay along the bottom and the heated effluent to stay on top—stratification.

Again, if the river is shallow, which most are, higher temperatures will reach closer to the bottom. The interest in this temperature profile is in the question to what extent the thermal plume might form a thermal block to various cold-water species of fishes, for example, salmon, shad, and trout. If the river is sufficiently deep and the plant does not use a significant portion of the flow, then the fish are able to stay in cool water and to pass under the plume as feeding and spawning habits necessitate. The Columbia River, with the Hanford Nuclear Plants, is an example of a large, cold river. Also of significance is the fact that the younger fish may not be able to escape being caught in the condenser flow or in the heated outfall (heated water discharge), and may die from fatigue or temperature shock. Other effects on the biotic community will be discussed later.

Lakes have been suggested as a possible dumping place for waste heat. However, many arguments have been raised against the use of lakes. Unlike rivers, the temperature of the water, the quantity of water, and flow patterns depend very much on the weather and the time of year. The turnover or flushing rate for many of the larger bodies of water is very slow, which negates any cooling by dilution. Also, the number of large lakes with water and flow capacity to handle thermal discharge with minimal ecological change or damage are relatively few. Furthermore, because many of our lakes have considerable recreational value, our citizenry feel that the construction of power plants on these lakes would be esthetically unpleasing. Few people imagine themselves wanting to swim and sun either in the shadow of a fossil-fueled plant belching forth sulfur oxides and fly ash or in the vicinity of a nuclear plant with *any* possibility of radiological contamination.

Perhaps the best known case in defense of keeping the lake shores clear and the waters cool and clean is that case made jointly by Cornell University scientists and concerned citizenry for Cayuga Lake. In 1969, a power company began digging the foundation for a nuclear generating station on the shores of New York's Cayuga Lake. A study by the Cornell professors yielded the following ecological picture of the lake (see Figure 13-4). During the warmer months, the lake is thermally stratified. Stratification or thermal layering of a lake occurs during the warmer months because the sun's heat penetrates mainly the upper layer of the water. Cold water is more dense than the warm water (a given volume weighs more) so that there is little or no mixing of the various layers. In Figure 13-4, a warm, constant temperature upper layer, **the epilimnion,** is observed whereas nearer the bottom a cold, constant temperature region, **the hypolimnion,** is observed. The transition region is called the **thermocline.**

Most of the biological activity (plant growth and reproduction) takes place in the upper layers where light is available for photosynthesis. The plants and animals that grow there eventually die and sink to the bottom, later to provide nutrients for further biological growth and reproduction. During the winter months, the upper layers cool. When they become the same temperature as the lower layers, the winds and currents cause mixing. Nutrients are swept to the upper layers for the next season's growth period and oxygen is mixed into the lower layers for the aquatic life existing in these regions. Then, as the air temperature begins to rise in the spring, stratification again occurs.

Figure 13-5 shows the predicted results of waste heat disposal. The power plant would withdraw cool water from the hypolimnion and dis-

Figure 13-4. Thermal stratification of a lake. The temperature induced layering effect is caused by the sun's light and heat mainly heating the upper layer, which then does not mix with the lower layers because of its lower density.

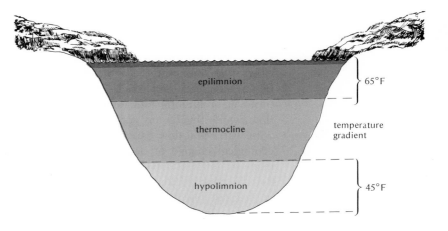

Figure 13-5. The effects on lake stratification of power plant operation. The effect basically is an increase in the thickness (and volume) of the epilimnion and decrease in the thickness (or volume) of the hypolimnion. Accordingly, mixing occurs later in the winter and stratification occurs earlier in the spring.

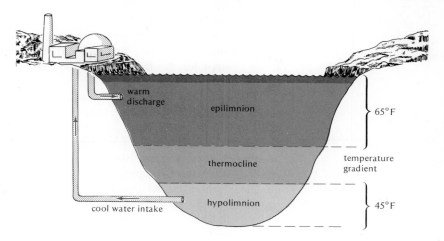

charge the heated water into the epilimnion. The epilimnion would remain at the same temperature, but its extent or thickness would increase with the other layers changing accordingly. The effect of this would be to extend the growing season for the plants and animals in the upper layers. Ultimately, as these plants and animals die and sink to the bottom, this means more nutrients and, hence, still more biological activity. Extending the stratification later in the fall and beginning earlier in the spring not only means a longer bioproduction period but means a shorter mixing period resulting in less oxygen being provided to the hypolimnion where game fish, such as trout, live.

The degrading effects of heat on the lake, for example increased eutrophication (see p. 401), are further intensified by a slow flushing period of about 6 years. This analysis indicating possible destruction of the natural ecological balance effectively delayed construction of the nuclear facility. As a result of the case for Cayuga Lake, further consideration of lake sites for possible power plant location, particularly on those lakes that have recreational value, is unlikely.

Several nuclear facilities have been sited on estuarine waters. The Humboldt Bay Nuclear Plant (172 Mw) in California was the first to be built on estuarine waters. With the exception of the Turkey Point plant in Biscayne Bay, Florida, in general, little damage to marine life has been reported as a result of the thermal discharge in estuarine waters. Considerable ecological change and biological damage has been observed at Turkey Point resulting in much controversy, particularly with the two additional nuclear stations (700 Mw each) proposed. The particular problem is that some bays are shallow and the turnover of water rather slow. The water withdrawn by the plant in a day or so is equivalent to the entire bay volume (in the case of Biscayne Bay). As a result, one would expect excessive heating of the water and consequent damage to the marine life. A similar situation would exist for the Oyster Creek facility (600 Mw), New Jersey, except for the flows and diluting effects of the Forked River and Oyster Creek.

The ocean can be considered as an immense **heat sink.** The world's oceans, covering 75% of the surface of earth, have a mass of 1.43×10^{21} kg (3.14×10^{21} lb or 3.77×10^{20} gal) and have an average depth of 3800 m (2.4 mi). Why not build our power plants off-shore? Only recently has there been much discussion and consideration of using the ocean for power plant cooling. What has been proposed is the building of a floating island with a nuclear generating station (1000 Mw) aboard (see Figure 13-6). The first proposal, to the authors' knowledge, is the controversial floating station to be built about 3.5 mi outside of Little Egg Harbor in southern New Jersey. The station presently is scheduled for completion in 1979.

One argument for using the ocean as a large heat sink considers the large quantity of water available for cooling, which should minimize the biological effects. An advantage not to be taken lightly is the reduction of the heat projected to be received by the inland waterways. However, it is not clear how fast the thermal plume will dissipate in the ocean—several feet below the surface—and to what extent biological damage might be incurred, even in the ocean. Much depends on the location and the ocean current conditions for the particular location. Public Service consultants forecast that the rise in temperature will be no more than 5°F over an

Figure 13-6. A proposed off-shore nuclear power generating plant. The plant would be sited about 3.5 mi off shore and would be little more than a dot on the horizon when viewed from the beach. [Courtesy of General Public Utilities Service Corporation.]

area of 5 acres, close to the plant's water outlet—outfall. Environmentalists confirm these advantages of the off-shore power concept. As in the case of runoff use, a detailed study of the ecology of any proposed plant site must be made and a complete estimate made of the environmental impact.

To estimate the gross effects on water temperature of the ocean, the authors have calculated the total impact of waste heat deriving from total fossil-fuel combustion. If the derived waste heat were to be uniformly distributed over all the ocean waters, the calculated temperature rise is only 0.04°F (Example 13-5). The simplifying assumption has been used that all of the fossil-fuel resources have been burned at 40% efficiency and the waste heat (60%) has been discharged uniformly throughout the entire depth which averages 2.4 mi. If we were to assume a mixing depth of only $\frac{1}{2}$ mile, the temperature rise would be about 5 times the above estimate or 0.2°F. This total temperature rise may or may not be significant.

In actuality the heat discharge would take place over a considerable length of time, so some of the heat would be lost to the atmosphere by evaporation, conduction, and radiation. In effect, then, our estimate of temperature rise is a maximum. However, as in the runoff problem, the concentration of heat from the newer 1000-Mw plants may still constitute a local hazard to the biotic community, that is, temperature rise in the immediate vicinity of ocean-sited power plants may have substantial ecological effects.

Another pro argument for the off-shore location (3 to 10 miles) is that power can be generated in the vicinity of the energy users, which is important for minimizing transmission losses. Inasmuch as the three heavy population growth areas in the United States, namely Megalopolis (the East Coast from Portland, Maine, to Newport, Virginia), central and south Florida, and California between San Francisco and San Diego are in areas not far distant from ocean coastlines, off-shore power plants (nuclear or fossil fuel) seem relatively attractive. For inland areas with substantial population growth, that is, the mid-west (from Pittsburgh to Chicago), other modes of waste heat discharge to the environment may be more attractive, although off-shore siting in the large Great Lakes is a possibility.

Some 14% of the nation's power is expended in the 150-mi wide strip

EXAMPLE 13-5

If 128×10^{18} Btu $(0.6 \times 214 \times 10^{18}$ Btu) is discharged uniformly into 3.14×10^{21} lb of ocean water, the entire volume, what is the anticipated rise in temperature?

$$\frac{128 \times 10^{18} \text{ Btu}}{3.14 \times 10^{21} \text{ lb}} \times \frac{\text{lb °F}}{1 \text{ Btu}} = 40 \times 10^{-3} \text{ °F} = 0.04 \text{ °F}$$

of land along the eastern shore between Boston and Washington—an area well within the delivery range for off-shore plants. Census figures further show that 42% of the United States' population lives within 200 mi of a coastline. Thus, one can imagine that a string of off-shore power plants located along the Atlantic, Gulf, and Pacific coasts, together with plants in the Great Lakes, could supply a vast proportion of the nation's electrical energy needs by the end of the century with a minimum thermal impact on the aquatic environment.

Perhaps the most serious argument against building large nuclear generating stations off-shore is the possibility of accidents caused by severe weather conditions. The public utility has guaranteed an adequate structure to withstand any foreseeable weather conditions. Others have worried about collisions with ships, aircraft, and so on. Many express concern about radiological contamination of coastal waters due to potential nuclear accidents. The esthetics of having a plant built 3.5 mi off-shore have been used extensively as a con argument by environmentalists. However, models show that the plant would be little more than a dot on the horizon. This should be compared with the prospects of sunning in the shadow of a 400-ft high cooling tower, or, perhaps doing with no power.

A Question for the Reader: When considering the trade-offs—benefits versus hazards, for off-shore siting of power plants, how would you rate this potential with the potential for continued inland siting?

Alternative Methods for Cooling Water

There are alternates for disposing of waste heat, but all represent a significant added cost to the kilowatt-hour of electricity. Furthermore, in considering the pros and cons of the various means, one can hardly be sure that he is not jumping from the frying pan into the fire. Varying degrees of technological sophistication are available for the cooling process. The simplest and least sophisticated is the cooling pond. **Cooling ponds** are being used mainly in the Southwest, although some smaller industries use cooling ponds where the volume of water required for cooling is not large, see Figure 13-7.

The advantages of cooling ponds are that they do not pollute the natural waters and the only water that is consumed is that necessary to replace the amount evaporated daily. Recall that 540 kcal of heat are needed to evaporate 1 kg of water; thus, this amount of heat is dissipated when 1 kg of water evaporates from the cooling pond. Anything that takes heat away, that is, the evaporation process, acts to lower the temperature of the water in the pond. In effect, the heat burden, instead of being borne just by the water environment, is being transferred to the air environment to be diluted over a large volume of wind-blown air.

Figure 13-7. A small industrial cooling pond utilizing spraying to increase cooling. [Courtesy of Bell Telephone Laboratories, Holmdel, New Jersey.]

The major disadvantage is the large land area required to contain the large volume of water necessary for cooling and the large surface area of the pond needed for effective cooling by evaporation. For example, a 1000-Mw plant would require about 2 mi² (about 1500 acres) of pond surface area; or, it is estimated that between 1 and 2 acres of pond surface is needed per Megawatt. In the eastern United States, with land increasingly at a premium, such ponds would be prohibitively expensive. As mentioned, this method is mainly used in the Southwest where land is less expensive and the low humidity conditions are favorable for more rapid evaporation. A considerable reduction in land area is obtained by spraying the water, but the increased cost in installation and maintenance makes this considerably less attractive.

In both the eastern and far western sectors of the United States (and also in Europe), an alternative that transfers the waste heat from the water environment (or before it gets to the aquatic environment) to the air environment, seems to be the use of various types of **cooling towers.** There are two basic types: the **mechanical-draft** tower (Figure 13-8) and the **natural-draft** tower (Figure 13-9). The mechanical-draft structure is somewhat less overwhelming than the natural-draft, since the chimneys are usually only about 30 to 50 ft in height. The natural-draft tower is best described as a huge hyperbolic-shaped chimney about the height of a 40-story skyscraper (400 to 500 ft high) and 300 to 400 ft across the base. A football field would easily fit in the base, perhaps along with stands for

Figure 13-8. The mechanical-draft cooling tower. The mechanical-draft towers use fans to force air circulation and hence evaporation of water. [Courtesy of the Morley Co.]

Figure 13-9. The natural-draft cooling tower. As the moist heated air rises in the hyperbolic shaped tower, cooler air is drawn in at the bottom to provide a natural draft, or flow of air. The natural-draft towers are about 300 ft across the base and 400 ft high. [Courtesy of General Public Utilities. Photo from Pennsylvania Electric Co.]

some 5000 spectators. One can hardly say that the esthetics of either cooling tower and, in particular, the natural-draft type, are pleasing.

The mechanical-draft tower in Figure 13-8 utilizes fans to force air through the warm water spray to effect rapid evaporation. The schematic diagram of the natural-draft wet tower, in Figure 13-10, shows how this tower utilizes the natural flow of air as in a chimney. The heated moist air moves upward, drawing cooler less moist air in at the bottom. The cooling of the water in both types of wet towers is effected by evaporation. The mechanical-draft type is mainly used with smaller power units since maintenance and operating costs are high. Initially, the capital expenditure for the natural-draft is higher but operating and maintenance costs are lower; it is also more efficient which makes it desirable for the larger

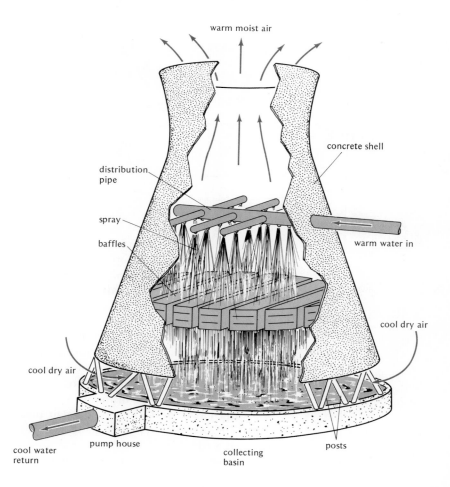

Figure 13-10. Drawing of a natural-draft wet cooling tower. [After a drawing suggested by General Public Utilities.]

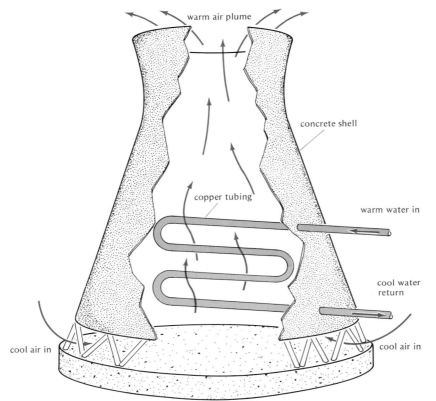

warm air plume

concrete shell

copper tubing

warm water in

cool water return

cool air in

cool air in

Figure 13-11. The natural-draft dry tower. The dry tower is necessarily larger than the wet-type tower since the heat capacity of air is considerably smaller. However, the advantage is that it is a closed system with no humidity problem. The additional expense of dry over wet, though, prohibits its consideration in the immediate future.

generating facilities—800 to 1000-Mw—being built today. A major advantage of both is that only a limited quantity of water is necessary and the power plants can be operated very near the mine (with fossil fuels) for a lower per kilowatt-hour operating cost.

In Europe, the natural-draft dry tower is sometimes (Figure 13-11) used. The circulating water is in a closed system, and the cooling is effected by conduction and convection. Because the specific heat capacity of air is only about one quarter that of water, much larger quantities of air are needed per kilowatt. The towers are then considerably larger. Because of the size and the approximately 600 mi of copper tubing necessary, the cost is prohibitive here in the United States.

A major disadvantage of the wet-evaporative type towers is the 1% to 3% water loss from the tower to the air that is distributed to surrounding areas. This moisture equates to about 30×10^6 gal/day, which is equivalent to about a 1-in. rainfall per day over about a 2 mi^2 area. Thus, the humidity conditions in the plant vicinity become a prime concern. For example, on cool days an extensive cloud is likely to be produced. If this cloud obscures vision at a critical place, this is a form of air pollution produced by the transferring of the waste heat burden from the water environment to the air environment. Many factories have operations that give off moisture as a cloud. Several years ago, such a plant was giving off a cloud that passed across a runway at the Kansas City Airport, interfering with flight patterns. Isn't this impairment of the environment? That is what was adjudged to be the case when operations of this plant were ordered changed.

Table 13-1 summarizes the relative ground area requirements and estimates per kilowatt costs for various cooling methods. The data for this table and our discussion on cooling is found in the references listed at the end of the chapter [2, 3, 4].

As we have seen, none of the above choices are exactly what society might consider ideal. However, the natural-draft wet cooling tower appears to be a sufficiently good alternative that we probably will see many of these dotting the landscape in the next 20 to 30 years at least. This brings up the question of esthetics. Living in the shadow of 40-story skyscrapers (Figure 13-12) is not exactly a bright prospect for the future, particularly when one considers that a 1000-Mw generating plant (the typical size today) requires two towers. Alas, it appears as though the commonplace 40-story skyscraper is a necessary evil since our society needs power. An old cliché seems to describe the situation very well; "You cannot have your cake and eat it too."

Again, the choice reduces to considering trade-offs: benefits versus hazards (costs). In order to have cheap, plentiful electrical energy, are we willing to forego the disadvantages of power production? It is the authors' opinion that off-shore facilities will minimize the undesirable biological

Table 13-1. Natural Cooling Media and Alternative Devices

Device	Relative Ground Area (mech. draft = 1)	Capital Cost Per Kilowatt ($)
ponds:		
cooling	1000	10
spray	50	10
mechanical draft tower	1	8
natural draft tower:		
wet	2–4	11
dry	2–4	25
river	—	5
bay-lake	—	6

Figure 13-12. Living in the shadow of cooling towers. T.V.A.'s Paradise steam plant in western Kentucky supplements the cooling capacity of the Green River. [Courtesy of Tennessee Valley Authority.]

effects that accompany society's demand for power and is surely one of the better alternatives, esthetically, for cities and metropolitan regions within a hundred or so miles of the shoreline.

A Question for the Reader: Are you willing to forego the disadvantages of "eye-sores" and pollution in order to have cheap, plentiful energy? Which disadvantages and to what extent?

Biological Effects of Thermal Pollution

Ultimately, the extent to which waste heat may be considered as thermal pollution depends on the adverse biological effects it produces. This, of course, means that we must have some understanding as to what biological effects are caused by excess heat and some way of discriminating between effects caused by natural phenomena and those caused by power plants. Unfortunately, research and study in this area is in its early stages because a general awareness of and vigorous concern for the problem has developed only since the mid-1960s. One of the more difficult problems is determining the "natural" state of things. Such determination means an extensive study of the aquatic biota[2] in the region proposed for plant siting and how the

[2]Aquatic biota is an all inclusive term of including plants and animals from the lower orders—phytoplankton and zooplankton—to the higher orders of aquatic life—fish, and so on.

biotic populations vary annually with seasons and years. Populations of various types of plants and animals are well known to vary in a cyclic way due to natural causes that are not fully understood.

Some of the undesirable biological effects [5] of thermal discharge are as follows. An increase in water environment temperature may

1. Kill aquatic life and render useless this natural resource for both recreational and commercial purposes (such as fishing).
2. Lower the dissolved oxygen content, perhaps to the point where aquatic organisms cannot survive.
3. Encourage massive development of aquatic plants that will not only be a nuisance but further lower the oxygen content.
4. Enhance the toxic effect of any toxic pollutants present in the water.

Any one of these effects alone or any combination could be extremely harmful to the aquatic environment (Figure 13-13). The extent to which any of these various conditions are already present and whether or not the body of water is already marginal for various types of aquatic life may play an important role in the final decision on siting the plant. In the following material, we will discuss some of the important considerations for each of the above effects, including some of the differing philosophies about the importance of the various contributing factors.

Figure 13-13. The direct effect of heat and chemical pollutants on aquatic life. Dead fish are removed from the Rhein River near Ehrenbreitstein (1969). [Courtesy of the German Information Center.]

Direct Effects of Heat on Aquatic Life.

Most of the effects of heat on aquatic animals may be attributed to increased rate of metabolism. A rule of thumb is that the metabolic rate doubles with each 18°F (10°C) rise in temperature. Remembering that the aquatic animals are cold blooded, their activity depends directly on the water temperature. As the temperature of the water increases, the animals' activity increases; this is accompanied by a greater need for oxygen and food. As we mentioned, the solubility of oxygen is lower in warm waters and, hence, may not be adequate to supply the animals' needs above some temperature point.

At some "high" temperature, the metabolic rate begins to slow, which signifies fatigue and reaction to stress. Further elevation in temperature leads to the point that we will call the **lethal temperature** or the temperature at which death occurs for the particular species. As might be expected, in addition to a lethal temperature, characteristic of each species, each species has optimum temperatures for growing well and for spawning. This is evident particularly in the salmon family of fish. The temperature for satisfactory growth is about 65 °F, whereas the optimum temperature for spawning is about 55 °F. Thus, salmon migrate inland to cooler waters in the spring for spawning. Herein lies the concern about plant discharges and the potential thermal blocks (Figure 13-3) in the Connecticut and Columbia Rivers (and others). In addition, warmed waters may encourage the hatching of eggs before food supplies have become adequate outside the nesting areas.

It should also be mentioned that aquatic animals are more susceptible to temperature extremes while they are in the egg or larvum stage, which occurs in the spring for many species. However, spring is a time when the river waters are cooler so that the heated effluent should have lesser effects. A list of various species of fresh and saltwater fish along with data on their temperature tolerances are shown in Table 13-2. As can be seen, the more popular game fish, the salmon and trout, function over a lower range of temperatures than some less popular species of fish, such as the bluegill and large-mouth bass, which function over a range at higher temperatures. However, the popularity of the fish often depends on the part of the country and its local inhabitants. In the southeastern part of the United States, large-mouth bass is a popular game fish, whereas in the more northern regions, the small-mouth bass is considered more desirable. Here, clearly, is a case of the environment (water temperatures) influencing the attitudes of people. How about your attitudes?

In the vicinity of plants utilizing estuarine waters for cooling, we are concerned also about various invertebrate animals composing the bottom fauna. Table 13-3 presents some data recorded at St. Andrews, New Brunswick, Canada, by Gunter [6]. The lethal temperature of these bottom organisms is relatively high, and it is expected that they will be highly

Table 13-2. Temperature Tolerances for Various Species of Fishes

Aquatic Biota	Spawning T (°F)	Satisfactory Growth T (°F)	Lethal T (°F)
salmon	50	55–65	77
lake trout	50	55–65	77
small-mouth bass	—	60–80	85
winter flounder	—	55–60	85
silverside	—	70–75	92
yellow perch	65	65–75	90
bluegill	—	85–90	93
striped bass	60	55–70	93
large-mouth bass	75	80–90	96
gizzard shad	79	75–93	98
gold fish	—	83	107

(Data in part from J.R. Clark, *Scientific American, 220,* 3, 1969, P. 24.

resistant to the increased temperatures caused by the heated effluent from the power plants. In fact, there are various reports that the heated waters have been very favorable for some of the native invertebrates near the Humboldt Bay plant in California. It is reported also that currently in Long Island Sound, oysters raised in power plant effluent reach market size 1 year sooner than those raised in other regions of Long Island Sound. Japanese shrimp have been found to grow larger and mature sooner in such heated effluent.

Also of concern is the effect of high temperatures on plankton. **Plankton** consists of all the free floating microscopic or near microscopic organisms that are transported by the water currents. The plankton are composed of both plants (mainly algae) and animals. The plants (**phytoplankton**) are

Table 13-3. Lethal Temperatures for Invertebrates and Algae

Organism	Lethal Temperature (°F)
oyster (invertebrates)	119
hardshell clam	113
clam (*Macoma fusca*)	108
softshell clam	105
edible mussel	105
periwinkle	108
diatom (algae)	68
green	83
blue-green	95

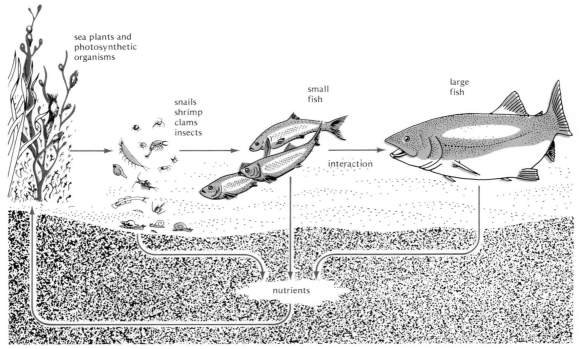

Figure 13-14. The aquatic food cycle—an example of the "balance of nature."

the chief agents in the fixation of energy and the animals (**zooplankton**) are major food items for other organisms in the aquatic food cycle (Figure 13-14).

The temperature tolerance of the various species is not well known, but they would be expected to tolerate summer temperatures in the vicinity of 80°F. Some specific data is available on three types of algae: diatoms, green, and blue-green. The diatoms are food for many species of organisms. Green and then blue-green algae represent lesser food sources, respectively. The lethal temperatures for the three species are 68°F, 83°F, and 95°F, respectively. The death of blue-green algae is evident when green slime is observed on rocks and is found floating in bodies of water. Lake Erie (which has lots of green slime) is an example of a body of water in the advanced stages of the process of **eutrophication,** the death of a body of water as a dynamic living system.

A final important factor in determining the direct detrimental effects of heat is the rate of thermal change. If aquatic animals are given time to acclimate, they can survive relatively large changes in temperature. The major point of disagreement between authorities is the length of time necessary for acclimation. Estimates range from a necessary time of days to acclimation in a matter of minutes. In view of the available data, it does appear that most animals can withstand thermal shocks of about 15°F for a short period of time without serious injury. However, longer periods at excess temperatures, as we have mentioned, will have the expected deleterious effect—death for the animal.

Oxygen Depletion

The second biological effect, that of inadequate oxygen supply, is perhaps an obvious effect. In general, gases, such as oxygen (O_2), are less soluble in warm water than in cold water. Put simply, warm water has a lower oxygen content than cold water. For example, the dissolved oxygen content falls from about 0.015 g/liter to 0.007 g/liter, as the temperature is increased from 0°C (32°F) to 30°C (86°F). Moreover, because the aquatic animals are cold blooded, their metabolic rate directly depends on water temperature. A higher metabolic rate means a greater oxygen demand. Thus, as mentioned earlier, with an increase in temperature, the animals need for oxygen increases while the available O_2 supply decreases. If the body of water is already marginal with regard to oxygen content, the effect is even more serious—it actually is critical. However, if the water is not marginal, there is lack of good evidence as to the seriousness of this particular point.

Algal Bloom—Eutrophication

Effect number three is one often cited because most people have observed a greenhouse and the growth rates obtainable at the higher temperatures (and, too, at higher carbon dioxide concentrations). However, to sustain an accelerated growth rate, the proper nutrients, actually excess nutrients, and adequate sunlight (for photosynthesis) must be present. Such an effect is observable when a stream or pond is polluted with excess nitrates (fertilizer). The excess growth, **bloom,** occurs in the warmed surface layers of the body of water. The plants and animals die and sink to the bottom where decay occurs, depleting the lower, cooler regions of its oxygen supply.

This is particularly pertinent to consideration of the stratification caused by the heated discharges of power plants which was discussed earlier in the chapter. Most fish prefer to feed on diatoms. Referring again to Table 13-3, we see that warmer water algae, green and blue-green, are the ones particularly abundant at elevated temperatures. Unfortunately, these species of algae are food supplies for relatively few species of fish and are even believed to be toxic to some. Also, the dead blue-green algae contribute an odor that is difficult to eliminate, which is particularly undesirable for drinking water supplies.

An additional effect of continued algae bloom (excessive growth of the higher temperature algae) is the gradual filling of the bottom of the body of water by partially decayed plant and animal remains. Eventually, the pond or lake becomes a bog or swamp, and ultimately a meadow. This is the process of **eutrophication** that we mentioned earlier. The loss in oxygen that accompanies the bloom (excess growth) comes from decay of the plants when they die—the process of oxidation.

Increased Chemical Activity

A fourth biological effect that does not specifically apply to power plants but is mentioned here for completeness, assumes the presence of various types of other pollutants, that is, sewage, chemicals, and so on. The higher temperatures activate (speed up reaction rates) the materials, which often are toxic to both plants and animals. The polluting materials may also foster rapid growth of organisms that kill the desirable plants and animals. The "red tide" that periodically appears off both the eastern and western coastlines is an example of such an effect. Furthermore, increased dissolved mineral content results from higher water temperatures. Iron and manganese in the form of oxides dissolve from the bottom materials and contribute a fouling odor to water that is difficult and costly to eliminate.

Uses of Heated Discharge

Because we seem destined to live with an increasing level of heated discharges from electrical power generation, some attention is being paid to prospects of benevolence from the waste heat. Steam and heated water have been used for heating buildings for some time. However, any large scale use necessitates building our communities around and in the vicinity of the plants to avoid the loss of the heat during transmission. Such a change in our present style of living raises questions as to our willingness to adapt—trade-off: benefits versus hazards (costs).

The heated discharge may be used to desalinate salt water to provide fresh water for human consumption and/or agricultural irrigation. Mineral-rich sea salt would be a by-product. Waterways that normally freeze during the winter months could be kept ice-free using waste heat. The marinas near Oyster Creek now stay open the year round with no worry about freezing. Fishermen, likewise, take advantage of the year-round fishing for certain species living in the heated effluent and not migrating to warmer waters in the winter. As we mentioned earlier, **aqua culture,** or **hyponic farming** is receiving increased interest. Since 1965, the Long Island Lighting Company has cultivated oysters in the discharge lagoon of its Northwest Power Station. Japanese shrimp and other crustacea appear to flourish in such warmer waters. In Oregon, warm water irrigation or running warm water through pipes in the soil to increase the length of the growing season is being investigated. Perhaps in the benevolence and utilization of these heated discharges lies the secret of increased food production for the world. However, research in these areas is just beginning, and the power crisis is upon us.

Some Results of Site Studies

Few results of the thermal effects on biota of various sites are available but some recent results follow [3, 7, 8].

1. Hanford Nuclear Plants, Columbia River (Washington). There has been no evidence of damage to salmonoid resource. However, you should note that the Columbia is a large, cool river.
2. Chalk Point (fossil-fueled, 770 Mw), Patuxent River (Maryland). No major detrimental effects have been observed. There have been some minor changes in population and production rate.
3. Oyster Creek (nuclear, 640 Mw) in Barnegat Bay (New Jersey). Reports some population changes have been observed, but no major detrimental effects have been reported during regular operation. However, Oyster Creek is the first power plant to report substantial fish kills, apparently resulting from thermal shock from sudden drops in temperature while shutting down for repairs. Of interest here is the fact that environmental groups have succeeded in the courts on imposing fines on the utility for such fish kill damage to the environment.
4. Contra Costa (1298 Mw) on the San Joaquin River (California), showed that young salmon could tolerate an instantaneous temperature increase up to 25°F for 10 min with no mortality.
5. Turkey Point (fossil-fueled, 864 Mw and nuclear, 1442 kw) at Biscayne Bay (Florida) reports substantial effects of heated effluent: reduction of population, kills to various species of vertebrates and invertebrates including algae.
6. Petersburg Plant (220 Mw) on the White River (Indiana), shows no evidence of thermal effects.

It would seem that with the exception of Turkey Point, that the thermal effects of electrical power production, to date, have not been large. However, the fear is that with rapidly increasing demand and the construction of larger single units (1000 Mw), that future impairment of the environment in the vicinity of plant sites is inevitable, unless alternate methods of cooling are employed.

Global Heating from Energy Consumption

Ultimately, all of the energy we consume is degraded into heat. The extent to which this degraded energy represents global thermal pollution depends on the environmental impact. As we have seen, globally, the impact is the possibility of large-scale effects on weather and climate, the climate being determined by the earth's radiation balance. From Chapter 3, we learned that the earth receives solar energy at the rate of about 178×10^{12} kw. Of this, 30%, the albedo, is reflected back into space mainly as short wavelength radiation. Most of the remaining 70% or 125×10^{12} kw is in the form of light (about 50%) and heat (about 20%). To maintain the radiation balance, 125×10^{12} kw must be reradiated to space. Globally, the reception, redistribution, and reradiation of this quantity of solar energy determines the gross features of the earth's climate. Small changes, of the order of 1% or less, in the rate due to changes in albedo (for example, changes due to carbon dioxide concentration or to particulates) or due to heat dissipation at the earth's surface are believed to cause drastic changes in climate.

From the Stefan-Boltzmann Law (equation 7-12 or 13-1),

$$P = \sigma A T^4 \qquad (13\text{-}1)$$

rate of radiation (power) = a constant $\times$ area $\times$ (temperature)^4

the earth is seen to appear as a black body radiating at a temperature of about 255°K. An increase of heat dissipation at the surface of the earth of 1% would cause an increase in temperature of about 0.64 °C or 1.2 °F. Such changes, or even smaller ($\sim$0.1%), have been recorded in our geologic past and have even been thought to cause major climate changes, such as the onset of ice ages (due to a decrease in temperature). Such an estimate would appear to set upper limits to heat dissipation due to energy consumption. If we assume that the ultimate population to be 20×10^9 people, all consuming at the rate of 20 kw, twice the present United States rate, the increase in average surface temperature of the earth would be about 0.64°C (see Example 13-6). Thus, there would appear to be no major global heating problem in the immediate future (see also Example 13-5).

Much more serious is the distribution or dispersion of heat from those vast islands of energy consumption, *the major population centers*. For example, the heat dissipation on Manhattan Island of New York City is about 7 times (700%) [9] the solar energy received by that area. If all that energy had to be radiated just from the Manhattan area, the temperature would rise to about 187°C or 372°F, well above the boiling point of water. In the Los Angeles area, the heat dissipation is 5% (this is a much larger area and the percentage is less than Manhattan) of the solar input, and such is the case for many of the industrial and population centers for the United States and the world. Apparently, heat dispersion and reradiation from larger areas prevents catastrophe in these "heat islands." As we discussed before, the atmospheric cellular flow patterns cause a large heat flux toward the polar regions.

EXAMPLE 13-6

Assuming energy dissipation at the surface of the earth is 1% of the solar input of 125×10^{12} kw, find the increase in average surface temperature of the earth. First, an increase of 1% is given as

$$125 \times 10^{12} \text{ kw} + 0.01 \times 125 \times 10^{12} \text{ kw} =$$
$$125 \times 10^{12} \text{ kw} (1 + 0.01) = 125 \times 10^{12} \text{ kw} (1.01).$$

The new equilibrium temperature (T_2) will be (using equation 13-1),

$$P = 1.01 \times 125 \times 10^{12} \text{ kw} = \sigma A T_2{}^4,$$

where A = area of earth, and σ = constant. Before the change, the effective temperature (T_1) of the earth of 255°K was obtained by,

$$125 \times 10^{12} \text{ kw} = \sigma A T_1{}^4$$

So, we construct the ratio

$$\frac{1.01 \times 125 \times 10^{12} \text{ kw}}{125 \times 10^{12} \text{ kw}} = \frac{\sigma A T_2{}^4}{\sigma A T_1}$$
$$T_2{}^4 = 1.01 \, T_1{}^4 \qquad \text{or} \qquad T_2 = (1.01)^{1/4} T_1$$

Because

$$T_1 = 255 \text{ °K}$$

we obtain

$$T_2 = (1.0025) T_1 = 255.64 \text{ °K}$$

which is an increase of 0.64°C or 1.2°F.

Note: To take the fourth root of a number, we simply ask ourselves what must we multiply by itself four times in order to get that number, 1.01.

The authors would like to make it clear that there are varying opinions of scientists, demographers, and so on, on the subject of the possible ultimate population of earth and the world per capita energy consumption. We do not choose to debate the issue here nor make value judgments, but we do feel that maximal differences in our figures of perhaps 50% do not substantially change our treatment or our conclusions [10].

Conclusion

From our studies, there is little doubt that excess heat can and will have a detrimental effect on our environment, particularly in certain locations. In the past, the thermal discharge from electrical power generating plants has apparently caused no major damage to the aquatic environment. However, in the future, with the need for cooling waters doubling by 1980, substantial damage must be expected. The most important question is how the environmental effects can be minimized. In so deciding, our citizenry (*you*), our politicians, our technologists and our scientists must put aside arguments based solely on emotion and replace them with arguments based on sound knowledge and fact.

As each power plant site is being planned, a thorough study must be made on-site to determine what type of aquatic environment is present and a careful assessment of the effects on this environment from waste heat. Do we want to live in the shadow of a 40-story cooling tower? Yet, do any of us want to be lightless, applianceless, or be cold in the winter and hot in the summer? Ultimately, our arguments and our decisions must reduce to the old question of trade-offs: benefits versus hazards (including costs). An example of benefits versus economic costs is that of the Vermont Yankee Power Corporation agreeing to construct a $6.5 million cooling tower to avoid disposing waste heat in the Connecticut River in order to preserve a marginal, at best, cold water fishery amounting to a $37,000 per year angling business. Can anyone say that such a decision is based on sound judgment? Are we willing to pay the price for similar decisions?

As we mentioned in Chapter 9, to reverse entropy–pollution requires energy, a term synonymous with money. Should we perhaps just admit that some rivers might better be used for cooling, some for municipal withdrawal (human consumption), and some for fishing? The answer to this question falls in the area of water management which has yet to be explored adequately. You will find it very easy to take an extreme position. That is very simple. But, you must understand fully all the consequences of such extreme positions. Any middle-ground position is necessarily more difficult to take. Where do you come down?

This question of how much heat our rivers can bear, entangled in the more complex question of water management, is but a part of a much larger question about our entire environment; how much energy can be consumed and how much of a heat burden can our environment withstand? Our power system will soon outgrow our rivers. Will the power system outgrow the cooling capacity of the total water environment? Will man soon outgrow spaceship earth?

Questions

1. Heat and, hence, temperature are considered to be the primary controls of life on earth. Under what circumstances is heat considered to be thermal pollution?

2. Give two reasons why you think *we are in hot water* (if not now, within the next three decades).

3. Explain the statement: "Pollution is entropy."

4. Using the concept of entropy, explain why it is costly to remove pollutants.

5. (a) List the various media available for the disposal of waste heat. (b) Compare present practices with alternatives giving the advantages and disadvantages of each.

6. Why are we concerned about the new large electrical generating facilities?

7. Discuss the pertinence of the word esthetics with regard to the locations of plant sites and methods used for cooling.

8. What is meant by thermal stratification?

9. Describe the plant-marine-food cycle that exists in dynamic equilibrium in a lake.

10. (a) Where are the greatest concentrations of population in the United States and hence the energy consumers? (b) What are the implications of part a with regard to siting power plants?

11. What do you consider as being the major argument for off-shore nuclear power plants? Against?

12. Explain the statement, "Dilution is the solution to thermal pollution."

13. What is the major disadvantage of cooling towers in colder regions?

14. Explain the operating principle of the natural-draft cooling tower.

15. (a) What is the difference in methods of cooling between the wet and dry towers? (b) Why are dry towers larger for the same quantity of heat disposal?

16. In what way might the warm moist discharge from cooling towers be considered as air pollution?

17. The dry-draft cooling tower is preferred to the wet-draft tower because no water is lost to the surrounding atmosphere. However, the heat must then be absorbed by the air. Can you suggest how this might affect the local weather conditions?

18. What is the difference between a warm blooded creature and a cold blooded creature with regard to its reaction with environmental conditions (for example, temperature)?

19. What are two direct effects of increased water temperature on aquatic animals?

20. Give one important indirect effect of increased water temperature on an aquatic animal.

21. What is thermal shock with regard to aquatic biota?

22. What is the relation of acclimation to temperature tolerance?

23. (a) What is the temperature range for existence of fish life in Table 13-2. (b) Does this verify one of our opening remarks about the very narrow temperature range within which life may exist on earth?

24. How does the temperature tolerance of the invertebrate (Table 13-3) compare with the vertebrate (Table 13-2)?

25. Which of the three species of algae is the most important source of food for aquatic life?

26. What does the term eutrophication mean?

27. How does the solubility of oxygen in water depend on temperature?

28. What is the relationship between thermal stratification and oxygen depletion?

29. What are the necessary ingredients for algal bloom?

30. Suggest three ways in which heated discharge may be potentially benevolent?

31. If the environmental impact of thermal discharge has been minimal to date, then why all the concern about it?

32. (a) Is there presently a global heating problem? (b) At what future time might world energy consumption present a global heating problem? (Use 1% of the solar input as your guideline.)

33. If heat dissipation on Manhattan Island is 700% of the solar input to that area, why is it not abominably hot?

34. (a) What is meant by heat islands? (b) What regions in the United States do we expect to become future heat islands?

Numerical Exercises

1. Show that 4.6×10^5 gal/min is equivalent to 2.3×10^8 lb/hr, using the conversion factor—1 gal weighs 8.3 lb.

2. (a) Convert 2.3×10^8 lb/hr to kilograms per hour. Remember that 2.2 lb = 1 kg.
   (b) Convert 4.4×10^9 Btu/hr to kilocalories per hour.
   (c) Do Example 13-1 in metric units.

3. Do Example 13-2 and 13-3 in metric units.

4. Do Example 13-4, parts a and b, in metric units. (Conversion factors you may need: 3.28 ft = 1 m, 1 m^3 of water has a mass of 10^3 kg.)

5. Perform the calculation in Example 13-5 using metric units. Refer to Appendix A for conversion factors.

6. Show that, if the mixing depth for heat in the ocean is only $\frac{1}{2}$ mi instead of 2.4 mi, waste heat from burning the fossil fuels will raise the temperature 0.2°F.

7. (a) Determine the amount of water evaporated per hour and per a 24-hr day from a cooling pond or tower in order to cool a 1000 Mw nuclear plant, assuming that all the heat goes into evaporation of water.
   (b) The quantity of water in part a is equivalent to 1 in. of rain over what area?

8. If land in a northeastern region of the United States is selling for $10,000 per acre, estimate the cost of a cooling pond for a 1000-Mw plant. Is the price competitive with cooling by cooling towers?

9. If population and power prospects on earth are 20×10^9 people and 20 kw per capita, what percentage of the solar input is this?

10. A power plant discharges 3×10^6 kwh (kilowatt-hours) of heat per week into a lake 1 mi^2 (28×10^6 ft^2) in surface area and 5 ft deep.
    (a) What is the volume of water in cubic feet contained in the lake?
    (b) If 1 ft^3 of water weighs 62.4 lb, what is the weight of the water in part a?
    (c) If there are 3412 Btu of heat in 1 kwh, how many British thermal units are in 3×10^6 kwh?

(d) If 50% (half) of the heat discharged in one week is lost by way of evaporation and radiation, what is the rise in temperature of the lake per week (that is, in 1 week)?

11. Assuming the present rate of population doubling, that is, about 40 years, and the present rate of power consumption doubling, that is, about 10 years, we have predicted that the power consumption by the year 2020 over the eastern coastal region as 52×10^{12} w. If the solar input is 173×10^{12} w over the same region,
    (a) find the percentage of solar input that this power consumption represents.
    (b) if a 1% change in solar input would cause a 2.0 °F change in temperature, what would be the predicted temperature change due to part a?

References

[1] The data for our example was provided by the courtesy of General Public Utilities.

[2] Cooling towers—a special report, *Power* (March 1973).

[3] P. A. Krenkel and F. L. Parker (eds.): *Biological Aspects of Thermal Pollution*. Vanderbilt University Press, Nashville, Tenn., 1969, pp. 44–50.

[4] P. A. Krenkel and F. L. Parker (eds.): *Engineering Aspects of Thermal Pollution*. Vanderbilt University Press, Nashville, Tenn., 1969, pp. 249–327.

[5] Unless otherwise noted, many of the biological effects presented here are from a General Public Utilities Environmental Report with material obtained for the report by consulting biologist, Dr. C. G. Wurtz, LaSalle College, and from [3] and [4].

[6] Gunter: Memorandum No. 67, Geological Society of America, Vol. 1, *Ecology*, reprinted 1963.

[7] A. A. Levin, T. J. Birch, and G. E. Raines: Thermal discharges: ecological effects. *Environ. Sci. Technol.*, **6**(3):225 (1972).

[8] *Environmental Reports*. General Public Utilities, New Jersey. (See [1].)

[9] D. R. F. Harleman: Heat—the ultimate waste. *Tech. Rev.*, **74**(2):44–51 (Dec. 1971).

[10] For further discussions refer to: (a) A. M. Weinberg and P. P. Hammond: Limits to the use of energy. *Amer. Sci.*, **58**(4):416 (1970). (b) R. S. Bryson: All other things being constant. . . . In *Global Ecology*. Harcourt, Brace, Jovanovich, New York, 1971, p. 78; (c) J. Harte and R. H. Socolow: Energy. In *Patient Earth*. Holt, Rinehart & Winston, New York, 1971, p. 276.

14

Our Air Is
Unclean—
Air Pollution

In the previous two chapters, we examined two important forms of environmental pollution—radioactive materials and heat. In fact, the previous four chapters broadly surveyed the scientific and technological essentials (including problems) of the generation of electrical power from nuclear reactions. Chapter 13 stressed that both the actual and potential thermal pollution of the environment in the vicinity of power plants is common both to nuclear power plants and to fossil-fuel fired power plants, but is greater with nuclear plants due to the 40% more thermal discharge to the water environment.

The problem of thermal pollution is simply the how and where to dissipate the large quantities of waste heat associated with the conversion of heat energy to electrical energy in the power plant. Most of the waste heat from all power plants is discharged to the water environment. Heat dissipation to the air environment is now being used on a limited but increasing scale by means of cooling towers. However, the tower effluent is not considered generally to be air polluting as is the stack effluent from burning fossil fuels.

Let us then, in this chapter, examine the more extensive pollution of the air environment by fossil-fuel fired power plants. More specifically, we will discuss the minor products of the combustion reaction of the fossil fuels. By way of introduction, we examine the nature of air as a gas. Then, smoke or particulate matter is studied as the visible air pollutant known for many centuries. The invisible air pollutants are then identified and are shown to be central to certain air pollution disasters that have occurred at times of unusual meterological conditions known as temperature inversions. How each of the primary, invisible air pollutants is formed in the combustion process and how these materials can affect your environment is laid before you. Our society today is concerned with the effects of small but significant quantities or concentrations of these materials on the quality of our life on this planet. The technology of the removal of these materials (pollution abatement) is discussed by reference to the scientific possibilities and limitations on this technology. Finally, the air quality standards deriving from the Clean Air Act of 1970 are set forth for your use.

Air Is a Gas

You have encountered clean air and unclean air. Your eyes and your nose sense uncleanness. But, these organs are not always accurate. The air that is "clean" to your eyes and to your nose may be contaminated with deadly, odorless carbon monoxide. The misty, overcast day may be either deadly noxious or pollutant-free. So then, what is meant by unclean air or air pollution? Let us begin by reviewing what is clean, unpolluted air. Table 14-1 records the principal constituents of clean dry air near sea level. The constituents present in larger amounts are listed both as percentages and as parts per million, whereas the constituents found in lesser amounts are listed only in parts per million (ppm).

The use of the unit, parts per million, is a convenient way to avoid both very large numbers and very small decimal numbers with several zeros between the decimal point and the numeral. If we mix 999,999 liters (1 liter equals about a quart) of clean, dry air with 1 liter of pollutant, such as the hydrocarbon, ethylene, we have a total of 1,000,000 liters of polluted air (as in Figure 14-1). The polluted air contains 1 liter of ethylene in 1,000,000 (1 million) liters of polluted air or 1 part in 1 million parts or 1 ppm. For gases at the same temperature and pressure, the volumes of the gas are directly proportional to the number of molecules therein. Thus, polluted air containing 1 ppm of ethylene actually means that one molecule out of every million is a molecule of ethylene (C_2H_4).

This amount may not sound or seem significant, but we will see later

Figure 14-1. Schematic illustration of 1 ppm level of air pollution.

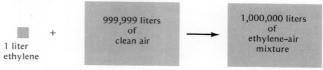

1 liter ethylene + 999,999 liters of clean air → 1,000,000 liters of ethylene–air mixture

1 ppm polluted air

that such small quantities of certain materials may have profound effects on us and parts of our environment. Thus, as we examine pollutants we will note some of the effects of rather small quantities of these materials.

Actually, our air atmosphere is a huge reservoir of the gases of Table 14-1 speckled with clouds and dusted with particles of almost every conceivable composition. Overall, the weight of our atmosphere is about 5.1×10^{15} metric tons. Each square inch of the earth's surface at sea level rests below a column of air that weighs 14.7 lb or 6.7 kg. Thus, we speak of sea level air pressure as 1 atm (atmosphere) of pressure or 14.7 lb/in.^{2} (pounds per square inch) or 10,333 kg/m^3. Furthermore, the atmosphere contains some amount of moisture. The amount of moisture varies with location and temperature so that the percentage moisture may be very small or as high as 3%. Most of the time the water content is somewhat less than 1%.

In any consideration of our air, particularly as we focus on air pollution, there are two important properties of gases that influence how gases move in our atmosphere. First, you all know from experience that most materials

Table 14-1. Gaseous Composition of the Atmosphere. (Clean, Dry Air Near Sea Level)

Substance	Formula	Percentage	Concentration (ppm)
nitrogen	N_2	78.09	780,900
oxygen	O_2	20.94	209,400
argon	Ar	0.93	9,300
carbon dioxide	CO_2	0.0318	318
neon	Ne	0.0018	18
helium	He	0.0005	5
methane	CH_4		1.5 (approx.)
nitrogen oxides	NO_x		0.5 (approx.)
carbon monoxide	CO		0.1 (approx.)
ozone	O_3		0.02 (approx.)
hydrogen sulfide	H_2S		0.01 (approx.)
sulfur dioxide	SO_2		0.002 (approx.)

Table 14-2. Variation of Density with Temperature for Air

Temperature (°C)	Density (g/liters)
−20	1.40
0	1.29
21	1.20
48	1.10
100	0.95

expand when heated and that hot air rises. In scientific terms, hot air is less dense than cold air *or* a given volume, say 1 liter, of a warm or hot gas weighs less than does the same volume of a cold gas. This variation of density with temperature is illustrated in Table 14-2.

Secondly, the vertical movement of gases (up or down) depends on molecular weight. You all know that a balloon filled with helium (molecules of He, molecular weight = 4.0) rises in air, even in warm air. Principally, air is made up of about 80% nitrogen (N_2 molecules, molecular weight = 28) and about 20% oxygen (O_2 molecules, molecular weight = 32), so we can consider that air has an average molecular weight of about 29. This number is a very useful reference point.

You can now see why helium (molecular weight = 4) rises, and it should be clear that gaseous molecules with molecular weights greater than 29 will tend to sink, move downward, or settle toward the ground. For example, gasoline vapors (on the average C_8H_{18} with molecular weight of 114) settle as do the constituents of dry cleaning fluids, primarily CCl_4 (molecular weight 154). In contrast, lighter molecules, for example, water vapor, move upward. Carbon monoxide (CO, molecular weight 28) would tend to remain in the area and altitude in which it is discharged, assuming no air movement, that is, winds. Thus, to repeat, the movement of air and the gaseous materials placed into the air is going to be influenced particularly by the temperature of these gases (relative to the air itself) and the molecular mass of the gaseous molecules (also relative to air).

Smoke (Particulates)

Smoke means different things to different people. The cottage with a curl of smoke wafting from the chimney for some suggests comfort and coziness. Smoke belching from factory chimneys for many means progress, often powerful progress (Figures 14-2, 14-3 and 14-4). Today, for many, smoke indicates desecration of the environment.

Since the introduction of fossil fuels, smoke has become a matter of importance in all centers of population. P. B. Shelley, as he referred to

Figure 14-2. Progress (Düsseldorf): 2 million tons of dust, 4 million tons of sulfur dioxide, and 6 million tons of carbon monoxide are passed into the air annually in Germany. [Courtesy of the German Information Center.]

Figure 14-3. A power plant located in Ohio, across the Ohio River from the city of Moundsville, West Virginia. The plant is not only a source of air pollution but also of thermal pollution, as it discharges heated water into the river. [USDA-SCS photography by Edward A. Gaskins.]

Figure 14-4. A smoggy-foggy day at an industrial complex in Cleveland, Ohio.

London, said "Hell is like a smokey city—London." However, the term smoke as used by most people is a very general term. Let us look at some specific smokes. The smoke from a coal or oil furnace not operating properly is black, due to uncombusted carbon. The smoke from a steel making, blast furnace tends to be red from particles of iron ore. The smoke from a cement plant is white from particles of limestone having been carried up the stack. From an air pollution point of view, concern for black smoke has been with us for many years. By the early 1930s, in the United States some 120 communities had smoke ordinances. Enforcement of these ordinances focussed on improvement in furnace operating conditions, particularly for furnaces and stoves using coal, with the objective of obtaining more complete combustion.

The smoke coming out of chimneys or exhausts consists largely of solid matter and gases. The solid matter is now known as **particulates** or particulate matter. Particulate matter is divided up into several subgroups according to the size of the particles. The size ranges are tabulated in Table 14-3 in microns (μ), where $1\,\mu$ equals 0.001 mm, or 0.000001 (1×10^{-6}) m, or about 0.025 in.

In general, the coarse dust particulates (larger than $10\,\mu$) and the fly

Table 14-3.
Sizes of Particulates

Particulate	Size (μ)
dust	1.0 to more than 1000
fly ash	1.0–800
raindrops	500–1000
smoke or fumes	0.1–1.0
mists	0.1–100
aerosols	0.001–0.8

ash, deriving from the ashes left after coal has been burned, settle out of the air quickly. Smoke (as defined in Table 14-3), fine dust particles and fine fly ash, travel rather far. Particles less than $1\,\mu$—usually referred to as aerosols because they are small enough to remain suspended in the air—move as easily in air as do gases. Particulates are prevalent, especially in the most heavily polluted parts of cities where 50 to 100 tons/mi² fall each month.

Figure 14-5 shows the results of many years of a building being exposed to a "smoky" environment and the transformation arising from cleaning by sand blasting. An average winter in New York City produces some 350 tons/mi² of particulate matter. In Kansas City, the dust fall in the

Figure 14-5. Cleaning Cincinnati City Hall—34 years of dirt, 1963. [Courtesy of Environmental Protection Agency.]

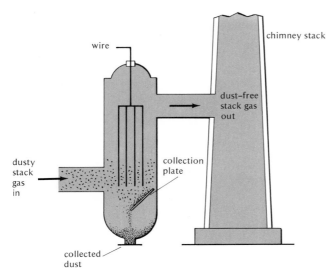

Figure 14-6. Schematic of an electrostatic precipitator (Cottrell). An electrostatic precipitator can remove small particles (smaller than 1 μ) from stack gases with efficiencies of up to 99.5%. The electric discharge between the central wires through the dusty stack gas causes the particles to be attracted to the collection plate. The precipitated dust is then periodically removed from the bottom of the unit. These units are used on a very large scale, particularly in reducing smoke (fly ash) from power plants that burn fossil fuels.

winter measures more than 67 tons/mi² per month. In Los Angeles estimates its aerosol emissions from gasoline powered vehicles at 40 tons/day.

Particulate matter can be very easily removed from large, stationary combustion sources such as fossil-fuel fired electric power stations, mills, factories, and so on. The Cottrell electrostatic precipitator (Figure 14-6)

Figure 14-7. **(A)** Stack with precipitator off. **(B)** Stack with precipitator on. Note the small amount of smoke remaining, which shows that the precipitators are not 100% efficient. [Courtesy of Eastman Kodak Co.]

(A)   (B)

was patented in the early 1900s, has been improved over the years, and is often used along with a mechanical collector. Cottrell precipitators are estimated to have removed some 87% of the particulates produced by utility power plants (see Figure 14-7). The guaranteed fly ash collection efficiency of new precipitators has risen from 96% in 1940 to more than 99% today.

Disasters

Where does the smoke go? All of you have observed smoke plumes from chimneys (Figure 14-8). Certainly, the direction of the wind determines who or what part of the landscape is to be affected. Sometimes meteor-

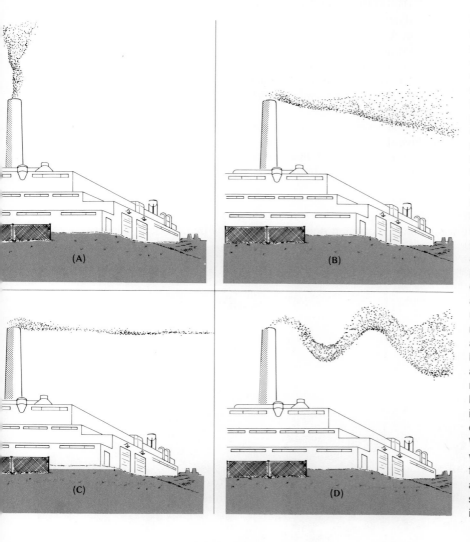

Figure 14-8. Smoke Plumes. The behavior of a plume of smoke coming out of a high chimney depends mostly on the strength of the wind and the stability of the atmosphere. Four common classes of plumes are shown. **(A)** In a stable atmosphere with no winds, the smoke plume continues to rise until it is dispersed by air movement at higher altitudes. **(B) Coning.** With winds greater than 20 mph and a stable (neutral) atmosphere, the smoke spreads out in a form resembling a cone with only small upward and downward movement of the plume axis. **(C) Fanning.** With low winds and a stable atmosphere, the smoke remains in a shallow layer but spreads out like a fan in this layer as it moves downwind. **(D) Looping.** With light to moderate winds and unstable air, the smoke plume loops up and down and the plume spreads out rather rapidly.

ological conditions lead to accumulation of wastes, that is, air pollution wastes in a particular region. Today, when air pollution is sudden and serious enough so that a lot of people die in a hurry, this is a disaster. Although the immediate death statistics are seldom very terrifying, at least compared to a plane crash or an earthquake, the full implications of the air pollution disaster often do not show up until long after the air pollution event has occurred. Most of the isolated catastrophies have involved the emanations from industries coupled with an unusual degree of **temperature inversion** wherein foul air (polluted air) is kept close to ground level for people to breathe. This effect is illustrated in Figure 14-9.

In 5 days, December 1 to December 5, 1930, a temperature inversion occurred in the Meuse River Valley in Belgium between the towns of Liége and Huy. In this valley, a large number of plants had been giving off significant amounts of gases such as sulfur dioxide and hydrofluoric acid. Sixty-three deaths have been attributed to the poisonous air and accumulated air-borne pollutants that quietly lay in the valley. Actually, in the Meuse River Valley, poisonous fogs have been recorded in 1897, 1907, and again in 1911.

A similar event, shocking to the United States, occurred at Donora, Pennsylvania, between October 26 and 31, 1948. Donora is situated some 20 mi south-southeast of Pittsburgh on the shore of Monongahela River. The town, lying near the bottom of a steep horseshoe-shaped valley, is about

Figure 14-9. Denver, Colorado, under smog (1963). The pollutants were trapped by a temperature inverted layer of air. [Courtesy of Environmental Protection Agency, photography by Charles E. Glover.]

Figure 14-10. Smoke and fog in Donora, Pennsylvania (1949). [Courtesy of Environmental Protection Agency. Photo of National Air Pollution Control Administration.]

500 ft below the surrounding terrain. These 5 days, when a temperature inversion capped the valley, the smoke and fumes from a steel factory, a sulfuric acid plant, and zinc plant (producing oxides of sulfur) were bottled up in the town (Figure 14-10). The air near the ground was very humid. Fog formed in the night, and in some low lying areas of western Pennsylvania it persisted during the day. At Donora, the visibility, cut by smoke and fog (**smog**), ranged from about 0.6 mi to only 1.5 mi. The sulfur dioxide concentration reached 2 ppm, 43% of the population of 12,000 were made ill, and 20 persons died, mostly on the third day. Old men were the principal victims.

The Donora and Meuse Valley episodes were very much alike; although visibility was not that poor (about 1 mi), each breath of that air carried into one's lungs thousands and thousands of tiny grains of matter, as **aerosols** (Table 14-3). The largest aerosols are captured in the nose and the throat. The smaller ones pass into the lung where many are deposited on the linings of the lungs, establishing the foundation for serious disease. In these two disasters, sulfuric acid—from sulfur dioxide (see p. 431)—aerosols are believed to be the poisonous "normal pollution" suddenly made deadly by inversion and fog.

English and Scottish cities have long suffered conditions arising from frequent fogs and temperature inversions. Prolonged smog seems to have killed several thousand people in London in 1872, but nothing much was done about it except to set up smoke-abatement committees, which abated no smoke. The "black fog" of December 5 to 9, 1952 brought the city to a standstill and caused 4000 deaths in 4 days. This smog occurred because moist, foggy air over the city stood stagnant while huge quantities of smoke were spewed into it. This was perhaps the most serious air-pollution disaster in history.

Today, although the prevention of air pollution disasters has been sought, society, especially in Europe and the United States, now recognizes that prevention of air pollution disasters alone is not enough. The quality of the air to which we are exposed every day may have profound effects on our lives. The subject of air quality standards is an important issue, will continue to be for the foreseeable future, and is discussed later in this chapter.

Temperature Inversions

The feature common to the air pollution disasters we just described and to any other near disaster is an accumulation of pollutants in the lower atmosphere. Normally, in most regions of the world, both vertical and horizontal air movement is sufficient to disperse the materials man and nature throw up into the air. Recall hot air is lighter than cooler air. Our exhaust gases are usually hot. Thus, normally the exhaust gases rise. But, as you well know as the altitude increases, the air temperature (Figure 14-11) gradually decreases. Thus, hot noxious emanations will rise and continue to rise even though they cool off somewhat until dispersed rather widely.

This vertical movement is aided by winds that blow materials along and away in horizontal direction, also resulting in effective dispersion. This is illustrative of getting rid of wastes by dilution—dilution is a commonly used solution to pollution. Mountain ranges interfere with the horizontal movement of air. Such interference may be uniquely observed in certain

Figure 14-11. Diagram of normal temperature conditions in which noxious fumes rise readily and are then dispersed in the large volume of air above the community.

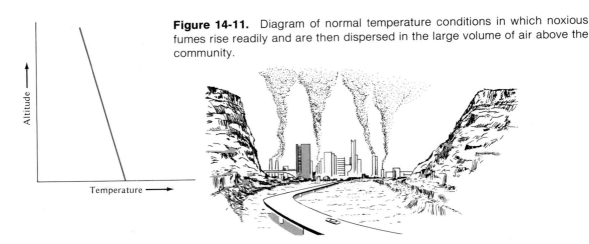

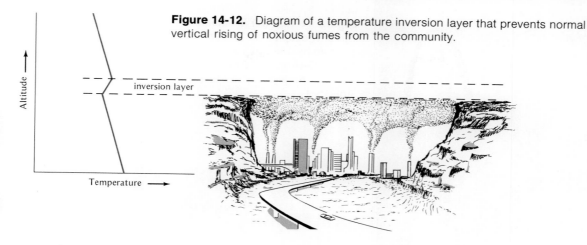

Figure 14-12. Diagram of a temperature inversion layer that prevents normal vertical rising of noxious fumes from the community.

valleys. When winds do not assist in horizontal air movement, then dispersion of pollutants depends entirely on vertical movement of the air.

Sometimes, vertical movement is blocked by the development of a layer of warm air at a low altitude, see Figure 14-12. The normal temperature decrease with increasing altitude is interrupted by a layer of air that is warmer and within this layer there is a temperature-altitude profile opposite to the normal. The warm, lighter polluted air rises through the heavier cooler air but, when it reaches the unusually warm air layer, the polluted air is not light enough to rise any further.

Thus, in this unusual meteorological situation, pollutants put into the air can rise only as far as the warm air layer which is known as an inversion layer. Thus, the pollutants are trapped in the lower layer of air and gradually recycle back towards the ground level. This is the air we breath. Such situations may remain unchanged for days. With an accumulation of noxious materials, an air pollution disaster may occur! Only a change in weather conditions effected by winds leads to the break up of the inversion layer with the consequent dispersion of the accumulated pollutants from the lower, trapped regions. See Figure 14-13.

Common Air Pollutants

The air pollution disasters and the weather conditions facilitating these disasters have led us to recognize the importance of the invisible materials (the minor products of combustion reactions) placed into the air by combustion processes. Four of these materials are gases invisible to the human eye but damaging to man and his environment. The five substances known as primary air pollutants account for more than 90% of the nationwide

Figure 14-13. **(A)** Midmorning smog in St. Louis, Missouri, 1966. The top of the Golden Arch stands high above the smog trapped by a temperature inversion. [Worldwide Photo, courtesy of Environmental Protection Agency.] **(B)** Smog formation in Los Angeles. [Courtesy of Environmental Protection Agency.]

(A)

(B)

pollution. They are carbon monoxide (CO), nitrogen oxides (NO_x), hydrocarbons (HC), sulfur oxides (SO_x), and particulates, which we have already discussed. Table 14-4 shows the sources and amounts of these materials produced in the United States during 1968.

The 214.2 million tons of primary pollutants (which would be much more without extensive smoke control) represent more than 1 ton/year per person or 6 lb/day per person in the United States. However, examination of a table of data such as Table 14-4 by itself is not sufficient to assess the relative importance of any particular source or the pollutants from that source. Data has been assembled by the state of California that takes into account the difference in human tolerance to these primary pollutants. Tables 14-5, 14-6, and 14-7 record the tolerance level and then assign relative toxicities to these pollutants. Combination of the relative toxicities with the amounts of the various pollutants and conversion to a 100% total basis, provides easy guidelines for assessing the relative impact on the environment of a particular source.

Conclusions based on the weighted mass calculations are not the same as those drawn just from the total masses of pollutants themselves. On the

Table 14-4. Primary Pollutant Sources and Amounts (millions of tons/year) 1968

Source	Weight of Pollutant Produced					Total Weight of Pollutant Produced by this source
	CO	NO_x	HC	SO_x	Particulates	
transportation	63.8	8.1	16.6	0.8	1.2	90.5
fuel combustion (stationary)	1.9	10.0	0.7	24.4	8.9	45.9
industrial processes	9.7	0.2	4.6	7.3	7.5	29.3
solid waste disposal	7.8	0.6	1.6	0.1	1.1	11.2
miscellaneous	16.9	1.7	8.5	0.6	9.6	37.3
total weight of each pollutant produced	100.1	20.6	32.0	33.2	28.3	214.2

SOURCE: Adapted from U.S. Department Health, Education & Welfare: *Nationwide Inventory of Air Pollutant Emissions,* 1968, p. 3.

Table 14-5. Pollutant Weighting Factors (Based on proposed air quality standards for California)

Pollutant	Relative Toxicity (weighting factor)
CO	1.00
HC	2.07
SO_x	28.0
NO_x	77.8
particulates	106.7

SOURCE: Adapted from L. R. Babcock, Jr.: *J. Air Pollut. Con. Assn.,* **20:**658 (1970).

Table 14-6. Weighted Masses of Pollutants (1968)

	CO	NO_x	HC	SO_x	Particulates	Total
transportation	63.8	630.2	34.3	22.4	128.0	878.7
fuel consumption	1.9	778.0	1.4	683.2	949.6	2414.1
industrial processes	9.7	15.6	9.5	204.4	800.2	1039.4
solid waste disposal	7.8	46.7	3.3	2.8	117.4	178.0
miscellaneous	16.9	131.2	17.6	16.8	1024.3	1206.8
Total						5717.0

SOURCE: See Table 14-5.

Table 14-7. Total
Emissions by Source.
(Mass versus Weighted
Mass Basis)

Pollutant Source	Total Emissions (%)	
	Mass Basis (From Table 14-4)	Weighted Mass Basis (From Table 14-6)
transportation	42.2	15.4
fuel consumption (stationary)	21.4	42.2
industrial processes	13.7	18.1
solid waste disposal	5.3	3.1
miscellaneous	17.4	21.1
Total	100.0	100.0

SOURCE: See Table 14-5.

basis of total mass, transportation would appear to be our major source of primary air pollutants; however, on the basis of the weighted mass analysis (please note the more significant items in Table 14-6), fuel combustion from stationary sources (primarily fossil-fueled electric power plants) appears to be the major air polluter. There is an element of correctness in both views.

In the Los Angeles region (Figure 14-14), the major source of air pollution is transportation based on the internal combustion engine. The ensuing Los Angeles smog, arising from nitrogen oxides, carbon monoxide, and hydrocarbon pollutants, is a different kind of smog than found elsewhere. In the New York metropolitan area (Figure 14-15), fuel combustion from stationary sources, industrial processes, and transportation combine to pollute the air. Although for general information purposes we discuss all these primary air pollutants, our discussion stresses those that come from stationary sources (fossil-fuel fired power plants). Air pollution from mobile units (transportation) will be referred to occasionally, but the reader should turn to other sources for more complete discussions of air pollution from transportation (mobile combustion units).

Let us look at the combustion of coal. The principal constituents of coal are carbon, hydrogen, and sulfur, with coal also containing certain assorted minerals. A typical anthracite coal might contain 92% carbon, 3% hydrogen, 3% sulfur, and 2% minerals. Thus, a metric ton of anthracite coal contains:

carbon	920 kg
hydrogen	30 kg
sulfur	30 kg
mineral	20 kg

A bituminous (soft coal) typically might show an analysis as follows: 72% carbon, 5% hydrogen, 1% nitrogen, 8% oxygen, 2% sulfur, and 12% minerals.

Figure 14-14. Smog in west Los Angeles, 1972. [EPA-DOCU-MERICA-Gene Daniels]

Figure 14-15. Thick smog over mid-Manhattan, New York, 1963. [Courtesy of Environmental Protection Agency. Photo by New York Journal American.]

427

EXAMPLE 14-1

How many kilograms of CO_2 are produced from 1 metric ton (M.T.) of anthracite coal containing 92% carbon?

$$C + O_2 \longrightarrow CO_2$$

From the balanced chemical equation, 12 g C yield 44 g CO_2 or 12 kg C yield 44 kg CO_2.

Thus, from this data and a unit analysis, we obtain

$$\frac{1000 \text{ kg coal}}{1 \text{ M.T. coal}} \times \frac{920 \text{ kg C}}{1000 \text{ kg coal}} \times \frac{44 \text{ kg CO}_2}{12 \text{ kg C}} = \frac{3370 \text{ kg CO}_2}{1 \text{ M.T. coal}}$$

In words, burning 1 M.T. of coal yields 3370 kg of CO_2.

The heating values of the two coals actually are not greatly different as shown in Table 6-5.

Coal molecules are vast interlocking networks of carbon atoms with hydrogen atoms attached to the networks. In the networks, a sulfur atom appears occasionally. For this chemical structural reason, sulfur is very difficult to remove from coal. Also, the burning of coal in a furnace occurs in air, not in pure oxygen. At the high temperatures of the combustion, some nitrogen reacts with oxygen. Thus, a ton of anthracite coal produces a number of materials, see Table 14-8 and Example 14-1.

In addition, over 10,000 kg of unreacted nitrogen also leaves the combustion chamber, so that about 14,000 kg or 14 metric tons of reaction products go up the smokestack for every metric ton of coal combusted. Let us now look at the primary pollutants more closely.

Table 14-8 Quantities of Combustion Products from One Metric Ton of Typical Anthracite Coal

Combustion Reactions	Quantity of Product from 1 Metric Ton of Coal
$C + O_2 \longrightarrow CO_2$	3370 kg
$4 H + O_2 \longrightarrow 2 H_2O$	270 kg
$S + O_2 \longrightarrow SO_2$	60 kg
$N_2 + O_2 \xrightarrow{\text{heat}} 2 NO$	14 kg
minerals $\longrightarrow$ smoke	10 kg
Total	3724 kg

Carbon Monoxide

Complete combustion of coal (principally carbon) gives carbon dioxide (CO_2). Similarly, complete combustion of oil (C_xH_{2y}) gives x CO_2 and y H_2O. Methane, CH_4, of natural gas simply gives 1 CO_2 and 2 H_2O. In discussing air pollutants, you should have noted by now that carbon dioxide is not considered an air pollutant, although, as discussed in Chapter 8, the greenhouse effect of carbon dioxide may have a significant influence on our lives on earth. Only the minor products of combustion reactions are known as pollutants.

Carbon monoxide (CO) arises principally from partial combustion of fossil fuels. Partial combustion is normally caused by insufficient air (or oxygen) or poor burning conditions. Briefly, the amount of carbon monoxide produced can be decreased and nearly eliminated by more efficient combustion (more air or oxygen and better mixing of fuel and air) by methods similar to those used to eliminate unburned carbon, which provides black particulate matter in the exhaust.

A simplified view of the combustion of carbon fuels is as follows:

$$2\,C + O_2 \longrightarrow 2\,CO \tag{14-1}$$
$$2\,CO + O_2 \longrightarrow 2\,CO_2 \tag{14-2}$$

Actually, the fuel combustion is considerably more complex, but these two equations are significant inasmuch as reaction (14-1), leading to CO, proceeds about 10 times faster than reaction (14-2), of CO with O_2 to give CO_2. Extra or excess oxygen must be present and good mixing of the fuel and the air must occur in order to assure rather complete conversion of carbon and carbon monoxide to carbon dioxide.

Carbon monoxide is a colorless, odorless, tasteless, nonirritating gas that boils (or liquifies) at $-192°C$. This gas is toxic to human beings. The effect on man of high concentrations, that is, greater than 100 ppm, have been well documented: 800 ppm leads to unconsciousness in 1 hr followed by death in 4 hr. Effects of quantities less than 100 ppm have been intensively studied in recent years. The Environmental Protection Agency has set forth National Air Quality Standards for carbon monoxide and other pollutants. These are discussed in the last section of this chapter.

Of concern, is the fact that atmospheric oxidation of carbon monoxide to carbon dioxide proceeds at a very slow rate. In sunlight, only about 0.1% of the carbon monoxide in the atmosphere is reacted in each hour. Fortunately, some soil microorganisms and some organisms in the sea are highly effective in removing carbon monoxide from the atmosphere. Thus, except in local regions where considerable carbon monoxide is being produced (remember that carbon monoxide does not rise to any extent due to its molecular weight) there is little likelihood of large concentrations of carbon monoxide persisting over large areas.

Nitrogen Oxides

From the coal combustion shown in Table 14-8, of the 14 metric tons of materials going up the smoke stack, some 14 kg (0.1%) is nitric oxide, an oxide of nitrogen. Nitric oxide (NO) continues to react rapidly with air after being formed, as shown in equation (14-3).

$$2\,NO + O_2 \longrightarrow 2\,NO_2 \tag{14-3}$$

The term NO_x, then, refers to a mixture of NO and NO_2 (nitrogen dioxide).

Nitrogen dioxide, a reddish brown gas, is quite toxic; that is, concentrations of nitrogen dioxide greater than 100 ppm are lethal to animals such as cats, guinea pigs, rats, mice, and rabbits with most of the deaths caused by pulmonary edema (strangulation caused by the filling the lungs with fluids). The effects on man range from unpleasant odor and mild irritations to serious lung congestion to death. Plants exposed for several weeks to nitrogen dioxide at concentration levels of only a few tenths of a part per million have had growths retarded by as much as 35%. At these very low concentrations, there are no acute symptoms on the leaves. Interestingly, the plants are more susceptible to the effect of nitrogen dioxide at night than during the daylight hours. To the best of our knowledge, nitrogen dioxide concentrations in polluted outdoor air usually are not high enough to produce serious effects on humans but may aggravate already existing respiratory ailments.

Hydrocarbons

Hydrocarbon pollutants deserve only a brief mention—brief because they are produced in only very small amounts by fossil-fuel fired utility plants. However, hydrocarbon air pollutants are very important components from automobile exhaust. These materials arise from partial combustion of oil (gasoline) and natural gas. Hydrocarbons, as the name implies, are chemical compounds containing only atoms of carbon and hydrogen. Many thousands of such compounds exist. For example, ethane (C_2H_6), ethylene (C_2H_4), acetylene (C_2H_2), and propylene (C_3H_8) are all found in the exhaust from automobiles in small quantities. In addition to simple hydrocarbons, the jargon of air pollution includes a variety of other organic chemical compounds put into the atmosphere by the handling of gasoline, vapors from spray painting or lacquering, and other related industrial operations.

All these organic substances, although not directly harmful themselves in small quantities, enter into a photochemical process involving nitrogen dioxide (see previous section) and ozone (O_3—formed in small quantities by effect of ultraviolet light on oxygen) to give Los Angeles-type smog. The products of these reactions found in the smog are toxic and irritating to people and cause damage to vegetation. More efficient combustion continues to be studied as one of the ways to reduce the tonnage of hydrocarbons put into the environment.

Sulfur Dioxide

Primary pollution from oxides of sulfur (SO_x) consists almost exclusively of sulfur dioxide (SO_2) with from 1 to 10% sulfur trioxide (SO_3) depending on the combustion temperature. The combustion of any sulfur containing material produces both these oxides of sulfur, as shown in Equations 14-4 and 14-5. Interestingly, the amount of sulfur trioxide produced is not particularly dependent on the amount of oxygen present. Thus, sulfur dioxide is the major product (90+%) even in excess oxygen.

$$S + O_2 \longrightarrow SO_2 \tag{14-4}$$
$$2\,SO_2 + O_2 \longrightarrow 2\,SO_3 \tag{14-5}$$

Sulfur is a common constituent of coal, oil, and natural gas. In coal and oil, the amount of sulfur may be as high as 7% with 3% sulfur being rather common. In the United States, no large deposit of coal containing less than 1% sulfur is found east of the Mississippi River. In 1964, the coal burned by utilities had the following sulfur composition: 21% had greater than 3% sulfur, 60% had sulfur between 1.1 and 3.0%, and only 19% had less than 1.1% sulfur.

Review of Tables 14-4 through 14-7 emphasizes the significance of sulfur dioxide being put into the air by fuel combustion in large stationary sources. Of the 24.4 million tons of sulfur dioxide produced in 1968 from these sources, 16.8 million tons came from power plants, 5.1 million tons from industrial plants, and 2.5 million tons from heating devices used in homes and businesses. The power plants accounted for over half of the annual sulfur dioxide pollution in 1968. Furthermore, high sulfur dioxide concentrations have been associated with each of the major air pollution disasters mentioned earlier.

Sulfur dioxide is a colorless gas that liquifies at $-10°C$. Sulfur dioxide has a characteristic suffocating and pungent smell. To humans, the main effect of sulfur dioxide appears to be irritation of the respiratory system by constriction of the air pathways with corresponding increases in resistance to air flow during breathing. Such effects are most severe in people already suffering from emphysema or similar impairment of lung function. Some responses of humans to sulfur dioxide are tabulated in Table 14-9.

Once in the atmosphere, sulfur dioxide is converted catalytically and photochemically to sulfur trioxide. One of the characteristics of sulfur trioxide is its high affinity for water. Sulfur trioxide and water combine rapidly to form droplets of sulfuric acid (H_2SO_4), equation (14-6).

$$SO_3 + H_2O \longrightarrow H_2SO_4 \tag{14-6}$$

Table 14-9. Effect of Sulfur Dioxide on Humans

SOURCE: Data adapted from R. E. Kirk and D. F. Othmer (eds): *Encyclopedia of Chemical Technology,* 2d ed., Vol. 19. John Wiley & Sons, Inc. New York, 1969, p. 417.

SO_2 (ppm)	Effect
3–5	Least amount detectable by odor
8–12	Least amount causing immediate throat irritation
20	Least amount causing immediate eye irritation or causing immediate coughing

Sulfur trioxide and sulfuric acid are referred to as secondary air pollutants in that they derive from sulfur dioxide, a primary or first-formed substance. Although insufficient data are available for the quantitative assessment of air-borne sulfuric acid as a health hazard, it is known that sulfuric acid is a much more potent irritant to man than sulfur dioxide. In the environment, much of the damage from SO_x pollution is caused by sulfuric acid solutions, droplets, or aerosols. For example, both in New England and in Sweden, the acidity of natural waters—lakes and ponds— has been measurably increasing in the past several years, most probably as a result of sulfuric acid added to these natural waters from the air.

Many metals such as zinc, iron, and chromium are pitted (Figure 14-16) and corroded by sulfuric acid droplets deposited from the atmosphere. The effect on plants varies from stunting of growth to leaf damage (Figure 14-17). Textile fibers such as cotton and linen lose strength and may be completely destroyed when exposed to acids. Paper becomes brittle and fragile. Canvasses and paintings on canvasses are damaged (Figure 14-18). Sulfuric acid aerosols from atmospheric pollution attack building materials

Figure 14-16. Corrosion of an automobile due to acid in the atmosphere. This corrosion of iron is just below the rear window of the automobile.

Figure 14-17. The effects of sulfur dioxide and ozone pollution. **(A)** Dwarf white pines at relatively low levels. Note how close the branches are as a result of very little annual growth. **(B)** Damage to white birch leaf. **(C)** Damage to tobacco leaf. [Courtesy of U. S. Department of Agriculture.]

(A)

(B)

(C)

433

(A)   **(B)**

Figure 14-18. Fifty years of exposure to polluted air (1920 to 1970). A painting from a small church near Milan. **(A)** 1920. **(B)** Fifty years later. [Courtesy of the Italian Art Landscape and Foundation.]

with carbonate-containing substances; limestone, marble, and slate are particularly susceptible (Figure 14-19). As examples, the oldest cemetery in the Western Hemisphere in the Netherlands Antilles has had its grave stones become illegible as a result of such reactions. The limestone buildings of Venice and the marble statues of Milan are being rapidly eroded as the result of the chemical reaction in equation (14-7).

$$CaCO_3 + H_2SO_4 \longrightarrow CaSO_4(s) + CO_2(g) + H_2O \qquad (14\text{-}7)$$
$$\text{limestone} \qquad\qquad\qquad \text{calcium sulfate}$$

where (s) = solid and (g) = gas.

The calcium sulfate formed on the surface is about twice as bulky as the carbonate of the stone from which it was formed. Thus, the stone appears leprous or diseased. Furthermore, the calcium sulfate is more soluble in water than the calcium carbonate, so the reaction products are gradually dissolved away leaving pits and weakening the stone.

Figure 14-19. Examples of limestone and marble erosion due to sulfuric acid. **(A)** A marble statue from the Cathedral in Milan. [Courtesy of the Italian Art and Landscape Foundation.] **(B)** A statue at the entrance to the Hirten Castle, near Rücklinghausen in the Ruhr, West Germany. [Courtesy of Engineering Technology, Inc. Photograph by Esber I. Shaheen.]

1908   1968

Pollution Abatement (SO_2)

Let us examine now the question of decreasing the amount of air pollution or pollution abatement by specific reference to sulfur dioxide. What has, or what can be done about these sulfur oxides and the derived sulfuric acid? How can the quantities of these materials be decreased? For many years, it has been believed that using taller and taller chimneys in utility plants, factories, and so on, was a way out. In other words, as we have said before, dilution is a solution to pollution.

The tall stack does not reduce the amount of pollutants emitted or produced; it only reduces the ground level concentrations of the pollutant in the immediate vicinity. For coal-burning power plants built in 1960, the average height of the stack was 243 ft; in 1969, the average was 609 ft. Stacks as high as 800 ft or more are being built (see Figure 14-20). Although the higher stacks certainly are needed for the larger power plants (500 to 1000 Mw or more), the tall stack merely is a device for spreading the pollutants out in a larger volume of air and thus over a larger area of ground surface.

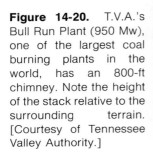

Figure 14-20. T.V.A.'s Bull Run Plant (950 Mw), one of the largest coal burning plants in the world, has an 800-ft chimney. Note the height of the stack relative to the surrounding terrain. [Courtesy of Tennessee Valley Authority.]

However, for real abatement (not just dilution) of SO_x pollution, we can identify four approaches to the actual reduction in the amount of sulfur oxides entering the air environment. These are

1. Better efficiency of use of the energy from the fossil fuel.
2. Remove SO_x after the fuel has been combusted.
3. Remove sulfur before the fuel is combusted.
4. Use alternate sources of energy not producing SO_x.

The first suggestion has already been discussed as we examined the laws of thermodynamics, particularly in connection with efficiencies of heat engines (Chapter 8). In the twentieth century, we have seen a remarkable improvement in the practical efficiency of fossil-fueled power plants rising from less than 10% to over 40%. Certainly, if our power plants operated today at only 10% efficiency we would be placing nearly 4 times as much sulfur oxides into the atmosphere. Unfortunately, our brief study of the Carnot engine and the consequent scientific limits to technology told us that only 60–62% was the best we could ever expect to achieve in a heat engine (a power plant) operating between certain temperatures. This limit is a law of nature. Put bluntly, any increase in efficiency of fossil-fueled power plants above 40% is going to be obtained with great difficulty. Thus, since substantial efficiency increase (that is, several percent) is not likely, considerable attention has turned to the other three possibilities.

What is the problem of removing sulfur oxides after fuel combustion?

EXAMPLE 14-2

Calculate the percentage of SO_2 appearing in the flue gas from the combustion of coal containing 3% sulfur.

From the balanced equation: $S + O_2 \longrightarrow SO_2$, 32 g of S yield 64 g of SO_2 (1:2).

$$1000 \text{ kg} \times 0.03 = 30 \text{ kg S} \qquad \text{which gives } 60 \text{ kg SO}_2$$

$$\frac{60 \text{ kg SO}_2}{14,000 \text{ kg gases}} \times 100 = 0.43\% \text{ SO}_2 \text{ in the flue gases}$$

Recall the figures determined previously that for every metric ton of coal burned in air some 14 metric tons (14,000 kg) of exhaust gases, mostly nitrogen, go up the chimney. Included in these gases is about 60 kg of sulfur oxides (as sulfur dioxide). This represents only 0.43% sulfur oxides as sulfur dioxide in the flue gases or 4300 ppm (Example 14-2). This rather small amount of sulfur oxides in the total stack effluent makes removing them difficult. Nonetheless, considerable study has been given to the development of sulfur dioxide removal processes.

How can the gaseous sulfur dioxide (or sulfur trioxide) molecules be taken out of the hot, gaseous mixture? One simply takes advantage of the same high chemical reactivity that causes these substances to damage the environment. Both sulfur dioxide and sulfur trioxide react with water to produce acids.

$$\underset{\text{sulfuric acid}}{SO_3 + H_2O \longrightarrow H_2SO_4}$$

$$\underset{\text{sulfurous acid}}{SO_2 + H_2O \longrightarrow H_2SO_3}$$

Furthermore, a characteristically fast chemical reaction is that between acids (bitter taste) and bases, (slippery feel).

$$\underset{\substack{\text{sulfuric acid}}}{H_2SO_4} + \underset{\substack{\text{sodium hydroxide} \\ \text{(a base)}}}{2 \text{ NaOH}} \longrightarrow \underset{\substack{\text{sodium sulfate} \\ \text{(a salt)}}}{Na_2SO_4} + \underset{\text{water}}{2 H_2O} \quad (14\text{-}8)$$

Similarly,

$$\underset{\text{sulfurous acid}}{H_2SO_3} + \underset{\text{lime}}{Ca(OH)_2} \longrightarrow \underset{\text{calcium sulfite}}{CaSO_3} + 2 H_2O \quad (14\text{-}9)$$

A number of bases, or substances that can be converted to bases [$Ca(OH)_2$, CaO, and so on] such as limestone (reaction 14-10) have been studied as to their effectiveness in removing SO_2 from flue gases.

$$CaCO_3(s) \xrightarrow{\text{heat}} CaO(s) + CO_2(g) \qquad (14\text{-}10)$$

Although many processes are in development, none is truly commercial. You should note that if ground-up limestone is used to remove the sulfur dioxide, reaction (14-11) would represent the process.

$$SO_2(g) + CaCO_3(s) \longrightarrow CaSO_3(s) + CO_2(g) \qquad (14\text{-}11)$$

where (g) = gas and (s) = solid.

You should note that removal of the sulfur dioxide gas results in the formation of carbon dioxide gas, not normally considered a pollutant (although a problem in respect to the greenhouse effect) *and* the formation of calcium sulfite ($CaSO_3$) solid—a white solid, which has to be disposed of. Fortunately, this particular solid can be usefully decomposed in a separate system (Equation 14-12).

$$CaSO_3(s) \xrightarrow{\text{heat}} CaO(s) + SO_2(g) \qquad (14\text{-}12)$$

The sulfur dioxide, now in concentrated form, can be used to produce sulfuric acid in a commercial plant. The calcium oxide can be reused as the base in the smokestack. This process, if commercially feasible, would be a particularly good choice. However, any sulfuric trioxide that is in the flue gas ends up as calcium sulfate ($CaSO_4$, gypsum) which cannot be similarly decomposed, so that this substance must then be disposed of somewhere. What is the point? Removal of one pollutant generates something else that must be disposed of. Where? How? We deal with profoundly interlocking problems!

Inasmuch as flue gas clean-up processes were not available in the late 1960s when the need to remove sulfur oxides was apparent and immediate, other solutions to the sulfur problem were sought, such as using fossil fuels that are low in sulfur content. For low-sulfur content fossil fuels, one has two choices.

1. Find a source of fuel that naturally has a low sulfur content.
2. Put a high-sulfur content fossil fuel through some chemical process that removes some of the sulfur so that a low-sulfur content fuel is produced.

New York City serves as a good example for the first choice. In 1966, Local Law 14 decreed that by May of 1967 only fuel with a maximum of 2% sulfur could be used and that by May of 1971 only fuel with a maximum of 1% sulfur could be used. The Consolidated Edison Company, by converting a number of its coal burning plants to oil *and* by shifting its source of oil from Venezuela ($>$ 1% sulfur) to African oil ($<$ 1% sulfur from Liberia and Nigeria) reached the 1% target of the Local Law by October 1967, ahead of schedule. What were the costs? At the time of change over, Con Ed's annual fuel bill had been $63 million. Fuel costs went up by $7.5 million, and $3 million was needed to build the fuel oil handling facilities. Ultimately, *you* must pay these additional costs. New York City turned to this costly solution both because low sulfur oil was available naturally and because methods of removing sulfur from crude oil have a reasonably developed technology.

As far as the naturally occurring low sulfur fuels are concerned, at present a considerable tonnage is being imported from underdeveloped nations, including many nations whose governments have undergone drastic changes and may not be particularly friendly to the United States. But, the political implications and the long term effects on our balance of international payments (economics) are beyond the scope of this book.

A Question for the Reader: How dependent are you prepared to be on low sulfur oil imported from unfriendly nations? Or, for that matter, on any oil imported?

What can be done to reduce the sulfur content of fuel oils, say from 3% down to less than 1%? The residual fuel oils used in power plants can be treated (catalytic treatment) to remove much of the sulfur content. This operation adds some 20–35% to the price of the oil, whatever the base price of high sulfur content oil may be. Con Ed turned away from coal because the technology for removing the sulfur from coal was not available nor is it expected to be in the foreseeable future.

What is the problem with coal? With coal, the sulfur present is found in three principal forms:

1. as pyrites or FeS
2. as sulfates (minerals)
3. as sulfur bonded into the network of carbons making up the coal

Coal can be finely ground up and, by appropriate flotation methods, the pyritic sulfur can be removed. The sulfate sulfur on combustion ends up either in the ashes from the pit or in the fly ash. However, in most coals these two forms of sulfur represent only about one half of the total. The remainder is sulfur bonded into the molecular structure of the coal. *No* method for removal of this type of sulfur has proved successful, even in the laboratory. Thus, at best with coal, our science and technology could remove only half of the sulfur content of the coal and this only at a substantial cost. Utility coal in 1970 had an average cost of around $6.00 a ton. One study of reducing the sulfur in coal, including recovery of the sulfur for sulfuric acid, which is used extensively in manufacturing processes, concluded that the cost of clean-up (50% of the sulfur removed) would be about $2.80 per ton—almost half the original price of the coal.

Thus, we turn to the fourth alternative: other sources of energy for manning the electric utility plants. One of the sources that is being intensively studied both by the government (Bureau of Mines) and by industry is based on coal, that is, the conversion of coal to a gaseous fuel. The various processes being studied are all referred to as **coal gasification.** A typical equation for the gasifying process might be as shown in reaction (14-13)

$$\underset{\text{coal}}{4\,C} + 3\,H_2O \longrightarrow CH_4 + H_2 + 3\,CO \qquad (14\text{-}13)$$

with the sulfur in the coal being removed as H_2S (hydrogen sulfide) by using the H_2 in the products of reaction (14-13). The formation of H_2S is shown in reaction (14-14).

$$H_2 + S \longrightarrow H_2S \qquad (14\text{-}14)$$

The advantages of these processes are that any uncombustibles (ash) are left as residue when the gas leaves. The H_2S can be removed by washing with a base (NaOH, $Ca(OH)_2$, CaO) to produce a sulfur-free gas. The manufacture of gas from coal actually was done in the United States for many years, prior to World War II. However, the advent of natural gas, with a higher heat value per pound, largely supplanted such processes in the United States in the 1940s. Variations of the older process are now under study in view of an imminent natural gas shortage.

At this point we have considered the four general ways of reducing the amount of sulfur dioxide entering the air environment as a result of oil and coal combustion. (Natural gas in general has a very low (near zero) sulfur content.) We have pointed out the scientific possibilities, both actually known and some based on present research. Again, the world of science identifies the limits of technology. Nonetheless, both the scientific facts and the technology, present and future, seem reasonably clear. So, here we are again faced with the prospects of trade-offs: benefits versus hazards. What do you think?

At this point, perhaps you will want to review the earlier discussion of nuclear reactions and compare the benefits and hazards therein to those of the combustion of fossil fuels for the production of electrical energy. Now, what does the future hold for us? The next chapter will briefly summarize some of the alternatives we face. However, before we proceed, there is one more aspect of air pollution of which you should be aware.

Air Quality Standards

In part, as a result of the air pollution disasters, but more importantly as a result of the recognition that extremely large quantities of primary air pollutants are being produced each year, the Environmental Protection Agency (E.P.A.) on April 30, 1971 set forth National Air Quality Standards.

On the basis of considerable data, ambient air quality standards were set that have been designed to be protective of human health with a margin of safety. These standards are tabulated in Table 14-10.

By July 1972, implementation plans for all the 250 federal air quality regions were to be completed. By July 1, 1975, the level of ambient air quality specified in Table 14-10 was also to have been achieved. Following the setting forth of these standards, considerable questions were raised in the early 1970s as to the feasibility of meeting the Act's deadlines. Conceptually, the Act's intent was, by 1975, to effect a 90% reduction in primary air pollutant emission from any given source. Such air quality standards are clearly a governmental step (a societal step) deriving from a consideration of benefits versus hazards or trade-offs.

Put another way, such standards set forth a hazard level beyond which the benefit is stated to be unacceptable. Of concern, here, is the fact that the standards do not eliminate air pollution, rather they set specific emission goals for air pollution abatement. The setting of such standards is of the utmost importance. Whether an energy-hungry or fuel-short society chooses to use these as guidelines is another matter. The debate ensues as to how

Table 14-10. National Air Quality Standards (as set forth by the Environmental Protection Agency on April 30, 1971)

Pollutant		Level Not to Exceed*
sulfur dioxide	80 μg/m^3	(0.03 ppm) as the annual arithmetic mean.
	365 μg/m^3	(0.14 ppm) as the maximum 24-hr concentration not to be exceeded more than once a year
Particulate matter	75 μg/m^3	annual geometric mean.
	260 μg/m^3	maximum 24-hr concentration as above
carbon monoxide	10 mg/m^3	(9 ppm) maximum 8-hr concentration not to be exceeded more than once a year
	40 mg/m^3	(35 ppm) maximum 1-hr concentration not to be exceeded more than once a year
hydrocarbons	160 μg/m^3	(0.24 ppm) maximum 3-hr concentration 6-9 A.M. not to be exceeded more than once a year
nitrogen oxides	100 μg/m^3	(0.05 ppm) annual arithmetic mean.
photochemical oxidants	160 μg/m^3	(0.08 ppm) maximum 1-hr concentration as above.

*μg = microgram or 10^{-6} g = 0.000001 g

these standards will be coupled with society's determination (or lack of) to enforce or monitor the guidelines.

Any guidelines must be based on the best available scientific information. Even though scientific data is used, a complexity arises when technological feasibility has not been demonstrated. Herein lies the necessity for additional research and development study of the problems. However, we wish to lay before you the thought that no matter what is done by way of legislation, such legislation represents a choice of man and not strictly a law of nature. As a choice of man, the standards can be changed. Furthermore, as a member of society, particularly a member of energy hungry society, you must consider benefits and hazards of your consumption of electrical energy—your way of life—our way of life.

Conclusion

Most pollution exists because we demand the goods and services of a technology which, for the most part, has been given little consideration as to the long range effects of its operations and its products. The 1970s could be considered the beginning of a real awakening of concern for the environment. The costs of abating pollution are already high, and they are likely to increase much more. Such costs of necessity will be passed on to the consumer, that is, increased electrical rates. Although, the price may ultimately be high, can we really say it is too high if in the bargain (trade-off) we live in clean air or at least cleaner air?

This chapter concludes a set of three chapters (12, 13, 14) that briefly lay before you the environmental impact of electrical power production. In this chapter, we have examined some of the lesser products of the combustion of fossil fuels. These products arise in straight forward simple chemical processes, but some of them are extremely noxious to our environment. The primary air pollutants are of particular importance because today and lasting through the twentieth century the major portion of our electrical power is to be generated in fossil-fuel fired power plants. This is reality. All the electricity you consume has left a "double" environmental impact at the site of the power plant in terms of heat plus an impact on the air environment by the generation of significant quantities of sulfur dioxide, nitric oxide, and particulates. We have looked at the chemical nature of these primary pollutants, how they are formed, and to some extent how the amount produced can be reduced.

This present reality now permits us to understand where we are now and what sort of tomorrow we can look forward to. The tomorrow consists both of those generating stations now already providing us electrical energy *and* of those generating stations to be built. Additionally, there has been considerable speculation as to other modes of power generation. Thus, "What Next?" is the subject of the next and final chapter.

Questions

1. What is air pollution? What are some air pollutants?

2. What are the natural components of air?

3. Air is a gas. Describe qualitatively the behavior of a gas on being heated.

4. What conditions are necessary for a thermal inversion?

5. What are the primary air pollutants?

6. What are the major sources of the primary pollutants?

7. What is a secondary air pollutant? Give two examples.

8. Define:
   (a) ppm   (e) incomplete combustion
   (b) microgram (μg)   (f) smog
   (c) micron (μ)   (g) aerosol
   (d) smoke   (h) an air quality standard

9. How are sulfur dioxide and sulfuric acid related? Which is the primary and which the secondary air pollutant?

10. Is air pollution only a recent problem? Discuss.

11. List four ways by which sulfur dioxide emissions from fossil fuel combustion can be reduced.

12. Why is natural gas (or synthetic gas from coal) considered to be a "cleaner" fuel than either coal or oil?

13. Gas turbines have been used to generate electrical power. The gas turbine engine has a combustion temperature higher than either coal or oil furnaces. Estimate how these combustions compare as to NO_x emissions.

Numerical Exercises

1. Which of the primary air pollutants will rise readily in air? Which will not? Why?

2. 1 liter of sulfur dioxide is mixed with 99,999 liters of air, without changing the temperature or the pressure. What is the concentration of SO_2 in the mixture in parts per million by volume? In parts per billion by volume?

3. If the atmosphere has 8 ppm of SO_2, how many liters of SO_2 are found in how many liters of air?

4. If a ton of coal has 87% carbon, how many tons of carbon dioxide will be produced on total combustion of 1 ton of coal? If 7.5×10^8 tons of coal are burned in the United States in a year, how many tons of carbon dioxide are emitted to the atmosphere?

5. (a) If a metric ton of soft coal has 72% carbon, how many metric tons of carbon dioxide will be produced on total combustion of 1 metric ton of coal?
   (b) If this soft coal contains 2.5% S (sulfur), how many kilograms of SO_2 will be sent up the smokestack after combustion of 1 metric ton of this coal?

6. By reference to the table below, you can estimate the approximate amounts of the primary pollutants emitted to the atmosphere by home furnaces and automobiles.

Pollutant	(lb/ton) Coal Burned	(lb/million ft³) Natural Gas Burned	(lb/1000 gal) Fuel Oil Burned	(lb/1000 gal) Gasoline Burned
carbon monoxide	50	0.4	2	2910
hydrocarbons	10	negligible	2	524
NO_x	8	116.0	72	113
SO_x	100	0.4	400	9
particulates	20	19.0	12	11

This table was derived from a compilation prepared by Martin Mayer, Public Health Service, Cincinnati, Ohio, May 1965.

(a) If a home burns 12 tons of coal per year, how many pounds of each of the primary pollutants are being put into the atmosphere by this home each year?

(b) If a home burns 2000 gal of oil per year, how many pounds of each of the primary pollutants are being put into the atmosphere by this home each year?

(c) If a home burns 2,500,000 ft³ of natural gas, how many pounds of each of the primary pollutants are being put into the atmosphere by this home each year?

(d) If a New Jersey community of 100,000 averages four people per dwelling, and if 10% of the homes are heated by coal, 40% of the homes are heated by natural gas, and 50% of the homes are heated by oil, calculate the number of pounds of each of the primary pollutants being put into the atmosphere by this community each year. Assume that each dwelling uses fuels at the annual rates noted in parts a, b, and c.

(e) Compute the quantity of fuel you burn in your furnace each year. Record your figure in the following terms: For coal use tons, for fuel oil use units of 1000 gal, for natural gas use units of 10^6 ft³.

(f) Compute the quantity of gasoline (in units of 1000 gal) consumed by your automobile in a year. (*Note:* If you have no idea of this number, estimate the number of miles you drive in a year. From this estimate and an estimate of the number of miles you get per gallon of gasoline, calculate the number of gallons of gasoline you use each year.)

(g) Calculate the pounds of primary pollutants put into the atmosphere by *your* home furnace and by your automobile each year by multiplying the results from part e and part f by the pounds of pollutants listed in the table.

15

What Next?

Well, there you have our situation. Up to this point we have surveyed the ways in which electrical energy is produced for your consumption. We have analyzed these methods in terms of the essential chemistry and physics. Within these specific areas of science, scientific man has grouped the laws of nature he has discovered. These laws of nature are fundamental to man's existence on earth in that these laws cannot be defied or circumvented. All that man does is controlled by these laws.

We have looked at the generation of electrical energy by the use of falling water to drive turbines. We have studied the generation of electrical energy by the use of hot steam to drive turbines. Our emphasis has been on the use of the chemical energy (chemical bonds) contained in fossil fuels to obtain the heat necessary to change water into steam. We have also considered the nuclear (binding) energy in uranium (or plutonium) as a source of heat energy. The reason for major emphasis

445

on our fossil fuels is due to the fact that fossil fuels (mainly petroleum) provide and, in the immediate future (next 10 to 15 years), will continue to provide more than 95% of our fuel energy. Although nuclear energy has provided only a small fraction of the thermal energy used by society in the early 1970s, we see it increasing significantly in importance by 1985 and perhaps growing to such an extent of our having an extensive (over 50%) nuclear electric economy shortly after the turn of the century.

However, we must caution that the extent to which our predicted trends continue will depend on the availability of domestic and imported petroleum supplies, the attitudes of society toward nuclear power (and its potential hazards), or, alternately, a return to an all coal economy with its "dirty power." We further suggest that the availability of more exotic sources of energy through advancing technology may play an important role, but probably not until the early years of the next century. In particular, a search for clean, pollution-free power may represent a substantial portion of the research and development effort by scientists and technologists in the next 50 years.

Put simply, if we are to continue to have the easily available energy upon which our "great society" has been built, especially in the face of decreasing petroleum supplies in the United States, we must pay the price. There is no such thing as a free lunch. As we have already said so many times, much of our development and our energy use in the past has been with little concern both for the finite energy supplies of our planet and for the environmental impact of energy use. Future industrial development will be considerably more difficult and more costly due to the debt that we already owe and due to the continuing concern for the future of spaceship earth. Thus, in brief, we have learned that there are at least two limits to energy consumption on earth and to our social institutions: resource depletion and pollution.

In this final chapter, we begin by reviewing patterns of energy consumption with some forecasting into the future. We then survey alternate energy sources including the potential of returning to "King Coal." Included in our discussion is a brief preview of how science and technology may be able to increase the efficiency of energy conversion. Because energy conservation is a new household word, we include a look at where we might save energy in the various user sectors. Then, we ask you to join us in a piercing glance into the future as we consider answers to the questions as to whether there are limits to growth and whether or not technology is really a salvation.

Inasmuch as we believe the answer to the first question is yes and to the second, no, we then offer for your consumption our biased but reasoned view as to what the future holds. At the outset we do not claim novelty, originality, or accuracy for our personal view. We then ask that you seriously consider the next step: developing your own view based on the facts and, of course, on the laws of science.

Future Energy Demands and Limitations

The pattern of energy consumption in the United States (and to a lesser extent the world) is presently characterized by three basic factors: an ever-increasing demand (exponential growth), a shift in the energy-power base from coal to petroleum, and burgeoning demands for electricity. Study of growth rates since 1900 has shown that our demand for energy (in the United States) has indeed grown exponentially, with the immediate past showing growth rates of 4 to 5% for gross energy consumption and 7% (or more) for energy consumed in the production of electricity. Even though the rest of the world may produce less electricity than the United States, the gross energy consumption by the world too has been growing at a rate of 5%.

Perhaps even more revealing of our energy consuming habits is the fact that in the last three decades the per capita energy consumption in the United States has doubled. Accompanying the rather phenomenal growth of our energy appetite has been an even greater growth (rate) of our gross national product. Thus, the actual unit cost of energy has been decreasing. A major factor in this decrease has been due to advancing technology resulting in raised conversion efficiencies, for example, the diesel engine and the increasing efficiency of electrical power production.

However, technological advances in the efficiency of steam–electric power generation have slowed. Plants have probably reached their optimum size, 1000 Mw, because materials technology limit boiler temperature and pressure. In the film *Oklahoma*, which depicted life in the early 1900s, there is the fascinating line, "We've gone about as far as we can go." As we look forward to 2001 A.D., have we gone as far as we can go, or are we where we were in 1900 with a future of great technological advancement and growth ahead of us?

Two rather crucial questions are on many of our minds: Will growth rates be maintained to 2000 A.D. and perhaps beyond? Will simple exponential extrapolations describe well the demand for energy? To continue along the way that we presently are going requires that we continue to produce energy at the necessary demand rate, maintain a low unit cost of energy, and apparently ignore the environmental impact of energy production and consumption. One of the few things that most environmentalists, scientists, technologists, and so on, agree upon is that none of these requirements is realistic.

When, how, and what will cause growth rates to decline remains to be seen. Unless something drastic occurs—particularly in the political arena—we feel that things will continue pretty much as they are, at least until 1985. Further, rather than an overall doubling of gross energy consumption, a somewhat more realistic estimate can be obtained by considering the growth in per capita energy consumption. Making the realistic assumption that our population (United States) will be 300×10^6 by the year 2000 and that the per capita consumption will double again in 30 years due to our energy-intensive economy, we estimate a total power requirement of 6×10^9 kw or an annual rate of energy consumption of about 53×10^{12} kwh or 180×10^{15} Btu by the year 2000.

This level of consumption reflects an average growth rate of about 3% subsequent to 1970, which is to be compared with forecasts of a 4.3% growth at least until 1985 by major oil companies. However, if we continue at 4 or 5% growth rate in per capita energy use, our per capita energy

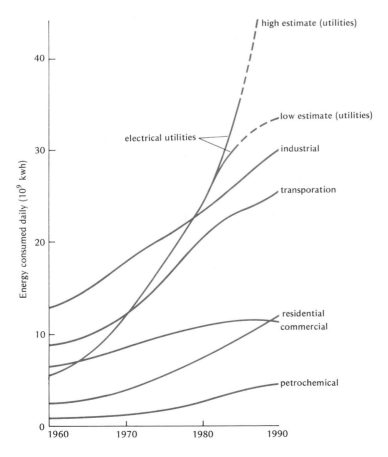

Figure 15-1. Individual market demands for energy. Electric power production is included in the market demands.

consumption will at least triple by 2000 A.D. The strain on our energy resources, not to mention the economic disaster that would be caused by having to import large quantities of fuels, would be unprecedented in our history since 1900. We further predict that as much as one half of the energy consumed would be for the production of electricity. But, here again we emphasize the need to consider the many various factors of demand, especially, cost limitations.

The major features of market demand are illustrated in Figure 15-1.

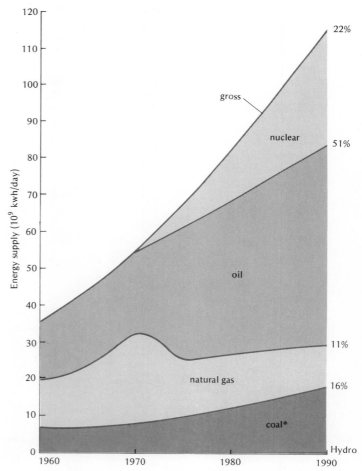

Figure 15-2. United States energy needs through 1990. Forecasts indicate that oil will continue to play a key role along with increasing roles of coal and nuclear energy.*

*Note: Potentially, coal could substantially replace the oil supply, depending on the availability of oil and the development of coal resources.

Three factors, population growth, increasing per capita income, and greater use of energy intensive processes, indicate a continuing increase in demand, unless, of course, population growth declines below the zero growth rate potential of about 1%. However, a significant increase in the cost of energy [1] could substantially moderate or even reduce the demand. Some of the factors contributing to greater cost are listed: the cost of environmental protection and the cost of obtaining new and greater supplies. Remember that in the United States the average oil well in 1972 cost about $95,000 and that discoveries occur at a much reduced rate. Both these cost factors lead us to forecast shrinking supplies and higher costs of energy. Evident from Figure 15-2 is that in the United States there is the lack of a broad-based energy economy.

Put simply, we presently depend heavily on petroleum and to a lesser

Figure 15-3. The energy gap. Our dependence on foreign sources of petroleum will continue to increase until an alternate source of energy is found here in the United States. This may, of course, require a different life style for us.

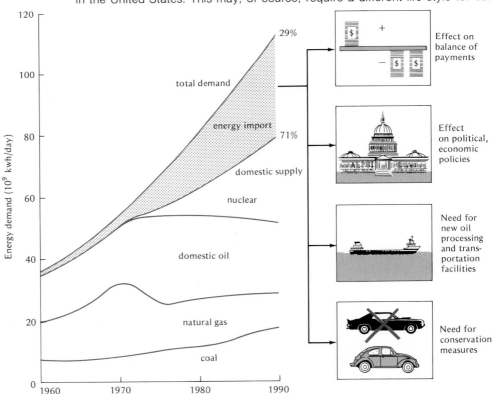

extent on coal. As we have discussed earlier, our large post-World War II increase in use of natural gas and oil has placed us in the position of turning increasingly to foreign (non-U.S.A.) sources of these fossil fuels. The United States, with its present consumption of oil and natural gas, must continue to increase its imports of these materials. Figure 15-3 illustrates our needs and the energy gap. As we have mentioned, the political and economic implications are very serious. All these resources are finite. When these resources are consumed, they are gone! With our outlook to 1990–2000 restated, let us survey potential supplementary sources of energy.

Nonrenewable Energy Resources

Power from King Coal

Recall that coal was the fossil fuel that fired the Industrial Revolution. In 1890, coal provided 87% of our energy supply, whereas in 1973 it only provided about 20%. "King Coal" yielded to the cheaper, cleaner fuels, oil and natural gas. Of the fossil fuels, coal has the greatest impact on the environment. Burning coal releases much of its sulfur content in the form of sulfur dioxide gas and also deposits large quantities of fly ash over surrounding areas. The pollution due to coal use is also acute in the regions being mined. Large quantities of acid-containing soils and rocks are deposited in the vicinity of the mine and these materials gradually wash into waterways, thus fouling them. By 1970, some 1.5 million acres of land had been devastated by strip mining with little attempt at reclamation (Figure 15-4).

Yet, even with these disadvantages, coal is our major fossil fuel resource, some 3×10^{12} tons of it! About 70% of our resource is in Montana, the Dakotas, Wyoming, and other western states. Much of this coal would be obtained by strip-mining. However, the ranchers in that area side with the environmentalists and stress that laying waste that open land area would seriously reduce the grazing area necessary for cattle and hence our beef supplies might decrease. It does seem evident that reclamation of strip-mined lands in these areas, which should be possible, would add substantially to the unit cost of energy provided by coal.

Due to the concern about environmental damage, most of any expanded coal production could go into synthetic gas and oil (syngas and syncrude). However, much of the technology to produce these synthetic substitutes is lacking. One technological objective in this area is to produce a low heat value "power gas" (for the utilities) about 180 Btu/ft^3 as compared to 1000 Btu/ft^3 for methane. A second objective is to produce a gas with high heat value, similar to methane, for residential and commercial use. The third objective is to produce syncrude to be processed into the many oil products—for the petrochemical industry and for gasoline. Actually,

(A)

Figure 15-4. (A) Dipper of a surface coal mining shovel as it lifts about 210 tons of earth and rock. [Courtesy of the National Coal Association, photograph Bucyrus-Erie Co.] **(B)** Devastation of land by strip coal mining in Wise County, Virginia, 1963. **(C)** The same land in Wise County after reclamation, 1971. [B and C, courtesy of the National Coal Association.]

(B)

(C)

in order to obtain these objectives, the United States will need to establish a crash program comparable in magnitude with the Apollo program. In making such commitments, we must carefully consider the trade-offs, benefits versus hazards (and costs), in developing this large energy resource [2].

Nuclear Power

The most immediate solution to our power problem appears to be the increased use of nuclear fission as a source of heat for operation of steam–electric power plants. As of 1972, 28 "nuke" plants were in operation, 50 under construction, and another 60 or so were planned. These plants would add some 100×10^9 watts (10^5 Mw) to the electricity-generating capability. All of this projected capacity would be provided by burner (light water) reactors.

This increase represents only about a 30% rise over the generating capability in 1972. Since electrical energy consumption has been expected to continue to double every 10 years, such projections represent a less serious total commitment to nuclear power as "our power for the future" than was expected in the early 1960s. Furthermore, the concern about radiation hazards and environmental impact (thermal) has delayed construction of many proposed "nuke" plants to such an extent that nuclear power may not actually become a truly substantial producer of electrical power.

Even more ammunition for the opponents (of nuclear power) has been added in the form of the Atomic Energy Commission's projection of a shortage of high grade, low cost uranium (supplying ^{235}U) by the turn of the century. In 1973, the breeder reactor program was just getting under way with prospects of a pilot plant being operable by 1980 to 1985. This project has been planned jointly by the Atomic Energy Commission and the Tennessee Valley Authority. With a time of about 10 years necessary to construct and put on-line a power plant, we foreseeably might have a few breeders on-line by 1990 to 1995, but only a few.

We also must consider that a finite amount of time is needed for breeding the plutonium fuel from ^{238}U. For example, it takes about 10 years to convert 1000 tons of the fertile ^{238}U into the fissile ^{239}Pu while burning ^{235}U. However, with the prospects of a serious shortage of fissile ^{235}U isotope by 2000 A.D., we ask the question, "Is it too little too late?" As we discussed earlier (Chapter 11), the rate of *consumption* of ^{235}U, that is, the number of burner reactors, may determine the length of the nuclear power age. Ultimately, though, if we assume that some national energy policy is established and that the breeder program is successful, nuclear fission could supply energy equivalent to about 100 times our present coal supply [3].

Fusion Power

Beyond the huge energy potential of the breeder reactor is the even more mammoth supply of energy potentially obtainable from controlled fusion. You recall from Chapter 3 that the sun obtains its energy from the fusion of hydrogen nuclei. Can we do the same and control it so as to produce electrical power? Lawrence M. Lidsky (Massachusetts Institute of Technology) in his article, "The Quest for Fusion Power," likens the search for economical controlled fusion power to the hunt for the Lost Dutchman Mine. "Only a few believers are absolutely certain that the goal exists, but the search takes place over interesting terrain and the rewards for success are overwhelming." [4]

The potential of fusion power depends on whether the deuterium-tritium ($^2_1$H-$^3_1$H) reaction is harnessed or whether the more difficult deuterium-deuterium ($^2_1$H-$^2_1$H) reaction is finally accomplished. The major difference in the two different fusion reactions is the temperature required to trigger them: deuterium-tritium about $50 \times 10^6\,°$K and deuterium-deuterium about $100 \times 10^6\,°$K. The tritium to sustain the deuterium-tritium reaction is obtained from lithium-6 ($^6_3$Li), which is relatively rare. If all the estimated lithium-6 reserve were to be used, the energy obtainable from the deuterium-tritium (actually $^2_1$H-$^3_1$H-$^6_3$Li) reaction is estimated to be of the same order as the world's fossil fuels. By comparison, the energy potential from the deuterium-deuterium reaction seems limitless. The fusion of the deuterium contained in 1 gal of water is estimated to provide energy equivalent to the energy obtainable from burning 300 gal of gasoline. Thus, if all the deuterium in the ocean were so harnessed, the energy equivalent would be about 100×10^6 times the world's initial fossil fuel supply [5]. Suffice it to say that the fusion process ($^2_1$H-$^2_1$H) represents the closest thing to an inexhaustible supply of energy for the inhabitants of spaceship earth.

Another attractive feature of fusion reactors is the rather small radiological hazard. In contrast to the large quantities of radioactive wastes produced from the fission reactor annually, the only radioactive material in the fusion reaction is tritium ($t_{1/2} = 12.3$ years). The hazard from tritium loss is expected to be negligible because any loss of tritium would actually be a loss of fuel. And, too, tritium has a relatively short half-life, 12 years, and thus decays relatively fast as compared to fission fuels and some by-products. Thermal pollution will present a problem, but the pollution should be considerably less because the fusion power plant is "guesstimated" to operate at an efficiency of about 50%, as compared to only 33% for fission reactors.

How close are we to fusion power? The scientific feasibility of fusion power is as yet to be shown. A mildly optimistic time table suggests the establishment of feasibility by about 1980. From there, one might expect a prototype reactor on-line by 2000 and reactors producing substantial quantities of electric energy by 2030 or so. Some are more optimistic. Others believe that commerical fusion power will never be achieved. One point on which there seems to be a general agreement is that the advances in technology necessary to achieve fusion electric power plants will be substantially greater and more complex than for any of the technologies developed thus far by any society or civilization.

Geothermal Power

The last of the nonrenewable, terrestrial resources is **geothermal** or heat from the earth. In Chapter 2 we discussed very briefly that the temperature of the inner regions (shells) of the earth is higher than the crust. The mantle, due to both its high temperature and pressure, is semifluid. Part of the core is thought to be molten. In principle, one should be able to use the difference in temperature between the earth's surface and inner regions to operate a heat engine (Figure 15-5). The deeper we are able to go, the higher will be the maximum temperature for operation and thus a higher maximum efficiency of engine operation should be possible.

As we discussed in Chapter 3 some heat from the interior does actually escape, constituting a very small part of the earth's radiation balance. The heat flow from the earth's interior takes place by conduction through the rock and by convection, that is, by the movement of the hot water or steam. The simplest and most direct way to use the earth's heat is by harnessing this steam from fumaroles (steam vents) and geysers, which occur along faults in the bedrock. The hot magma (molten rock) flows up from the mantle to the upper crustal regions and carries with it large quantities of heat. Normally, the crust beneath the land region of the continents is about 25 mi thick, which is too far to drill through to reach the hotter regions in the mantle.

Unfortunately, in the United States, the only region where steam from the earth is available is in northern California: The Big Geysers in the

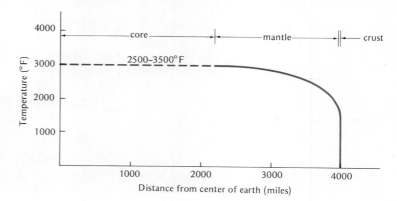

Figure 15-5. Suggested temperature profile of the earth's interior, (assuming that the interior temperature is 3000°F). Notice the rapid increase in temperature in going from the surface through the crustal regions (3 miles to 30 miles in depth) into the mantle.

Imperial Valley north of San Francisco (Figure 15-6). The Pacific Gas and Electric Company expects to have a generating capacity of 500 Mw by the mid-1970s using geysers. Steam-rich fields are to be found in other parts of the world, such as in Iceland, Greenland, Italy, and New Zealand. For example, the Larderello fields southwest of Florence (Italy) have been used for generating electricity since 1913. These "hot spots" are to be found in geological settings where the great continental plates (according to plate tectonics) have broken apart or perhaps have collided causing rifting of the crust. The regions in New Zealand, the California coast, and Iceland display these types of geological structure. In general, however, geothermal resources seem to be very definitely local ones.

Present technology utilizes the steam from geothermal fields directly to run turbines. This requires the water to be already present in the trapped hot water reservoir mentioned earlier. A novel idea has been proposed that suggests sinking a well into hot rocks several miles below the surface and then pumping water down into it. The cold water will fracture the rock. Nearby, a second well is drilled to allow the heated water to escape back to the surface as steam, which can be used directly to generate power. The water system can be a closed system by recirculation, or the water can be disposed of in nearby runoff (which is not desirable). Donald White [6] of the U. S. Geological Survey has estimated that the geothermal energy potentially usable is 4×10^{20} J (joules) (1.1×10^{14} kwh) or about one hundredth of actual world consumption in 1970. Even this figure is deceptive, though, because with it goes an assumed total withdrawal time of about 50 years [7].

Figure 15-6. Steam vents in northern California are potential geothermal sources. [Courtesy of J. R. Jackson, Jr., Exxon Company.]

As we have mentioned, so long as the cooled steam or water is returned to producing reservoirs, there is little problem. However, systems using runoff for cooling would indeed be both thermally and contaminant polluting. The steam in geysers contains small quantities of hydrogen sulfide (H_2S) gas, methane, and ammonia as well as some mineral content which could be undesirable.

Renewable Energy Resources

Tidal Energy

The tides, along with the winds, waters and direct solar energy, may be considered as eternal in that they will last as long as our solar system and our sun. These are our *renewable* energy resources. It has been the desire of man for ages to harness these renewable sources as his slaves; but, aside from dreams and wishes, we must ask the important question, "How much of this energy potentially can be harnessed?" Surely you can imagine the impracticality of attempting to capture all of the wind, or all tidal movement, or all of the sun's light for electrical power production. The reality of the situation suggests that only small fractions of any one of these resources may be harnessed ultimately. Perhaps even more sig-

Figure 15-7. Tidal power plant, Rance River, Brittany. [Courtesy of the French Embassy, Press and Information Division.]

nificant is the very complex technology necessary to utilize these resources to any great extent.

The energy inherent in the flow of the tides is normally dissipated in the form of heat via friction between the waters and the continental shelves (waves on the beaches or shores). The average tidal height (difference between the water level at high tide and low tide) is only about 3 ft, but there are specific areas around the world that have tidal heights as great as 50 ft. The largest such region in North America is the Bay of Fundy which has an estimated potential operating power of 29×10^6 kw (or 2900 Mw). The world's largest tidal generating facility in operation is the one in La Rance, France, operating at 240 Mw of a total 349 Mw capacity (Figure 15-7). Other small experimental power units have been installed in Russia.

Ultimately, the estimated world tidal power potential is 64,000 Mw, or about the same as the nonrenewable geothermal resource described above. In other words, the world's tidal resources have an ultimate capacity of about one-hundredth (10^{-2}) of the present world use of energy [8]. The major advantages are that power from the tides is nonpolluting and produces minimum ecological disturbances. An apparent disadvantage is the fact that the regions of great tidal heights are very distant from major population centers where electric energy consumption is the greatest.

Hydropower

The use of runoff for doing man's tasks represents the largest actual use of solar energy because runoff derives ultimately from the heat portion of the solar spectrum. From the energy balance sheet (Table 3-1) in Chapter 3, we can see that the greatest storage on earth of solar energy is the evaporation of water. Thus again, we must ask, "What is the ultimate potential for man's use in this renewable resource?" The world capacity for hydroelectric power has been estimated at 3×10^{12} w, which is about equal to the present total power capacity in the United States. Of this, only about 8.5% (210,000 Mw) worldwide had been developed by 1965. In comparison, the estimated potential of the United States hydropower is 161,000 Mw and, as of 1970 an installed capacity of about 60,000 Mw (37%) existed.

By far, the greatest potential concentration of water power is in the coal-poor regions of South America and Africa which have 20 and 27% of the total, respectively. At present, the people of these underdeveloped areas need it the least. In other words, the unused hydropower sites are located far from regions of high demand for electric power. Significantly, the ecological impact of building dams for hydropower usually has been minimal. One major exception is the Aswan Dam in Egypt, where the natural cycles have been severely interrupted.

Wind Power

Windmills, like water wheels and turbines, have provided power for many centuries. Before the rapid growth of electric power from hydro and fossil fuel sources, power for pumping water was provided by windmills. Denmark had several thousand windmills producing electricity in the early 1900s. Furthermore, most of us remember a picture of a little Dutch boy or girl standing in front of a windmill in Holland. Before the Rural Electric Administration (1930s), the farms and ranches on the western plains of the United States used windmills for pumping water and doing other tasks. In fact, windmills still dot the landscape near Lancaster, Pennsylvania, where the "Pennsylvania Dutch" have settled and built their farms (see Figure 15-8).

Several investigators [9] have predicted the rebirth of the windmill as a nonpolluting, power-producing device that would have a minimum ecological impact. A number of people have recently developed more efficient machines [10]. For large quantities of power, it has been suggested that huge windmills be built on floating platforms in the oceans to utilize steady, strong ocean winds. And further, to utilize the steady, high winds that blow in the upper troposphere, the towers could be several miles high. However, a tower 1 mile high would be equivalent in height to about five

Figure 15-8. Windmills still dot the countryside in Lancaster, Pennsylvania.

Empire State buildings stacked upon one another. We can only conclude that our technology is insufficient to make any large utilization of wind power. However, for local use where the winds blow rather steadily, the windmill could be an important addition to the power network. This would be particularly significant if an economical way could be found to store electrical energy for use during calm periods.

Direct Solar Power

The potential of the sun as a direct source of energy is tremendous. About 50% of the sun's energy is given off in the form of visible light. Visible light is the principal form of the sun's energy reaching the earth's surface. Presently, the chemical process of photosynthesis is the only process that utilizes significant quantities of this light energy, and, even this process only affixes about 2% of the total incident energy on plants in the form

Figure 15-9. An experimental all solar house at the University of Delaware. The glassed panels act as a heat trap via the greenhouse effect. Beneath the panels are solar cells that convert light to electricity, which in turn is stored in batteries located in the building to the right. The black-body radiation (heat) from the cells is then transferred to a coolant, which stores the heat for later use. [Courtesy of the Institute of Energy Conversion, University of Delaware.]

Figure 15-10. Large mirrors (with large surface areas) can be used to gather light and heat energy and focus it in a small area to obtain higher temperatures, analogous to the use of a magnifying glass to concentrate the sun's energy. The mirror shown, constructed by students at Monmouth College, concentrates the energy in the small heat trap at the focal point of the mirror. The heat is absorbed by circulating water and is then either stored or is used to power a freon cycle with a small gas turbine as the energy converter.

of sugar and plant fibers. However, if we were able to convert substantial quantities of light directly to electricity or to heat for steam electric generation, direct solar energy could provide the large quantities of energy that our society needs and desires.

Two methods are presently being considered. One direct way is to collect the sun's energy via a heat trap (Figure 15-9). Some use mirrors to collect the energy and focus it in a small area for higher collection temperatures (Figure 15-10). Another way provides the direct conversion of light into electric energy via photovoltaic cells, a by-product of space age technology. Sometimes called **solar** cells (Figure 15-11), these tiny devices (usually about 1 in. by 1 in.) are presently used to provide electric energy for operation of satellites and space probes. Unfortunately, two distinct disadvantages limit the large scale use of solar cells as energy converters: an efficiency of conversion of only 10% and a high cost of obtaining pure, single-crystal silicon, which is the major constituent of these devices. The estimated cost of electricity from solar cells is about 500 times the present cost of electricity [11]. The cost comparison is more easily seen when placed on a per watt basis: about $300 per watt for solar cells as compared to about $0.50 per watt for fossil fuel power.

Much of the problem with the direct use of solar energy is its diffuse nature, which means that large areas are required to collect sufficient quantities. The 178×10^{12} kw (30,000 times present rate of use) of solar power intercepted by the earth's diametric plane (Chapter 12) is spread over the large area of 1.275×10^{14} m² (which gives us our solar constant, 1.39 kw/m²). Furthermore, to use solar energy in steam–electric power production, large area mirrors would be required to concentrate the energy enough to obtain high enough temperatures and sufficient heat flow.

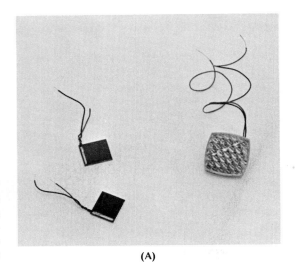

Figure 15-11. **(A)** Solar cells. The cell absorbs light energy and converts it directly to electrical energy. **(B)** A view of the Skylab space station as photographed from the Skylab 4 Command Module. The single panel on the right is an array of solar cells. [Courtesy NASA.]

(A)

(B)

An estimate of the collection area required for a solar power plant can be made again using our discussion in Chapter 3. The solar constant is 1.395 kw/m^2 of which only about half (visible), 0.7 kw/m^2, reaches the earth's surface under ideal atmospheric and weather conditions and with the sun's rays perpendicular to the collecting surfaces. If the entire 0.7 kw/m^2 could be collected and converted entirely into usable heat (under the best conditions, probably only about 0.5 kw/m^2 can be collected) then at least $4.3 \times 10^6 \text{ m}^2$ or 4.3 km^2 are necessary to generate electricity at the rate of 1000 Mw (10^6 kw) at 33% plant efficiency. See Example 15-1 for the details of this calculation. However, we should further note that since we receive significant quantities of energy only about 8 hours per day, approximately two times our estimated land area, or 8.6 km^2, would be needed to supply energy at the rate of 1000 Mw for a 16 hour electrical usage day. This estimate compares favorably with the Meinel and Meinel [12] estimate of 8 km^2 of collection area necessary for a 1000 Mw facility.

From Chapter 8 we have seen that electric demands are doubling every 10 years (7% per annum growth); thus, based on the production of 1.43×10^5 Mw in 1969, the yearly production needed by 2000 A.D. will be 1.12×10^6 Mw. Put simply, we will need about an additional generating capacity consisting of 1000 additional 1000-Mw generating facilities! To provide this power with *only* solar power would thus require 8000 km^2 or 3000 mi^2 (50 mi × 60 mi) of land area, or about 0.1% of the United States' land area (lower 48 states). The most suitable land areas are those in the western United States: Arizona, New Mexico, Nevada, Utah, Colorado and eastern California, and the 3000 mi^2 would amount to about 5% of suitable land area of that region.

Besides the large land area requirement there are other disadvantages to the potential use of solar power. An obvious one is that we only receive substantial quantities of light about 8 hr/day. If we plan to supply energy at the rate of 1000 Mw for at least 16 hr/day, we must have the equivalent of two, 1000-Mw facilities as compared to one conventional steam electric

---

EXAMPLE 15-1

Using the visible solar energy density as 0.7 kw/m^2, determine the minimum land area required to operate a modern, 1000-Mw (10^6 kw) electrical generating facility operating at 33% efficiency.

An efficiency of 33% indicates that the rate of heat input must be three times the rate at which we plan to produce the electricity (see equation 7-1). Thus, to produce 10^6 kw of power, heat must be supplied at 3 times this rate, or 3×10^6 kw. Thus

$$3 \times 10^6 \text{ kw} \times \frac{1}{0.7 \text{ kw/m}^2} = 4.3 \times 10^6 \text{ m}^2 = 4.3 \text{ km}^2$$

Figure 15-12. The Blenheim-Gilboa pumped storage power project (1973), located about 40 mi southwest of Albany, is designed to help meet peak consumer demands for electricity. [Courtesy of the Power Authority of the State of New York.]

facility. This also requires that we store the energy for the 8 hrs of early morning and evening. Unfortunately, there is presently no satisfactory way of storing very large quantities of electrical energy. It has been suggested that a pumped storage facility be used. In such a facility, water is pumped into an elevated basin (usually a dammed valley) during off-peak hours and then released to provide hydroelectric power during peak power requirements. Unfortunately, to store enough energy overnight for the United States would require the filling and emptying each day a lake the size of Lake Mead or a set of lakes equivalent to this lake! This, nonetheless, appears to be one of the best considerations for the near to intermediate future (Figure 15-12).

Another major disadvantage is the transmission of the energy from the more suitable areas of the western United States to the major consumers located along the western and eastern coastlines. The energy lost as heat using conventional transmission lines is far too great. It has been suggested that water (H_2O) be separated into hydrogen (H_2) and oxygen (O_2) and the hydrogen used as a fuel much the same as methane is used. The shipping and use of hydrogen as a fuel brings on a number of other technical problems (to be discussed later with Fuel Cells).

Nonetheless, in spite of the disadvantages of solar power, the prospects of this large renewable energy supply, potentially 30,000 times present use, make it an extremely worthwhile area for research and development. In the past, funding of solar power research has been negligible, but the future should see an increasing emphasis on developing this large, relatively nonpolluting energy source.

Solar Power from the Sea

As early as 1881, the French physicist Jacques D'Arsonval suggested that someday man would utilize the heat energy in the seas to power society. In the January, 1973 issue of *Physics Today*, the American physicist, Clarence Zener, described his proposal for solar sea power. The sun heats the surface layers of the seas and, as we described earlier, the body of water becomes thermally stratified. In principle, then, one can use the warmer upper layers as the high temperature reservoir and the colder lower layers

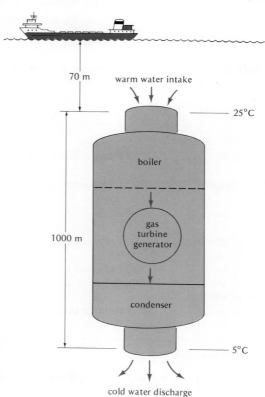

Figure 15-13. A schematic of a proposed power plant utilizing the temperature difference in the various ocean layers.

as the cold temperature reservoir. To obtain a significant temperature difference, one must work between about 70 m and 1000 m in depth where the temperatures (in the tropical oceans) are 25°C and 5°C, respectively.

A typical schematic is shown in Figure 15-13. In actuality, the working substance, ammonia or freon, works between the temperatures of 20°C and 7°C. Using these temperatures, we determine that the maximum theoretical efficiency (Carnot efficiency) is about 4%, and one might expect the practical efficiency or conversion factor to be about 2%. In either case, the efficiency is extremely low, and one could seriously question whether or not the plant could operate with that small a margin for success.

However, Zener suggests that with the cost of conventional power by 1980, such solar sea power plants could be economically feasible and provide energy for the 6 to 7 billion people of the world in 2000 A.D. at the present United States rate of consumption of 10 kw/yr. The distinct advantage is the large quantity of energy available which is relatively nonpolluting. Zener predicts that such plants would decrease the ocean temperature by 1°C. Whether or not this is a benefit or a hazard remains to be determined. furthermore, practical utilization of this concept of obtaining solar energy seems to be very far in the future, if at all possible.

Table 15-1.   A Summary of Power Prospects

Source	Power Potential	Technology	Environmental Impact
Nonrenewable			
coal	1000 times present use*	partly available (except gasification)	pollution—air and thermal
fission	2000 times present use (2 × coal)	available	thermal pollution; radiological waste problem
fusion	infinite	not available	thermal pollution; slight tritium hazard
geothermal	1/100 present use	partly available	probably small
Renewable			
tidal	1/100 present use	available	none
hydro	present (U.S.) use small†	available	some‡
wind	small	not available	none
solar	30,000 times present use	not available	none

*Present world use of 6 × 10⁹ kw.
†About 3 × 10⁹ kw.
‡Potential changes in area ecology. The environmental impact varies from location to location—loss of valleys, silting up of the dammed lakes.

Power for the Future—A Summary

Our discussions on power prospects are shown summarized in Table 15-1. First of all, these results represent, to a great extent, a scientific consensus on the power problem. Second, the table clearly shows where future research and development efforts (and hence, funding for same) might yield the greatest benefits for the time, effort, and money invested in the short, intermediate, and long terms. Finally, Table 15-1 provides you—the citizen, the voter and policy determiner—some necessary and useful insight into our energy needs and potential in order for you to voice an intelligent and informed opinion.

In summary, for the immediate present and short term future, our energy will come from oil and/or burner nuclear reactors. Increased exploration, drilling and production for oil will be necessary. The intermediate power prospects appear to be for coal as a principal energy source (including the developing technologies of coal gasification) and/or for breeder nuclear reactors. The decision of which *or* how much of which *and* the consequences, air pollution, mountains of ashes, land devastation, or chance of radiological contamination must be yours. Solar (direct, wind, and hydro) and fusion appear to be long term (50 years) power potentials. Within these long term prospects there *may* be found that unlimited power that our society and future societies will be seeking with ever-increasing fervor.

Future Energy Conversion Technology

Certainly, one solution (at least a partial solution) to both the energy "crunch" and the pollution problem is to utilize our fuels more efficiently. Significant technological advances in energy conversion over our present steam–electric technology will be required. In considering the direction in which changes in our present steam–electric technology might prove fruitful, first recall the important aspects of the Carnot engine in Chapter 7. The maximum theoretical efficiency (which disregards any engine details) of the engine depends on the temperature of the hot and cold reservoirs.

$$\eta(\%) = \frac{T_{\text{hot}} - T_{\text{cold}}}{T_{\text{hot}}} \times 100\%$$

Any potential increases in efficiency by lowering the temperature of the cold reservoir (T_{cold}) are necessarily small because temperatures of the atmosphere and/or runoff water are the cold reservoirs. On the other hand, the principal limit to the operating temperature of the boiler (T_{hot}) is the materials from which the boiler is constructed. Can materials technology develop new materials, that is, develop materials to withstand even higher temperatures and pressures? Even though the present metallurgical (metal material) limit for steam boilers appears to be about 1000°F (530°C), jet engine combustion chambers commonly operate at temperatures several times this, for example 4000 to 5000 °F. Further, the largest loss of efficiency occurs at the steam turbine, which employs the conversion of kinetic energy of steam to mechanical energy. Alternate methods include the magneto-hydrodynamic (MHD) generator, and the fuel cell (another by-product of space age science and technology).[1]

Magnetohydrodynamics (MHD)

Power generation by MHD involves the conversion of the kinetic energy of a conducting gas in the presence of a magnetic field to electricity. Figure 15-14 illustrates schematically the MHD principle. Gases are combusted at high temperatures in a chamber, similar to a jet engine, and allowed to expand through a nozzle. The high velocity gas is then "seeded" with a metal, such as sodium (Na) or potassium (K), which ionizes (loses electrons) easily at these high temperatures. The magnetic field exerts a force (called the Lorentz force) on the moving ions, causing the positive ions (the nuclei minus electrons) to move in one direction and the negative ions (the electrons) to move in the opposite direction, as shown. The deflected ions are collected on electrodes and, when a load, for example, a light bulb, is connected, an electrical current flows. This is a direct current (DC) system which contrasts to alternating current (AC) systems now widely employed. However, either the DC can be converted to AC or the DC can be used

[1] Other energy converters that may prove useful are thermoelectric and thermionic generators. However, a discussion of these is outside our scope.

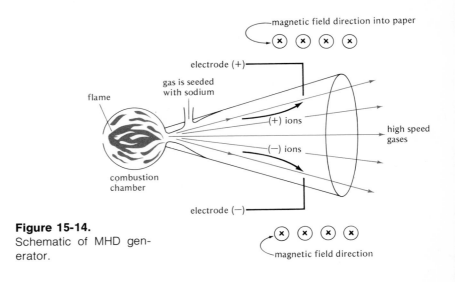

Figure 15-14.
Schematic of MHD generator.

directly as it is in many camper trailers and mobile homes containing battery electrical systems.

The advantages of MHD are a potentially high efficiency (at least 60%) due to the high operating temperature and probably low maintenance because there are no moving parts. In the intermediate term, it has been suggested that MHD generators be used in conjunction with fossil fuel plants as a "topping" cycle. The high velocity gases could be used to run a high speed, high temperature gas turbine, and then the hot exhaust gases from this system used as a source of heat for steam–electric generation. It has also been proposed to insert a low temperature, low pressure gas turbine as a "bottoming" cycle to use the cool steam before the condensing cycle.

The efficiencies of such a combination of devices could be as high as 60% with the obvious advantage of producing less waste heat. The advantages of adding "topping" (MHD) and "bottoming" cycles to steam–electric power facilities would in effect be to increase the temperature of the hot reservoir and to operate with less of a temperature and pressure drop between stages. This combination approaches a more reversible overall operating cycle. Russian engineers are presently putting on-line a pilot MHD generator with a 25 Mw output. The development of MHD power in the United States has been severely retarded due to the lack of funding, according to MHD expert Dr. Arthur Kantrowitz, Avco, Everett Research Laboratory. With adequate funding, the United States could put MHD "on-line" within the next decade.

Fuel Cells

The fuel cell converts the chemical energy stored in fossil fuels (hydrocarbons) in gaseous form or the chemical energy in pure hydrogen directly to electricity. Because the fuel cell avoids the chemical to heat energy conversion (combustion), it is not subject to the limitations of the Carnot efficiency of a heat engine. Figure 15-15 shows a schematic of a fuel cell. The operation actually is very similar to that of a storage battery. The fuel to be oxidized and the oxidizer are fed into the cell, as shown, and oxidation of the fuel takes place at the negative electrode. As a result, electrons flow through the external circuit to the positive electrodes where reduction of the oxidizing agent takes place. In contrast to the storage battery, the electrolytes and electrodes are not used up because the fuel and oxidizers are fed in continuously while the combustion products (for example water and carbon dioxide) are removed continuously.

The manned space probes use hydrogen for the fuel and oxygen as the oxidizer with the production of water. Using fuel cells, a conversion efficiency of up to 60% has been obtained. Fuel cells are small, quiet, and maintenance free, so one could foresee their use in homes to provide power

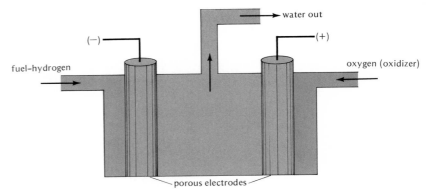

Figure 15-15. Schematic of a fuel cell. The fuel cell is similar to a storage battery with the exception that the reactants are fed in and the products removed continuously.

without the large transmission losses now experienced by the power industry. A given home really does not consume very much power, so that what was necessary could be generated on the spot economically. There is also no pollution, thermal or otherwise. Perhaps the future may see us stopping at the corner station for a tank of hydrogen rather than gasoline.

A Hydrogen Economy

The use of fuel cells as energy converters has led some [13] to suggest using hydrogen as a multipurpose nonpolluting fuel because water is the only product. The merit of such a plant is that, as fossil fuels become in short supply, other energy sources such as nuclear, solar, and so on, must supply our needs with the concomitant move toward an all-electric economy. For example, an off-shore nuclear power plant could use its energy, particularly during off-peak hours, to separate water molecules into hydrogen and oxygen by passing an electrical current through water. This process is known as electrolysis. The hydrogen gas can then be collected, stored, and transported to user locations for reconversion to electricity via fuel cells when power is needed. The necessary oxygen for the fuel cell would come from the air.

Perhaps the greatest advantage is the elimination of transmission losses. Such a plan also allows us to make maximum use of generating facilities because, at present, they operate at about 50% of capacity due to the lack of storage facilities (pumped storage) and technology (for storing electrical energy). It has been estimated that the costs of shipping hydrogen over 250 miles are considerably less than the cost of transmitting electricity by conventional power lines.

Beware though, this plan is no panacea. A major problem to be solved is the transportation of hydrogen. The hydrogen molecule is the smallest molecule and is very hard to contain. If there were any leaks in the pressurized tanks containing hydrogen and if there were a small spark, a violent explosion would result. Natural gas and bottled gas (propane) present similar hazards, but the hydrogen gas hazard is recognized as considerably worse.

Solar-Microwave Energy to Earth

If we were able to capture energy from the sun outside the earth's atmosphere, potentially we could get almost 10 times as much energy to a given area. The large difference in energy received is due to the day-night variation, earth and atmospheric reflections, weather and seasons. The 10:1 ratio is even higher in the more northern latitudes. To overcome the energy-reducing effects, Dr. Peter Glaser [14] proposed placing solar cell arrays in space in synchronous orbit with the earth (Figure 15-16). The cells would receive the sun's energy continuously and remain fixed over that location on the earth's surface which was to be the receiver. A microwave transmitter would then beam the energy at $\lambda = 7.5$ cm back to earth where it would be converted into DC (direct current) power. The total area of the arrays for a 10^4-Mw (10,000-Mw plant) [15] would be about 80 mi^2 and use an antenna about 1 km (0.6 mi) in length. The beam diameter on earth would be about 7 km (4 mi). The efficiency of such a system is estimated to be about 68% of the solar cell output.

Such a system would cause considerably less thermal pollution, due to

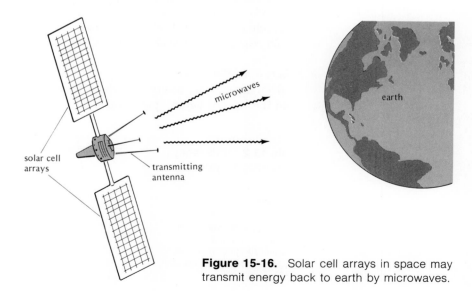

Figure 15-16. Solar cell arrays in space may transmit energy back to earth by microwaves.

its high operating efficiency. The greatest advantage, of course, is that the supply is eternal. There is some concern about biological effects on birds (and people) flying through the beam or contacting it in other ways (some of you may be familiar with the rapid cooking power of microwave ovens), and these effects are under study. The known major disadvantage, however, is cost. Nonetheless, as our need for energy increases and our earth-bound supplies dwindle, the next decade or two may see the economics of solar-microwave power becoming competitive.

Extending Our Resources—Energy Conservation

Earlier we learned that society is about 50% efficient (1969–1970) in its utilization of energy for producing goods and services. That is, a full 50% of our energy is directly discarded as waste heat. Furthermore, with our present patterns of consumption, we appear to be headed for an even more energy intensive–less efficient economy. An example of our increasingly energy intensive ways is the substitution of air and truck freight transportation for the substantially more efficient rail transportation. These modes of transportation are compared in Table 15-2.

The building construction industry is also using the more energy-intensive concrete and steel construction, particularly concrete, while paying little attention to more efficient ways of heating and reducing heat losses (and cooling costs, too). In the private sector, homes are inadequately insulated. We have changed to the convenience of electricity as we use energy-expensive frost-free refrigerators, self-cleaning ovens, electric heat (only 33% efficient use of fuel at the power plant) and the very costly air-conditioning. Specifically, our energy consumption has doubled in the last 20 years, while little attention has been paid to trying to "save-a-watt."

Where Can We Save?

Even though we (society as a whole) waste 50% of our energy, we know from the laws of nature (physics and chemistry for example) that society

Table 15-2. Efficiency of Transport Systems

SOURCE: E. Hirst and R. Herenden: A diet guide for chronic energy consumers. Sat. Rev. **55**:64 (Oct. 28, 1972).

Mode	Relative Scale	Energy Demand (Btu/ton-mi)
water	1	670
rail	1	680
truck	5.6	3,800
air	62	42,000

can never be 100% efficient. However, a saving of only 1% of the 19×10^{12} kwh consumed in 1969–1970 represents a significant gain in energy; this is equivalent to 100 million barrels of oil[2] or about 3 days total supply of energy for the United States, which had a total daily energy consumption equivalent to about 30 million barrels of oil in 1970. From Chapter 8 remember that there are five major energy consuming sectors: industry, 30%; utilities, 22%; transportation, 26%; residential and commercial, 22%.

To estimate potential savings, we must consider the individual sectors. In the home (residential) about 88% of a family's energy budget goes into heating, cooling, and refrigeration. The remaining 12% goes into lighting and operation of the many appliances. Thus, when looking for places to save in the private (and the commercial) sectors, we must look closely at present heating and cooling practices. Certainly our use of appliances, for example, electric clothes dryers, should not be forgotten. In the transportation sector, we look for the development of smaller automobiles and a move towards mass transit. In the aircraft transportation industry, load factors could perhaps be increased from 55% to 80%. In the industrial sector, more efficient operation and, in particular, the use of large quantities of the now-wasted process heat. There has also been a move in industry, for example, in steel mills, to use the less efficient electrical energy.

The potential increases in efficiency of electric energy production by the utilities could provide a large savings. Some of these potential savings for the future are summarized in Figure 15-17. Included is an estimate based on simple exponential projections of requirements for 1980 and 1990. In either case, the potential savings due to conservation measures by the different sectors seems to be about 10% of the gross projected consumption. Although these potential savings may seem small at first thought, nonetheless, in terms of daily amount of energy consumed, they represent about 1 month total supply for the country. This amount can mean the difference in having a normal operation of our economy or our living with an "energy crunch." Thus, even small savings are worthwhile.

Some Specific Recommendations for Energy Savings

We think that it would be safe to say that the initiative for energy consumption must come first from us—the individual energy consumer. Because the greatest portion of our (including commercial establishments) energy consumption goes for heating and cooling, we can immediately effect a savings by reducing our thermostat settings to the range between 65 and 68°F. In addition, the installation of weather stripping and storm windows can effect another substantial savings. A longer range goal dictates

[2]*Note:* One barrel = 5.8×10^6 Btu = 1710 kwh.

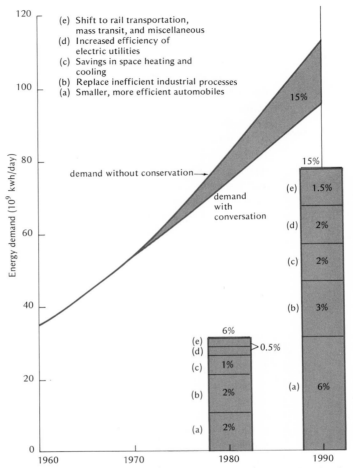

Figure 15-17. Summary of potential energy savings for the future.

a change in building practices to include optimum thickness of insulation to reduce heat loss by conduction through walls and ceilings.

In the immediate future, or until the utilities can raise their operating efficiency, a return to the direct use of fuels for space heating would save that large initial loss in electrical energy production. The middle 1970s may see extensive use of the newer technology of **heat pumps** for heating and cooling. In the past, manufacturers have paid insufficient attention to producing really efficient electrical appliances. For example, an increase in the efficiency rating of air conditioners from 6 to 10 Btu/w-hr could result in a savings of 16×10^9 kwh/year, or about 0.1% of the 1970 use.

Altogether, it has been suggested [16] that savings in space heating and cooling could save as much as 7% of the national energy use.

The American automobile is another prime example of gross waste. The average gas mileage is about 12 mi/gal as compared to about 25 mi/gal for the smaller European cars. Perhaps the time of the one-man, 2-ton, 400-in.³ engined automobile will pass away as gasoline prices increase. By 1990, about a 5% fuel savings (about 3×10^6 barrels of oil per day), could be effected by a change to smaller automobiles giving 15 to 20 mi/gal as compared to today's 10 to 12 mi/gal. This assumes that about half of the total automobile fleet is made of smaller cars. Interestingly, in 1972, 2 million more bicycles were sold than automobiles in the United States (about 11×10^6 bicycles as compared to 9×10^6 cars).

We further suggest that special attention be paid to research and development in both the industrial and utility sectors to eliminate the large quantities of waste heat. Specifically, for the electric utilities this means developing the MHD "topping" cycle and the low temperature-pressure turbine "bottoming" cycle that we mentioned earlier.

These are only a few specific recommendations for future fuel savings. Even with general awareness of impending fuel shortages, and a commitment to energy conservation, it would seem inevitable that some change in our highly energy-intensive life style is necessary. This would seem even more evident when we look around and see the rest of the world desiring our standard of living.

Our View of the Energy Future

All that has been set before you represents what the authors believe has been necessary for you to understand our subject of energy: energy as a resource, energy as a slave, and energy as a pollutant. It is tempting to stop, as most technical people do, with the presentation of the facts and the laws governing these facts. By now, as you have read with us, you have probably come to expect something more. We do not pretend to be prophets. The energy problem, however, we do believe is central to any consideration of the future of our society and, more broadly, the future of man and woman on earth.

From what we have presented earlier, we hope that we have given you the idea that there is a limit to growth. That limit, whether it be imposed by resource depletion or by pollution will be reached more quickly if growth continues at an exponential rate. The limit will be reached eventually by any pattern of increase. Although the concept of a limit has been discussed for centuries, most previous limits, for example those of Malthus, have been circumvented or found not to be limiting as a result of new technological developments. Will technology be our salvation too?

In consideration of the energy sources available to us and of the large quantities of energy that we consume, we believe that we have identified limits to man's existence on earth. Some, in particular, Paul Ehrlich, say that population growth will determine the future of man. Others subscribe to the Barry Commoner philosophy that certain highly energy-intensive technologies will determine the future of man. A group at Massachusetts Institute of Technology [17] has made some reasonable, but arbitrary, assumptions about future developments in population, energy resources, and pollution, and, from a computer analysis, came up with several forms of Hubbert's principle (Chapter 8). Any one or any combination of the three could represent the limits we have been discussing.

Any choice that we, the members of a scientific-technological society, make must ultimately be based on the concept of trade-offs: benefits versus hazards and costs. In choosing to develop cheap, accessible nuclear power, specifically power from breeder reactors, we must accept the immediate, short term hazards of the discharge of small quantities of radioactive waste into the biosphere and a potential (albeit remote) chance of accident with melt-down. We must accept the long term hazard of radioactive waste storage. With regard to the long term hazard, Albert Weinberg, Director of Oak Ridge National Laboratory, compared the nuclear situation to a "Faustian bargain."

The proponents of fusion power—a potentially limitless source of "clean" energy—say its technology is our salvation. We believe that the scientific and technological complexities of a fusion power plant surpass those of the Apollo project. Will science and technology come through with commercial fusion power? We believe we are being realistic in saying that we do not know, but, if fusion becomes a reality, it will not be before the turn of the century.

The eternal and almost limitless solar power may be the basis of our energy future. Most of the solar power technology that we need has been developed. Perhaps the remaining techniques can be fully developed by at least 1980. The greatest uncertainties with the extensive use of solar power lie in the cost of energy from this source and the transportation of electrical energy to areas where the prospects of building solar farms are not good, such as the northeastern sector of the country. An optimistic estimate suggests that the cost of solar electric power eventually may be about 5 times present costs. Are we willing to pay this cost for this potential pollution-free energy source? Certainly any use of solar power, whether it be in large solar farms or whether it be on a local basis with heat traps on our roofs and solar cookers in our back yards, may require a change in life-style!

Insofar as the other smaller sources of energy, geothermal, tidal, and so on, all of them represent a good potential for specific regions of the United States and the world. However, no particular one or combination appear to be sufficiently large to be any total answer to our future energy needs. Moreover, it will take time for scientists and technologists to develop ways to utilize them to any large extent.

In summary, our immediate energy future appears dependent on continued development of our fossil fuel resources: namely, the search and development of natural petroleum reserves and the expansion of coal production both for direct use and for producing syncrude and syngas. Our nuclear power resource should be developed, with due caution for the potential hazards, to fill the gap created by shortages of fossil fuels.

Our long range goal should be the investment (time, manpower, and money) necessary to develop the more exotic forms of energy in hopes that in them will be found means to meet the energy hopes of man and woman on earth. Any economy dependent mostly on any one fuel (as we in the United States are) can only expect trouble. We suggest a diversified energy economy and one that will make us independent of any one source (such as oil) and as independent as possible of foreign energy sources. To obtain such will require an interest and willingness on your part to see that future energy policy includes various aspects of the views we have presented.

In conclusion, let us recapitulate some of our facts and our philosophies. Underdeveloped nations, teeming with people, aspire from their very low standard of living to a rather higher one. Our standard of living is based on inexpensive, readily available energy (principally fossil fuel), which translates simply into large numbers (over 400) of "energy slaves" for every man, woman, and child in the United States. Without being specific, we are convinced that change in the growth pattern of energy use in the United States must take place, preferably sooner than later. How this will occur—voluntarily or by necessity—you will determine by your own life style. One delightfully extreme viewpoint is "abolish slavery." Such a view practically suggests the absence of energy consumption. This is impossible, short of the elimination of all life. Life by its very existence consumes energy. "Minimize slavery" sounds dull and uninteresting by comparison, but on reflection should have real meaning for each of us.

Some people, including many scientists and technologists (engineers), in an almost worshipful way suggest that technology is man's solution. "Technology has done it before; technology will do it again." Do you really believe this after reading the sections earlier in this chapter? We do not. We are not prepared to worship at technology's altar. Worshipping is a very different activity from working. We certainly must work at, or encourage others to work at, supplemental and alternate sources of energy and improvements in what we have. Some success in these will certainly extend the time in which we may reach a limit to growth. As long as people have good ideas, society should encourage their pursuit with as much keenness and vigor as possible.

So where are we? We certainly have found that energy considerations influence both what we say and do. Our study of energy has made us aware of many things. But we, like you, are members of American society. We believe that our life styles have been modified by our study in many subtle and, perhaps, not very dramatic ways, that is, walking instead of riding, bicycling instead of using a car, turning out unneeded lights, more insulation in our homes, more efficient rather than less efficient appliances, recycling materials especially aluminum.

We would like to end our presentation to you with something dramatic—particularly from either science or technology. We simply cannot. Perhaps our greatest pleasure here has been the quest for insight and understanding. We have gained much, and hope that you have also. As members of and believers in a democratic way of life, we remind you then that "the price of freedom is eternal vigilance." Come join us in being vigilant about energy—its resources, its service or slavery to mankind, and the minimizing of its destructive pollution to the environment around us.

References

[1] For further discussion, see D. Chapman, T. Tyrrell, and T. Mount: Electricity, demand, growth and energy crisis. *Science*, **178**(4062):703 (1972).

[2] For a more extensive discussion of coal use, refer to W. D. Metz: Power gas and carbon cycle. *Science*, **179**:54–56 (1973); W. Greenburg: Chewing it up at 200 tons a bite. *Tech. Rev.*, **75**(4):46 (Feb. 1973).

[3] M. K. Hubbert in *Resources and Man*. W. H. Freeman Company, San Francisco, 1969, p. 218.

[4] L. M. Lidsky: The quest for fusion power. *Tech. Rev.*, **74**(3):10 (Jan. 1972).

[5] See reference [3], p. 228.

[6] D. E. White: *Geothermal Energy, Publication No.* 519. U. S. Geological Survey, 1965.

[7] R. G. Bowen, E. A. Grok: Geothermal earth's primordial energy. *Tech. Rev.*, **73**(1):42 (Oct. 1971).

[8] The data for this section was obtained from Hubbert, reference [3] pp. 207–209.

[9] Dr. W. E. Heronemus, Massachusettes Institute of Technology; Dr. T. Sweeny, Princeton University; Dr. K. Bergey, University of Oklahoma; and others.

[10] S. Kidd and D. Garr: Pollution-Free Electric Power from Windmills. *Popular Sci.*, **201**(5):70 (1972).

[11] G. T. Seaborg and W. R. Corliss: *Man and Atom*. E. P. Dutton and Co., New York, 1971, p. 55.

[12] A. B. Meinel and M. P. Meinel: Physics looks at solar energy. *Phys. Today*, **25**(2):48 (1972).

[13] D. P. Gregory: The hydrogen economy. *Sci. Amer.*, **228**(1):13 (1973).

[14] P. E. Glaser: Power from the sun: its future. *Science*, **162**:857–861 (1968).

[15] W. C. Brown: Satellite power stations: a new source of energy? *IEEE Spectrum*, **10**(3)38–47 (1973).

[16] A. L. Hammond: Energy needs: projected demands and how to reduce them. *Science*, **178:**1186 (Dec. 15, 1972).

[17] D. L. Meadows, et al.: *Limits to Growth.* Massachusetts Institute of Technology Press, Cambridge, Mass., 1972.

Questions

1. Describe the three basic features of the energy consumption pattern in the United States since 1900.

2. Do you feel that the growth pattern that we have experienced in recent years will continue? Why or why not?

3. Describe the outlook for petroleum as a major energy resource for the United States and the world, through 2025.

4. Describe the outlook for coal as a major energy resource for the United States and the world.
   (a) Through 2025.
   (b) Post-2025.

5. What scientific and technological advancements would you foresee as desireable in order for coal to be used as a major source of energy?

6. What are the prospects for our having a nuclear all-electric (fission) economy?

7. What scientific and technological advancements are necessary for a bright "nuclear" outlook?

8. What features of fusion power make the rewards for success overwhelming?

9. List the nonrenewable resources and rate them as to their fuel potential and pollution potential.

10. List the renewable resources and rate them as to their fuel potential and pollution potential.

11. What is meant by a "hydrogen economy" and what technologies are necessary for its implementation?

12. Describe the areas in which energy conservation may produce the greatest savings.

APPENDIX A

Scientific Notation

As a member of our complex, scientific, technological society you are constantly being bombarded with numbers. On the one hand, you may be personally concerned about the mileage you drive and the cost of gasoline. Perhaps your latest experience at the supermarket was an unpleasant one, when you paid out a large sum of money for seemingly few groceries. On the other hand, you pick up the newspaper and learn that last year's gross national product (GNP) "topped" 1000 billion dollars and the Federal budget is measuring in hundreds of billions of dollars, a significant part of the GNP.

Eventually, you ask what this means to you, because the cost of running society comes from the taxpayer's pocket. You learn that the Apollo Project cost about $40 billion over 10 years, and that other governmental supported research and development efforts, namely energy research, are costing millions of dollars. Does this represent adequate funding or too much funding? Answers, unfortunately, to these types of questions are often relative, but the important point here is that you need to develop a familiarity with numbers and some ease in manipulating them.

Numbers and units are even more essential to a scientists' way of life. Not only does he live with the moderate range of numbers (tens to billions) that you most often experience, but he commonly thinks in terms of very small numbers from one ten-billionth (in meters, the size of an atom) to very large numbers like one hundred billion trillion (in miles, the size of the visible universe). As modes of communication become more elaborate, satellite, radio, and T.V., and as the scientific, technological community attempts to achieve a favorable public consensus on their efforts, you, too, are increasingly being exposed to large ranges of numbers and the units that give meaning to these numbers. Furthermore, as energy becomes a common household word, you must become accustomed to the very large numbers necessary to describe the phenomenal quantities of energy we (in the United States) consume, 20 trillion kilowatt-hours, 65 quadrillion British thermal units, or 12 billion barrels of oil equivalent. This requires that you master to some degree scientific notation: powers-of-ten notation, units, and conversions between units.

Powers-of-Ten Notation

Table A-1 illustrates the use of multiples of ten, and some prefixes commonly used to describe multiples are given in Table A-2. The number of times that ten has been used as a multiple—either as a multiplication

TABLE A-1.

Decimal	Power of Ten	Operation
0.000001	10^{-6}	1/1,000,000
0.00001	10^{-5}	1/100,000
0.0001	10^{-4}	1/10,000
0.001	10^{-3}	1/1000
0.01	10^{-2}	1/100
0.1	10^{-1}	1/10
1	10^{0}	1/1
10	10^{1}	1×10
100	10^{2}	1×100
1,000	10^{3}	1×1000
10,000	10^{4}	$1 \times 10,000$
100,000	10^{5}	$1 \times 100,000$
1,000,000	10^{6}	$1 \times 1,000,000$

TABLE A-2.

Multiple	Prefix	Symbol
10^{12}	tera-	T
10^9	giga-	G
10^6	mega-	M
10^3	kilo-	k
10^{-2}	centi-	c
10^{-3}	milli-	m
10^{-6}	micro-	μ
10^{-9}	nano-	n
10^{-12}	pico-	p

factor or as a divisor—is written as a superscript, as shown in Table A-1, which is called an exponent. These multiples of ten are collectively referred to as **powers of ten.** The minus sign preceding the number (call it x) in the exponent simply means that a number preceding the power of ten is to be divided by ten x times. For example, $2.0 \times 10^{-4} = 0.0002$. A simple rule says that, in order to convert a number and its power of ten to decimal form, start at the decimal (2.0) and count x times to the left (for negative exponents). For positive powers of ten, we do just the opposite. Start at the decimal and count x times to the right, namely $2.0 \times 10^4 = 20,000.0$. Generally, when expressing numbers in powers of ten, the digits are expressed as a decimal number between one and ten followed by the appropriate power of ten. This procedure is illustrated in Example A-1.

Multiplying and dividing numbers expressed in powers of ten is a little more complicated, but the procedure can be expressed by a simple rule: **when multiplying powers of ten, add the exponents algebraically, and, when dividing, subtract the exponents, algebraically.** The numbers are also multiplied or divided. The use of this rule is illustrated in Example A-2.

Example A-1

(a) $6.0 \times 10^{-3} = \dfrac{6.0}{10 \times 10 \times 10} = \dfrac{6.0}{1000}$

$\qquad = 0.006.0 = 0.006$

(b) $0.00006 = 0.00006 \times 10^{-5} = 6 \times 10^{-5}$

(c) $6.0 \times 10^3 = 6.0 \times 1000 = 6.000. = 6000$

(d) $600,000 = 6.00,000. \times 10^5 = 6 \times 10^5$

Example A-2

(a) $(2 \times 10^2) \times (3 \times 10^3) = 2 \times 3 \times 10^2 \times 10^3 = 6 \times 10^{2+3} = 6 \times 10^5$

(b) $(2 \times 10^{-2}) \times (3 \times 10^3) = 6 \times 10^{-2+3} = 6 \times 10^1 = 60$

(c) $\dfrac{4 \times 10^4}{2 \times 10^2} \left(= \dfrac{4}{2} \times 10^4 \times 10^{-2} \right) = \dfrac{4}{2} \times 10^{+4-2} = 2 \times 10^2$

(d) $\dfrac{4 \times 10^4}{2 \times 10^{-2}} \left(= \dfrac{4}{2} \times 10^4 \times 10^2 \right) = \dfrac{4}{2} \times 10^{4-(-2)} = 2 \times 10^6$

System de Units Internationale— Metric System of Units

Most of the world has already gone to a metric system of units and those nations that have not, including the United States, are expected to convert to the international system in the next 10 to 15 years. A selected list that is sufficient for our purposes here is shown in Table A-3.

TABLE A-3. Metric Units and Some Equivalents

Measurement	Unit	Symbol
length	kilometer (10^3 meters)	km
	meters	m
	centimeters (10^{-2} meters)	cm
time	seconds	sec
mass	kilogram (10^3 grams)	kg
	gram	g
energy	joule (newton $\times$ meter)	J
	kilocalorie (10^3 calories)	kcal
	kilowatt-hour (3.6×10^6 joules)	kwh
power	kilowatt (10^3 watts)	kw
	watt (joule/second)	w
temperature	degrees Celsius scale	°C
	degrees Kelvin scale (°C + 273°)	°K

Conversion Factors

We often find it necessary to convert between various systems of units, particularly, because we live with a mixed system, English and metric. A list of selected conversion factors follows in Table A-4.

TABLE A-4.
Conversion Factors

Length
English to Metric

      mile = 1.61 kilometers
      foot = 0.305 meters = 30.5 centimeters
      inch = 2.54 centimeters

Metric to English

      kilometer = 0.621 miles = 3280 feet
      meter = 39.37 inches = 3.28 feet

Area
English to Metric

      square mile = 2.59 square kilometers
      acre (43,560 square feet) = 4047 square meters

Volume
English (U.S.)
      cubic foot = 7.48 gallons

Metric

      liter = 1000 cubic centimeters = 0.001 cubic meters

English to Metric

      gallon = 3.785 liters
      quart = 0.946 liters
      cubic foot = 0.0283 cubic meter

Metric to English

      liter = 1.057 quarts
      cubic meter = 35.3 cubic feet

Weight/Mass
English

      ton (short ton) = 2000 pounds

Metric

      metric ton = 10^3 kilograms

English to Metric

      ton = 0.907 metric tons = 907 kilograms
      pound = 0.454 kilograms = 454 grams

Metric to English

      metric ton = 1.1 tons = 2205 pounds
      gram = 0.0022 pounds = 0.0352 ounces

Table A-4.
Conversion Factors
(Continued)

Energy
English

> foot-pound = 0.00129 British thermal units (Btu)
> Btu = 778 foot-pounds

Metric

> kilocalorie = 4185 joules
> kilowatt-hour = 860 kilocalories = 3.6×10^6 joules

English to Metric

> Btu = 0.252 kilocalories = 2.93×10^{-4} kilowatt-hour

Metric to English

> kilocalorie = 3.97 Btu
> kilowatt-hour = 3412 Btu

Energy Equivalents of Fuels*

> cubic foot of natural gas = 1000 Btu
> = 0.293 kilowatt-hour
> barrel of oil = 5.8×10^6 Btu
> = 1700 kilowatt-hours
> † pound of coal = 10,000 Btu = 2.93 kilowatt-hours
> ton of uranium oxide (U_3O_8) = 210×10^9 Btu
> = 61.5×10^6 kilowatt-hour

Power
English

> horsepower (hp) = 550 foot-pounds/second
> horsepower (hp) = 2546 Btu/hr

Metric   kilowatt = 1000 watts = 0.238 kcal/second

English to Metric

> horsepower = 746 watts = 0.746 kilowatts
> Btu per second = 0.106 kilowatts

Metric to English

> kilowatt = 1.34 horse-power = 0.948 Btu per second

Temperature

> $T(°F) = 9/5 \ T(°C) + 32°$
> $T(°C) = 5/9 \ [T(°F) - 32°]$
> $T(°K) = T(°C) + 273.15$

* SOURCE: National Petroleum Council, 1972.
† In Chapter 8 we used the coal equivalence of 7400 kwh/metric ton which gives for a pound of coal, 11,464 Btu or 3.36 kwh. Both figures are widely used in the literature and the discrepancy is of no real consequence to us. The disparity relates to the fact that different types of coal from various parts of the United States have different heat values, which means that either figure is an average.

APPENDIX B

Density, Pressure, and an Ideal Gas

Density

In our discussions of bodies (for example, the planets) and matter in general, one of the most characteristic features of a particular type of substance was its density. **Density** is defined as the *mass (or weight) of a body or substance divided by the volume.*

$$\text{density} = \frac{\text{mass}}{\text{volume}}$$

or
$$\rho = m/v \qquad \text{(B-1)}$$

The most commonly used units for density are grams per cubic centimeter (g/cm^3), kilograms per cubic meter (kg/m^3) and pounds per cubic foot (lb/ft^3). One of the most familiar substances is water which (at $4°C$) has a density of $1 \, g/cm^3$, $10^3 \, kg/m^3$ or $62.4 \, lb/ft^3$. In words, a small cube of water, 1 cm (0.4 in.) on a side has a mass of 1 gm. Or, the weight of $1 \, ft^3$ of water

TABLE B-1.

Substance	Density (ρ) (g/cm³)	Density (ρ) (kg/m³)	Weight Density (ρg)* (lb/ft³)
mercury	13.6	13,600	850
lead	11.3	11,300	706
iron	7.8	7,800	490
aluminum	2.7	2,700	169
salt	2.15	2,150	135
sugar	1.59	1,590	99.5
water (4°C)	1.000	1,000	62.4
ice	0.92	920	57
alcohol (ethyl, 20°C)	0.79	790	49.4
gasoline	0.68	660	42
oak wood	0.8	800	50
maple wood	0.49	490	30.6

Note: The weight density (ρg) is the mass density, ρ, multiplied by the acceleration of gravity, g.

(7.5 gal) is 62.4 lb. Table B-1 compares the densities of various common substances.

Pressure

A second physical quantity that we have used in a rather vague way is that of pressure. Specifically, we have referred to the interior-core of the sun as being at very high temperatures and pressures and to the steam in a boiler as being at a high temperature and pressure. The definition of pressure is simple but needs some interpretation depending on the circumstances involved. **Pressure** is defined as a *force divided by the area over which it is exerted.*

$$\text{pressure} = \frac{\text{force}}{\text{area}} \quad \text{or} \quad p = F/A \quad \text{(B-2)}$$

The common units of pressure are newtons per square meter (nt/m^2), pounds per square inch (lb/in^2 = psi) or atmosphers (atm). When equation (B-2) is applied to the concept of air pressure (at sea level), it simply means the force or weight of a column of air above a given area at the surface. For example, if we choose an area of 1 in.² at the surface, the weight and hence the force of the column of air (a column extending throughout the various atmospheric layers) is 14.7 lb. The pressure is then the force/area or 14.7 lb/1 in.² = 14.7 lb/in.². This is equivalent to 10^5 nt/m^2, the pressure of a column of mercury 76 cm high, or, a column of water 34 ft high.

EXAMPLE B-1

Show that the pressure of a column of water 34 ft high exerts a pressure at the base of 14.7 lb/in².

From Table B-1 we see that the weight density (ρg) of water is 62.4 lb/ft². So,

$$p = \rho g h = 62.4\frac{\text{lb}}{\text{ft}^3} \times \frac{34 \text{ ft}}{1} \times \frac{1 \text{ ft}^2}{144 \text{ in}^2} = 14.7\frac{\text{lb}}{\text{in}^2}$$

The interested reader may show that the pressure of a column of water 34 ft high (and similarly for mercury) is equivalent to 1 atm or 14.7 lb/in.² by noting that equation (B-1) may be rewritten as

$$\text{pressure} = \text{weight density} \times \text{height}$$
$$p = \rho g \times h \tag{B-3}$$

You should note that equation (B-2) relates pressure and density. Example B-1 shows this equivalence in detail.[1]

Pressure and the Ideal Gas

For our purposes, equation (B-1) relates more directly to the interpretation of the pressure of a quantity of gas contained in a given volume, V, for example, ionized hydrogen in the sun's core or steam trapped in a boiler. In general, when a gas is trapped in a container, the atoms or molecules collide with the walls, exerting a force over every part of the area; this according to equation (B-2) is what we call pressure. For a given volume, the average number of collisions of molecules with container walls will depend on the average speed of the molecules or, more appropriately, the average kinetic energy of the molecules. If we further note that the product of pressure and volume, pV, has units of energy, nt/m² × m³ = nt × m = J (joules), we may intuitively suggest that the pV product is directly related to the average kinetic energy of the gas atoms or molecules.

If we use the **kinetic model** of a gas, as discussed in Chapter 7, we know that the average kinetic energy depends on the absolute or Kelvin temperature, $\overline{KE} = \text{constant} \times T$. Thus, the pV product for a gas described by the kinetic model must depend directly on the temperature. The relationship between p, V, and T is called the **ideal gas equation.**

$$pV \propto \overline{KE} = \text{constant} \times T \tag{B-4}$$

---

[1] A more detailed discussion of pressure may be found in, V. Booth and M. Bloom: *Physical Science*. Macmillan Publishing Company, New York, 1971.

Equation (B-4) is usually written as

$$pV = mRT \tag{B-5}$$

where m is the mass of the gas-system in moles (see Chapters 5 and 6), and R is the ideal gas constant, 8.3×10^3 J/mole °K. Equation (B-5) also expresses a relationship between pressure and density for a gas, $p = \rho RT$ $\left(= \dfrac{m}{V} RT \right)$. We should note that the mass in equation (B-5) is usually given in moles which, when used with the joule unit, represents the molecular weight of the material in kilograms and Avogadro's number as 6×10^{26} atoms or molecules. In the chemistry of Chapters 5 and 6, we used the traditional mole unit used by chemists which represents the mole-weight of the material in grams and Avogadro's number as 6×10^{23}.

Equation (B-5) now allows us to understand, at least in a qualitative fashion, the relationship between pressure and temperature. For a given volume (and mass), as in a boiler, a high temperature means a high pressure. Example B-2 illustrates this principle and the use of the ideal gas law.

EXAMPLE B-2

A given volume and mass of gas in a boiler is initially at atmospheric pressure (10^5 nt/m²) and room temperature (300°K). What is the pressure of the gas in the boiler when it is heated to 800°K (about 1000°F) assuming that it behaves as an ideal gas?

Initially, we have

$$p_1 V_1 = m_1 R T_1 \qquad \text{or} \qquad \frac{p_1}{T_1} = \frac{m_1 R}{V_1}$$

Finally, we have

$$p_2 V_1 = m_1 R T_2 \qquad \text{or} \qquad \frac{p_2}{T_2} = \frac{m_1 R}{V_1}$$

Thus

$$\frac{p_1}{T_1} = \frac{p_2}{T_2}$$

and rearranging

$$p_2 = p_1 \times \frac{T_2}{T_1}$$

$$= 1 \text{ atm} \times \frac{800 \text{ °K}}{300 \text{ °K}} = 10^5 \frac{\text{nt}}{\text{m}^2} \times \frac{800 \text{ °K}}{300 \text{ °K}}$$

$$= 2.67 \text{ atm} = 2.67 \times 10^5 \text{ nt/m}^2$$

APPENDIX C

Basic Electricity

Even though a discussion of basic electricity is not necessary for our overall theme of energy, the concepts are both interesting and useful. The basis of electricity is the flow of electric charges or electrons. Very simply, when electrons move, they have kinetic energy that can be used to produce work. In fact, we take advantage of this principle every time we plug in an electric appliance. Hereafter, all the arguments that we have presented to establish the concepts of transformation of energy, via work, the concept of power and the appropriate units for each follow as before. There is, however, some terminology we need to establish so that the basic knowledge you already have may be put into practice.

Electric Force and the Flow of Charge

The supply of charge for electricity is provided by the electronic structure of the atom. You recall that the atoms contain protons (positively charged particles) in the nucleus and electrons (small negatively charged particles) orbiting about the nucleus (in the simple model). Just as an attractive gravitational force holds a planet in orbit about the sun, an attractive electric force (called a **Coulomb force**) between the protons and electrons holds the electrons in orbit about the nucleus. The electric force between protons and electrons is the third basic force in nature, the other two being gravitational and (short range) nuclear. Much the same as a body, having been lifted above the earth's surface, has gravitational potential or potential energy, an electron separated from a proton has electric potential or potential energy. Again, in analogy with the gravitational case, an electron moving from a region of high potential (energy) to a region of low potential transforms this potential energy into kinetic energy.

Although electric charges can be made to flow through air as in the case of lightning, practically, we use a material called a **conductor,** usually a metal. Copper and aluminum (wire) are two such examples. The way in which a conductor, such as copper, differs from an **insulator** (non-conductor) like air or glass is that in a conductor the outer electrons are rather loosely bound and can be made to move or flow when a small electric force is created in the wire. The way one creates this force is to cause a pile up of negative charge at one end of the wire and of positive charge at the other end. Practically, a **battery** or a fuel cell (see Chapter 15) does this for us.

A chemical reaction provides a pile up of electrons at the negative pole while there are fewer electrons than are necessary to maintain electric neutrality at the positive pole of the battery. When a wire—a conductor—is attached between the poles, electrons flow from the negative pole, a region of high electrical potential, to the positive pole, a region of low potential. The flow of electric charge is called **electric current** and is measured in **amperes (amp).**

The chemical reactions in the battery maintain the supply of current according to the capacity of the battery, which is measured in ampere-hours (amp-hr). For example, a 60 amp-hr battery could ideally supply 60 amp for 1 hr. It is interesting to note that it takes 6.25×10^{18} electrons passing a point in the wire every second to constitute 1 amp of current. The electric potential is called **voltage** and is measured in **volts.**[1] A typical wet cell (lead-acid) storage battery is rated at 12 volts. The wire or some appliance that we connect across the poles or terminals of the battery is called a **load** or a **resistance** and is measured in **ohms.**

---

[1] In electrical jargon, we speak in terms of potential rather than in terms of potential energy, which is used in the gravitational case. The subtle difference between the terms is that electric potential energy is the electric potential multiplied by the unit of charge existing at that potential. However, there is no further need for us to distinguish between them here.

A second way to produce a flow of charge is to take a wire loop connected to a device that measures flow—a galvanometer or an ammeter—and pass one side of it between the poles of a magnet. The interaction of the moving wire and the magnetic field (a magnetic force field) exerts a force (called the Lorentz force) on the charges in the wire, causing them to move. We say we have **generated** an electrical current and such a device is called an **electrical generator.** In practice, the motion or mechanical energy is provided by a steam turbine or a water turbine. The conversion from mechanical to electrical energy is done by our electrical generator.

Ohm's Law

The relationship relating current, voltage, and resistance is called **Ohm's law.** It states that the amount of current in an electrical circuit (a source of charge and a load) is proportional to the voltage and inversely proportional to the resistance.

$$\text{current} = \frac{\text{voltage}}{\text{resistance}} \qquad \text{or} \qquad I = \frac{V}{R} \qquad \text{(C-1)}$$

Equation (C-1) tells us that, if a potential of 1 volt is impressed across a load having a resistance of 1 ohm, then 1 amp of current will flow. A typical 100-w light bulb has a resistance of about 100 ohms. A typical iron or toaster has a resistance of about 20 ohms. From Equation (C-1), we see that, when the bulb is connected to a 120-volt household wall plug, about 1.2 amp of current will flow, whereas, when a toaster is connected, about 6 amp will flow. The transformation in either case is from electrical energy to (mainly) heat energy with some light. The details of this determination are shown in Example C-1.

Electric Power

When a charge moves in a circuit, it does work. The results of this work usually are heat or some motion, as in turning a motor. We know from the text that the rate of doing work is what we call **power.** Electric power

---

EXAMPLE C-1

A 100-w light bulb having a resistance of 100 ohms is connected to a 120-volt household supply (wall plug). How much current does this load draw?

$$I = \frac{V}{R} = \frac{120 \text{ volt}}{100 \text{ ohms}} = 1.2 \text{ amp}$$

is given simply as,

$$\text{power} = \text{voltage} \times \text{current} \quad \text{or}$$
$$P = V \times I \tag{C-2}$$

As is apparent, the unit of power is volt-amperes, which is equivalent to a watt ($=$ joule of work per second). Equation (C-2) illustrates the more practical view of electric energy.

Electric appliances are rated according to their power requirements in watts or kilowatts. Also, usually given (or sometimes understood) is the **working potential** of 120 volts for normal household use. Some appliances, such as large air conditioners, use 240 volts. With the power rating and the operating voltage given, we can determine the current that the appliance "draws" and the resistance of the appliance using equations (C-1) and (C-2).

Another practical use of equation (C-2) is in the determination of the maximum number of appliances that one can safely operate from a single household circuit. As a matter of explanation, usually, all of the wall plugs in a room are on the same circuit with the exception (in newer homes with updated electric service) of dining rooms and kitchens which usually have two circuits since they nominally have higher power requirements. The maximum power on each circuit usually is 1800 w or 15 amp at 120 volt. Some circuits, usually in kitchens, have 20-amp service or a 2400-w capacity.

In deciding on the number of appliances that can be safely operated on a circuit, we simply add the individual power requirements of each device and stay under the maximum allowable for that circuit, that is, 1800 watt. As an example, we could operate 18 100-w bulbs, or the equivalent in other appliances, on a single circuit. Any more than that, all operating at the same time, will blow the fuse or open the circuit breaker located in the fuse or breaker panel. Usually, it is recommended that the number of appliances operating on a single current require no more than about 80% of the maximum allowable power, that is, 15 100-w bulbs.

Fuses and breakers are safety devices that keep the household wiring operating within safe "heat limits." The wiring, itself, is a device that consumes electrical energy and gives off heat. Excessive heat due to overloading circuit wiring is a common cause of house fires. Knowing your household circuits and following this simple fact could save much grief to many consumers. A second cause for personal injury is electric shock. Normally, the body's resistance (with dry skin) to current is high, about 10^6 ohms. Even when your hand is across 120 volts, the current is low (about 0.0001 amp) and the hazard is minimal. However, if hands are damp or wet, the skin may have a resistance of only a few hundred ohms, resulting in a current of as much as 1 amp. Unfortunately, currents as low as 0.05 amp can be fatal and 0.1 amp through the heart is most always fatal. This is the basis for the frequent electrocution of persons standing in the bathroom or tub when the hands and feet are usually wet and plugging or unplugging electric appliances.

The Cost of Electric Energy

As we emphasize in Chapter 8, we consume energy not power. Work or energy is just the product of power and time, watts × hours or kilowatts × hours. Presently, the cost of electric energy to the consumer varies between about 1 cent and 5 cents/kwh depending on location, time of year, and quantity consumed. To evaluate the cost of running an appliance, we multiply the average use in hours, daily, and so on by the power rating in kilowatts. The resulting unit is the kilowatt-hour and the cost is about $0.03 per kwh. For example, ten 100-w bulbs, used for an average of 12 hr daily, would consume 12 kwh of electrical energy and cost about $0.36 or 36 cents (about $11.00 per month).

In conclusion, the authors have attempted to provide a layman's approach to electricity with some hints that could be helpful with respect to safety and to the conservation of electric energy. We have not tried to circumvent any rules established by professional, electric societies, and any specifics on use should be obtained from such organizations and your local power companies.

Glossary

Acceleration The rate at which the velocity of a body changes with time; the change in velocity may be in magnitude (speed) or direction or both.

Albedo The short wavelength portion (30%) of the sun's energy received by earth, which is immediately reflected back into space.

Alpha particle An energetic helium nucleus ($+2$ charge) emitted from the nucleus of a heavy atom during radioactive decay.

Astronomical unit A unit of distance equal to 9.3×10^7 miles.

Atmosphere The gases above the earth's surface; a unit of pressure equal to 14.7 pounds per square inch, which is the pressure of our atmosphere at sea level (abbreviated as atm).

Atom The basic constituent of matter which, when in combination, comprises molecules.

Beta particle A negatively charged particle (an

497

electron) ejected from the nucleus during radioactive decay.

Binding energy   The energy equivalent of the mass defect (according to Einstein's equation, $E = mc^2$); the energy needed to break up an atom totally into protons and neutrons.

Biosphere   The earth's entire biological community, including man, which is confined largely at or near the earth's surface.

Biotic   Biological; refers to the living components of the environment.

Breeder reactor   A nuclear reactor that produces more fuel (from ^{238}U) than it burns (^{235}U).

British thermal unit (Btu)   The amount of heat necessary to raise the temperature of 1 pound of water by 1 Fahrenheit degree at 59°F.

Burner reactor   A nuclear fission reactor which uses uranium-235 as the fuel without generating additional fuel from uranium-238. (See Breeder Reactor.)

Calorie   The amount of heat needed to raise the temperature of 1 gram of water by 1 Celsius degree at 15°C.

Carnivore   An animal that uses other animals as a food source.

Carnot engine   An ideal, reversible heat engine, in which the maximum amount of work output can be obtained for the heat energy input; the Carnot efficiency depends only on the temperatures of the heat reservoirs, $\eta = (T_H - T_C)/T_H$, where the subscripts H and C denote the hot and cold reservoirs respectively.

Chemical bond   The force or linkage holding one atom to another in a molecule.

Chemical change   The process whereby the arrangement of atoms in molecules (reactants) is changed to form different molecules (products).

Chemical energy   The energy stored in chemical bonds, which may be released upon combination or recombination of atoms and molecules.

Closed system   A thermodynamic system that exchanges energy but not mass with its surroundings, for example, the earth.

Coal gasification   A process by which the solid fossil fuel, coal, is converted to a gaseous fuel, such as, methane.

Conduction, electrical   The flow of electrons in a conductor, for example, a metal; electricity.

Conduction, thermal   The transfer of heat from one particle to another through direct contact.

Convection   The transfer of heat by the movement of matter, that is, the currents in gases or liquids caused by differences in densities.

Conversion factor or conversion efficiency   The observed or real efficiency of a heat engine.

Cooling tower   A chimney-like device used to transfer heat from water to the atmosphere via evaporation.

Demographer A person who studies populations and their statistics, such as births, deaths, health, and geography.

Density The mass of a substance divided by its volume.

Density of water In the metric system, 1 gram of water (at 4°C) has a volume of 1 cubic centimeter. Other substances are compared to water.

Diastrophism Geological processes involving movements of the crustal plates; the movements cause earthquakes, volcanic action, produce mountain ranges, and so on.

Doppler effect (Hubble red shift) The change in frequency or pitch of sound waves as the source and listener are moving towards or away from each other; a similar effect is observed with light from moving stars, the Hubble red shift.

Ecliptic plane An imaginary plane in space described by the earth's orbit; most of the planets' orbits lie within a few degrees of this great plane.

Ecology The study of the interactions of man, animals, and plants with the environment.

Ecosystem An environmental unit consisting of living organisms and their surroundings; the system may be very small or very large, as the earth.

Electricity (electric current) See Conduction, electric.

Electrolysis The process by which electrical energy causes atoms of a molecule to be separated into elements or other materials, for example, causing water to be converted to hydrogen and oxygen.

Electromagnetic spectrum The span of wavelengths from the short wavelength—high energy gamma rays down to the long wavelength—low energy radio waves.

Electromagnetic waves Coupled electric and magnetic disturbances that travel out from accelerated electric charges; the waves carry energy and travel at the speed commonly referred to as the speed of light, 3×10^8 meters per second.

Electron A subatomic particle, found in atoms, whose charge (negative) is equal and opposite to the proton and whose mass is about $\frac{1}{2000}$ of the proton mass.

Endothermic reaction A chemical reaction that does not release heat, but, in fact, requires energy to be supplied for completion.

Energy Commonly defined as the property a body has that gives it the capacity to do work; for example, kinetic and potential energy.

Energy cycle The reception, the use, and the discarding of energy as waste.

Energy slave A metaphor that expresses the per capita energy consumption in terms of the energy a man (slave) can produce in an 8-hour working day; because this energy provides the high standard of living that the average American enjoys, we humorously equate it to a number of slaves working for us, about 400 in 1970.

Energy spectrum The electromagnetic waves emitted by the sun; see Electromagnetic spectrum.

English system A system of measurement in which length, weight, and time are recorded in feet, pounds, and seconds.

Entropy The term that expresses the concept of energy becoming unavailable to produce work; it is also referred to as a measure of the disorder in a physical system.

Epilimnion The warm, constant-temperature, upper layer of a thermally stratified lake.

Eutrophication A process by which a lake becomes more productive, which ultimately results in a filling-in by partially decayed matter; the lake becomes shallower and eventually becomes a bog or swamp and finally a meadow.

Evaporation The process by which a liquid becomes a vapor (gas).

Exothermic reaction A chemical reaction that releases heat energy.

Exponential decay A nonlinear type of decay or decrease characteristic of the radioactive decay process of unstable atoms; it is characterized by a halving-time or half-life.

Exponential growth A highly nonlinear type of growth (change) that is characterized by a percentage annual increase and a corresponding doubling time (the inverse of exponential decay).

Fertile atom An atom that, after capture of a neutron by its nucleus, decays in some number of steps to another kind of atom which will fission when bombarded by a neutron.

Fissile atom An atom that, after capture of a neutron by its nucleus, undergoes nuclear fission, breaking apart into two approximately equal fragments.

Fission (also fusion) See Nuclear fission and Nuclear fusion.

Food chain A sequence of organisms (producers, herbivores, and carnivores) through which energy and materials move within an ecosystem.

Fossil fuel A substance, capable of combustion, that derives from plant life laid down in the earth in past geological ages; specifically, coals, petroleum liquids, and natural gas.

Galaxy An aggregate of stars separated from one another by much greater distances than those between member stars.

Gamma ray High energy x-rays (or electromagnetic waves) given off by the nuclei of some radioactive atoms.

Geiger counter A device for the detection and measurement of radioactive decay from unstable atoms.

Geothermal heat The heat energy conducted from within the earth's interior.

Gravity (gravitation) The attractive force found to exist between two bodies; this force depends on the masses of the bodies and the distance separating the bodies.

Greenhouse effect   The trapping of long wavelength heat energy by atmospheric water vapor, carbon dioxide, and ozone; the energy penetrates the atmosphere as light, but is reradiated as terrestrial heat, which is trapped.

Half-life   The time period needed for one half of the atoms originally present in a radioactive substance to be transformed to atoms of another element by the mode of decay characteristic of the substance (see Exponential decay).

Heat conduction   See Conduction, thermal.

Heat energy   Energy that manifests itself in the rapid motion of molecules, that is, in kinetic energy of molecules.

Heat engine   A device that takes in (uses) heat energy and converts some of this energy into work while discarding the balance as waste.

Heat of combustion   The amount of heat released by burning 1 mole of a substance.

Heat of fusion (also melting)   The amount of heat removed in order to transform 1 kilogram of liquid to 1 kilogram of solid at the same temperature (80 kilocalories per kilogram for water to ice).

Heat of reaction   See Heat of combustion.

Heat of vaporization (also condensation)   The amount of heat needed to convert an amount of a liquid completely to a gas at the same temperature; the heat of vaporization of water is 540 kilocalories per kilogram.

Heat sink   Any mass, for example an ocean, that is large enough so that when heat is added to it there is a negligible effect on the temperature of the mass.

Hubble red shift   The shifting in frequency or pitch of the hydrogen-alpha line (0.656 micron) in a star's energy spectrum due to the star's motion relative to us; see Doppler shift.

Hypolimnion   The bottom, cold, constant-temperature layer of a thermally stratified lake.

Inertia   The property of matter by which it remains at rest or continues its motion unless acted on by some external force.

Insolation   The portion of the sun's energy received at the earth's surface.

Insulator   A material such as wood, cork, wool, or glass fiber, that is a poor conductor of heat and electricity.

Ion   An atom (or molecule) that has gained or lost one or more electrons and consequently has become a charged particle; an electrically charged particle.

Ionization energy   The energy required to remove an electron from an atom, when the atoms of the element are in the gaseous state.

Irreversible process A process that cannot be reversed with an infinitesimal change in conditions; any spontaneous process; see Reversible process.

Isolated thermodynamic system A thermodynamic system that exchanges neither energy nor mass with its surroundings, for example, the universe.

Isotope An atom of the same element (same atomic number) but differing in mass number.

Joule A metric unit of work; a force of 1 newton applied through a distance of 1 meter (1 newton-meter).

Kelvin scale The temperature scale on which 273 degrees is defined as the ice point of water whereas 373 degrees is the normal boiling point of water; this scale is a direct measure of average molecular kinetic energy.

Kilocalorie The amount of heat needed to raise the temperature of 1 kilogram of water by 1 Celsius degree at 15°C. (Sometimes written Calorie.)

Kilowatt A unit of power in the S.I. system (and metric) equal to 0.238 kilocalories per second; 1 kilowatt is equivalent to 0.948 British thermal units per second or 1.34 horsepower.

Kilowatt-hour A unit of energy in the S.I. system (or metric) equal to 3.6×10^6 joules or 860 kilocalories; 1 kilowatt-hour is equivalent to 3412 British thermal units.

Latent heat The amount of heat that is absorbed or liberated when a change in state occurs; when water evaporates this amount of heat (540 kilocalories per kilogram) is stored in the vapor in the form of molecular potential energy and will be released upon condensation.

Lethal temperature The temperature at which death occurs for a particular species.

Light That part of the visible electromagnetic spectrum lying between 0.7 and 0.4 microns in wavelength.

Light year The distance light travels in one year; 6×10^{12} miles.

Lignite A brownish-black variety of coal intermediate in heating value between peat and bituminous coal.

Lithosphere The solid matter making up the earth's crust plus the upper mantle.

Mass The quantity of matter in a body (chemical definition); the measure of a body's inertia (physical definition).

Mass defect The difference between the mass an atom would have if it were made up of free electrons, neutrons, and protons and the mass the atom actually is observed to have.

Matter Anything that occupies space and has mass.

Maximum theoretical efficiency The Carnot efficiency of a heat engine or the best that can be done by an ideal engine operating between two specific temperatures.

Metric system A system of measurement in which length, mass, and time are recorded in meters, grams, and seconds, respectively. This system utilizes a counting system in units of 10.

Moderator Substances made up of lighter atoms that are used to slow the fast, fission neutrons via collision; the new slower neutrons are more efficient in producing more fission reactions. Materials such as light water (H_2O), heavy water (D_2O), graphite, and the lighter hydrocarbons are good moderators.

Mole The amount of any substance which contains Avogadro's number (6.02×10^{23}) of atoms or molecules or molecular formula units, depending on the nature of the substance; thus, 1 mole of material has a mass equal to its atomic or molecular mass in grams. For example, 1 mole of carbon has a mass of 12 grams.

Molecule The simplest component of a pure substance that still has all the chemical properties of that substance.

Natural background radiation The radiation that exists in our environment and is uncontrollable by man.

Neutron An electrically neutral particle found in the nucleus, possessing 1 atomic mass unit.

No-growth (linear projection) That situation in which the consumption rate per person would necessarily decline if there is any growth in population; No-growth implies that each year hence, the annual consumption rate is the same as the base period, for example, the annual rate in 1970. Based on this assumption, a graph of annual consumption versus year and the extension to future years would be a linear projection in contrast to the exponential growth projection, which incorporates growth.

Nuclear fission The splitting of the nucleus of a heavy atom (such as uranium-235) into two fragments by a slow neutron, accompanied by a release of energy and several neutrons.

Nuclear fuel The fissionable substance, containing sufficient uranium-235 or plutonium-239, used in nuclear reactors.

Nuclear fusion The joining together of two light atoms (such as deuterium and tritium) under conditions of high temperature and pressure to make helium, accompanied by a release of energy and a neutron.

Nuclear reactor A machine for containing and controlling the release of heat and other products from a nuclear chain reaction.

Nucleus A very small region of an atom near the center containing all the atom's protons and neutrons and nearly all the mass of the atom.

Particulates (particulate matter) The solid matter exhausted into the air from combustion of fossil fuels.

Periodic law Arranged in order of their atomic numbers, atoms exhibit a periodicity of properties, both chemical and physical.

Photosynthesis   The process by which plants convert light energy, water, and carbon dioxide into plant fiber and sugar; chlorophyll is a catalyst for this process.

Physical change   A change in state or appearance of a substance wherein no new substance is produced.

Pitchblende   A brown to black massive mineral (ore) with pitch-like luster, which is a good source of uranium and radium.

Plate tectonics   A geological theory of earth structure and changes thereof, in which the earth's crust is considered to be made up of a few solid plates floating on the semifluid interior.

Pollution   The contamination of the natural environment with substances that are toxic or detrimental to a life cycle or a food chain.

Power   The rate of energy consumption; the rate of doing work; the work done per unit of time.

Pressure   A force divided by the area over which the force is exerted; for example, 1 atmosphere of pressure is 14.7 pounds per square inch.

Primary pollutant   A substance produced directly by a process discharging into the environment.

Proton   A positively charged particle found in the nucleus of atoms, possessing one atomic mass unit and one atomic number; an element is identified by its number of protons (atomic number).

RAD (radiation absorbed dose)   A unit of radioactive energy corresponding to absorption by living tissue of 2.4×10^6 kilocalories of energy per kilogram of tissue.

Radiation   The transfer of energy by means of electromagnetic waves.

Radiation balance   The equivalence of the sun's energy intercepted by earth *and* of the energy returned into space by earth.

Radioactivity   A phenomenon peculiar to some atoms that spontaneously change by emitting particles and/or rays with tremendous energies from the nucleus.

RBE (relative biological effectiveness)   A numerical evaluation of the biological damage effected by the various kinds of radiation.

REM (radiation equivalent to man)   The amount of radiation of any type that, when absorbed by man, has an effect equal to the absorption of 1 rad of gamma rays or x-rays.

Reserve   The portion of the total substance (stock) in the earth that man can obtain with current technology and under current economic conditions.

Resource   The portion of the total substance (stock) in the earth known to exist that man can obtain under some technological or economic conditions different from those at present.

Reversible process   A process that can be made to go in the reverse direction by an infinitesimal (very, very small) change in operating conditions.

Roentgen (R) A measure of radiation dose used only when considering gamma or x-rays; see rad.

Science The coordinated and systematized body of knowledge obtained by methods based on observation, including careful experimentation and creative thinking.

Scientific method A technique used frequently in the practice of science consisting of four steps: gathering data, setting an hypothesis (model) on the basis of the data, taking more data usually experimentally to check on the predictions of the hypothesis, and formulating a reasonable theory encompassing the data.

Scrubbing (as in pollution abatement) The removal of water-soluble pollutants from effluent gases by water spray.

Secondary pollutant A substance deleterious to the environment and derived from reactions involving primary pollutants; for example, photochemical smog, sulfuric acid aerosols, and so on.

Silicates The major constituent of soil and rocks, composed of silicon, oxygen, and minerals.

Smog A mixture of smoke (gaseous and liquid air pollutants) and fog.

Solar constant The rate at which energy from the sun crosses the diametric plane, that is, 1.395 kilowatts per square meter.

Specific heat capacity The quantity of heat per unit mass required to raise the temperature of a substance by 1 Celsius degree; measured in units of kilocalories per kilogram per degree Celsius in metric units or British thermal units per pound per degree Fahrenheit in English units.

Speed The rate of covering distance; distance/time.

Stratification The existence of layers of different temperatures in a nonturbulent body of water caused by the absorption of heat in the upper layer with the upper warmer layer being less dense.

Subatomic particle A particle found in atoms whose size is less than that of the smallest atom.

S.I. (Systeme Internationale) A system of measurement in which length, mass, and time are recorded in meters, kilograms, and seconds, respectively.

Technology The means of controlling or manipulating nature for man's use, such as, in the production of goods and services.

Temperature The degree of hotness or coldness of a body; a measure of the average kinetic energy per molecule in a body; measured in degrees Celsius, Fahrenheit, or Kelvin.

Temperature inversion A meteorological condition in which a stable layer of warm air prevents the normal upward movement of warm air due to convection; a temperature profile characterized by an increasing temperature with height rather than decreasing.

Thermal pollution A deleterious effect caused by the raising of the temperature of some body of water by the discharge of heat from a man-made source into the water.

Thermocline The transition region of a stratified lake between the epilimnion and the hypolimnion.

Thermodynamic cycle See Energy cycle.

Thermodynamics A study of the laws governing heat flow.

Translational motion Movement from one point to another point; not including rotational or vibrational motion.

Troposphere The atmosphere of the earth extending from the surface out to about 10 miles above the earth's surface; the site of most weather events.

Turbidity Dustiness of air or water resulting in the scattering of light and a reduction in visibility.

Volcanism—Volcanic Action The modification of the earth's surface resulting from the flow of molten material to the surface; the heat for volcanic action via the theory of plate tectonics, is thought to be due to the friction between the great continental plates as they slide upon each other during collision.

Waste heat The heat discarded in order to complete a thermodynamic cycle; exhaust heat from an engine; discarded process heat.

Watt The metric unit of power equal to 1 joule per second.

Weight The force of gravity (attraction) on a body; the attractive force between bodies, that is the earth and a body.

Work The force applied to an object multiplied by the distance the object moves in the direction of the force.

Index

507